SEISMIC DESIGN OF BUILDINGS TO EUROCODE 8

SECOND EDITION

SEISMIC DESIGN OF BUILDINGS TO EUROCODE 8

SECOND EDITION

EDITED BY

AHMED Y. ELGHAZOULI

Imperial College, London

CRC Press
Taylor & Francis Group
Boca Raton London New York

CRC Press is an imprint of the
Taylor & Francis Group, an **informa** business

CRC Press
Taylor & Francis Group
6000 Broken Sound Parkway NW, Suite 300
Boca Raton, FL 33487-2742

First issued in paperback 2018

ISBN-13: 978-1-4987-5159-9 (hbk)
ISBN-13: 978-0-367-13905-6 (pbk)

Library of Congress Cataloging-in-Publication Data

Names: Elghazouli, Ahmed Y. (Ahmed Youssef), editor.
Title: Seismic design of buildings to Eurocode 8 / [edited by] Ahmed Elghazouli.
Description: Second edition. | Boca Raton, FL : CRC Press, Taylor & Francis Group, 2016. | Includes bibliographical references and index.
Identifiers: LCCN 2016042114 | ISBN 9781498751605
Subjects: LCSH: Earthquake resistant design. | Buildings--Earthquake effects. | Standards, Engineering--Europe.
Classification: LCC TA658.44 .S3996 2016 | DDC 693.8/52--dc23
LC record available at https://lccn.loc.gov/2016042114

Visit the Taylor & Francis Web site at
http://www.taylorandfrancis.com

and the CRC Press Web site at
http://www.crcpress.com

Contents

Preface

This book aims at providing a concise and clear introduction to the principles of seismic design for building structures and their foundations, with a particular focus on linking these concepts to the provisions of the European seismic code, Eurocode 8. It addresses the underlying approaches related to seismic hazard, ground motion models, basic dynamics, seismic analysis, siting considerations, structural layout, and design philosophies. Eurocode 8 procedures are applied in most parts with the aid of walk-through design examples which, where possible, deal with a common case study. As well as an update throughout, this second edition of this book incorporates three new and topical chapters dedicated to specific seismic design aspects of timber and masonry buildings as well as base-isolation and supplemental damping. The primary audience for this book are practicing structural and geotechnical engineers as well as post-graduate and senior undergraduate civil engineering students.

The introductory chapter, Chapter 1, offers an overview of the background to the development and implementation of Eurocode 8 together with an outline of this book. Chapter 2 provides a detailed review of the methods used in determining seismic hazards and earthquake actions. It covers seismicity and ground motion models, with specific reference to the stipulations of Eurocode 8. Chapter 3 presents a review of basic dynamics including the response of single- and multi-degree of freedom systems and the use of earthquake response spectra, leading to the seismic analysis methods used in Eurocode 8; this chapter also introduces an example building, which is used throughout most of this book to illustrate the application of Eurocode 8 in practical design. The provisions relating to general considerations for the design of buildings are dealt with in Chapter 4; the selected case study is then used to provide examples for the use of Eurocode 8 for siting as well as for assessing structural regularity.

The application of seismic design principles, as adopted in Eurocode 8, for buildings of different materials and configurations are described in subsequent chapters of this book. Chapter 5 focuses on the design of reinforced concrete structures, and a design example for a dual frame/wall lateral resisting system is presented and discussed. The design of steel structures is dealt with in Chapter 6, which includes design examples for moment and braced steel frames. Chapter 7 focuses on highlighting key design aspects for composite steel/concrete buildings, and includes a design variation for the example building. The fundamental concepts of the seismic behaviour of timber structures are outlined in Chapter 8, and the example building is used to illustrate the design of a laminated-timber shear-wall system. Chapter 9 deals with the seismic design of masonry structures, and includes a specific design example, which is employed to highlight the benefits and limitations of available analysis methods.

Seismic isolation and supplemental damping can offer low-damage solutions, which are being increasingly used in practice. In Chapter 10, background concepts on seismic isolation, types of devices available, numerical results, and design criteria from Eurocode 8 are discussed and an illustrative example using the case study building is presented. Moreover, although Eurocode 8 does not specifically cover supplemental damping devices, an outline of recommended design criteria for viscous and viscoelastic damping systems are included in this chapter. The final two chapters, Chapters 11 and 12, are concerned with the design of shallow and deep foundations, respectively, and cover issues related to liquefaction and settlement as well as static and dynamic performance requirements. Design examples to Eurocode 8 including pad, raft and pile foundations for the case study building are also included.

It is important to note that this publication is not intended as a complete description of Eurocode 8 requirements nor as a replacement for any of its provisions. The purpose of this book is mainly to provide general background information on seismic design, and to offer discussions and comments on the use of the code in the design of buildings and their foundations. It is also worth noting that although all the Eurocodes have undergone a rigorous process of development, drafting and revision before publication, it is only through their application by practicing engineers on real projects that areas for improvement become apparent. The current revision cycle for Eurocode 8 is likely to take at least five more years before it is approved and made available to practicing engineers. Although the main emphasis in this book is on the existing code provisions, the scope of several chapters extends beyond this in order to address specific aspects that are currently being dealt with in the revision process; these aspects are highlighted, where relevant, within the individual chapters.

This book stems primarily from practical short courses on seismic design, which have been run over a number of years and through the development of Eurocode 8. These short courses have been organised by the Department of Civil and Environmental Engineering at Imperial College London, as part of Imperial's School of Professional Development Programme, in collaboration with the Society for Earthquake and Civil Engineering Dynamics (SECED). The courses were also supported by the European Association for Earthquake Engineering (EAEE) and the Eurocodes Expert initiative of the Institution of Civil Engineers (ICE).

The contributors to this book are either specialist academics with significant consulting experience, or leading practitioners who are actively engaged in large projects in seismic areas. The unfailing enthusiasm of all the contributors and their employers, as well as their dedication towards SECED's activities including practical short courses, have been vital to the completion of this book. The editor would additionally like to acknowledge the constructive advice offered by the eight reviewers to the book proposal, including four anonymous reviewers as well as Professor Christoph Butenweg (RWTH Aachen), Professor Peter Fajfar (University of Ljubljana), Dr Gregory Penelis (Penelis Consulting) and Professor André Plumier (University of Liège). The expert assistance of the staff at Taylor & Francis publishing group is also gratefully acknowledged; particular thanks are due to Tony Moore for his constant support and encouragement throughout the development of the two editions of this book.

Ahmed Y. Elghazouli
Professor of Structural Engineering
Imperial College London

Editor

Ahmed Y. Elghazouli is a professor and head of structural engineering at Imperial College London. He is also a director of the post-graduate programme in earthquake engineering. His main research interests are related to the response of structures to extreme loads, focusing on the areas of earthquake engineering, fire engineering, blast assessment and structural robustness. He has led numerous research projects and published more than 200 papers in these areas, and has worked as a specialist consultant on many important engineering projects worldwide over the past 20 years. Ahmed is a fellow of the Royal Academy of Engineering, a chartered engineer, and a fellow of ICE and IStructE. He is a member of BSI and ECCS committees on seismic design. He is also a past chairman of SECED and UK National Delegate of IAEE and EAEE.

Contributors

Philippe Bisch is a specialist in structural analysis and currently a professor at the Ecole des Ponts ParisTech (ENPC) in Paris. In 1976, he joined Sechaud & Metz (S&M), Consulting Engineers, as a technical director and is now an international expert with the EGIS group to which S&M belongs. He is a honorary president of the French Association for Earthquake Engineering (AFPS) and he was formerly president of the European Association for Earthquake Engineering. He has long been involved in standardisation, in France and in Europe. He is now chairman of CEN TC250/SC8 on EC8. He was also involved in CEN committee TC340 for the anti-seismic devices standard. He is an active member of the French governmental group (GEPP) related to seismic safety.

Julian J. Bommer is a senior research investigator in the Department of Civil and Environmental Engineering at Imperial College London, specialising in ground-motion prediction and the assessment of seismic hazard and risk due to natural and induced earthquakes. He has published more than 100 widely cited journal papers on these topics. He served as SECED chairman from 2000 to 2002. His consultancy experience includes the Panama Canal, major pipelines and dams, and nuclear power plants in Abu Dhabi, Brazil, Romania, South Africa, Spain and the United States; he is a member of the Expert Panel on Seismic Hazard of the UK Office for Nuclear Regulation.

Edmund Booth, after 15 years of keeping his technical interests as broad as possible, took the specialist route into earthquake engineering in 1982. He founded his own practice in 1995, and undertook the seismic design, analysis and assessment of a very wide range of structures worldwide. Edmund is a fellow of the Royal Academy of Engineering, and has taught at Imperial College London and Oxford University as a visiting lecturer. He is author of the textbook *Earthquake Design Practice for Buildings* published in 2014 and chairs the BSI technical sub-committee for EC8.

Andy Campbell is an independent consultant, having spent the bulk of his career in the UK nuclear industry, dealing with abnormal or extreme environmental loading with seismic effects frequently forming the dominant action. He has many years of experience of seismic design and appraisal with a particular interest in reinforced concrete structures and performance-based design. He has been a member of the SECED Committee since 2003, serving as chairman from 2014 to 2016.

José Miguel Castro is an assistant professor at the University of Porto and an invited professor at the University of Coimbra. He earned a PhD in civil engineering from Imperial College London, in 2006. He is an expert in steel and composite structures subjected to extreme loading conditions and is a full member of Technical Committee 13 on Seismic Design of the European Convention for Constructional Steelwork (ECCS). His current research interests are on the development of novel seismic design methodologies as well as on the collapse and seismic risk assessment of existing buildings. He is author and co-author of more than 50 publications in peer-reviewed journals, book chapters and international conference proceedings.

Matthew DeJong is a senior lecturer in structural engineering at the Department of Engineering, University of Cambridge, and a fellow of St Catharine's College, Cambridge. He leads a research group focused on earthquake engineering and masonry structures, and has particular research interests in analytical dynamics, computational modelling, seismic collapse, heritage structures and the use of rocking in modern design. He is currently a SECED Committee member.

Damian N. Grant is an associate in Arup's Advanced Technology and Research practice, and leads the practice's earthquake engineering team. Since joining Arup in 2005, Damian has been responsible for the seismic design and assessment for a diverse range of projects, including buildings (from single-storey dwellings to super-tall skyscrapers), industrial facilities, offshore platforms, storage tanks, bridges and other special structures. He specialises in performance-based earthquake engineering, the practical implementation of non-linear finite element techniques in design and assessment, and the application of seismic protective systems such as isolators and supplemental damping devices. Damian serves on the SECED (Society for Earthquake and Civil Engineering Dynamics) committee, and chairs the BSI (British Standards Institution) subcommittee on Actions and Analysis in Eurocode 8, including the design of base-isolated structures.

Mário Lopes earned his PhD in earthquake engineering from Imperial College, London, and is an assistant professor at the Civil Engineering Department of the Technical University of Lisbon. His main research interest and design activity concern the seismic design of reinforced concrete buildings and bridges. He has experience in field investigations following destructive earthquakes and he is a member of the Portuguese Society for Earthquake Engineering.

Zygmunt Lubkowski is an associate director at Arup, where he is also the seismic business and skills leader for Europe, Africa and the Middle East. He has more than 20 years of experience working in civil, geotechnical and earthquake engineering projects. He has carried out seismic hazard and geo-hazard assessments and seismic foundation design for a range of structures in various sectors. He has acted as the seismic specialist for major projects such as offshore platforms, LNG plants, nuclear facilities, major bridges, dams, immersed tube tunnels and tall buildings. He has been the chair of SECED and Earthquake Engineering Field Investigation Team (EEFIT). He also participated in post-earthquake field missions worldwide and the development of a field guide for the survey of earthquake damaged non-engineered structures.

Gopal S. P. Madabhushi is a professor of civil engineering and the head of Geotechnical and Geo-Environmental Engineering Group at the Department of Engineering, University of Cambridge. He is also a fellow of Girton College, Cambridge and the director of the Schofield Centre that houses the centrifuge facility with earthquake modelling capability. He leads the research of the Earthquake Geotechnical Engineering Group that focuses on soil liquefaction, soil structure interaction, pile and retaining wall performance, dynamic behaviour of underground structures and performance of earthquake remediation strategies. He has an active interest in field investigations following major earthquakes and was the past chairman of the EEFIT.

Christian Málaga-Chuquitaype is a lecturer in the Structures Section of the Department of Civil and Environmental Engineering at Imperial College London. He earned his PhD from Imperial College, London, and his Master's from the University of Pavia (ROSE School) and the University Joseph Fourier – Grenoble 1. His current research interests relate to structural timber engineering and emerging building technologies, earthquake engineering and the response of structures to extreme loads.

Robert May is currently the chief geotechnical engineer with Golder Associates in Australia. He was previously with Atkins and has 30 years of experience in geotechnical engineering with a particular interest in seismic design. His recent seismic design projects have included major retaining walls, foundations and slopes. Dr. May is a former SECED Committee member and has chaired the 12th European Conference on Earthquake Engineering in 2002. He was a former member of the Géotechnique Advisory Panel.

Alain Pecker is chairman and managing director of Géodynamique et Structure, a French engineering consulting firm in earthquake engineering. He is also a professor at Ecole Nationale des Ponts et Chaussées and at the European School for Advanced Studies in Reduction of Seismic Risk (University of Pavia, Italy). His professional interest lies in soil dynamics, liquefaction, wave propagation, soil structure interaction and foundation engineering. Alain Pecker is past president of the French Society of Soil Mechanics and Geotechnical Engineering, and honorary president of the French Association on Earthquake Engineering. He has been elected to the French National Academy of Technologies in 2000.

Andrea Penna is an associate professor of structural engineering in the Department of Civil Engineering and Architecture at the University of Pavia and a fellow of the EUCENTRE Foundation, Pavia, Italy. His main research activity focuses on the experimental and numerical study of the seismic behaviour of masonry structures. He has been involved in many research projects focused on both existing masonry structures and innovative masonry technologies. He is a co-author of the TREMURI computer program, widely used for the non-linear seismic analysis of masonry buildings. He is also a member of international working groups on seismic aspects of monuments preservation and the revision of masonry design codes.

Peter J. Stafford is a reader in the structures section of the Department of Civil and Environmental Engineering at Imperial College London. His research interests primarily relate to engineering seismology and its interface with earthquake engineering. In particular, his work focuses upon the specification of loading for seismic analyses within probabilistic frameworks. He has significant experience working on high-level probabilistic seismic hazard and risk studies in various parts of the world. His broader research interests cover aspects of applied structural dynamics and structural reliability theory.

Indrasenan Thusyanthan is currently the principal geotechnical engineer at Saudi Aramco. He was previously a consultant with KW Ltd. and a lecturer in geotechnical engineering at the University of Cambridge, UK. He earned his PhD, BA and MEng from the University of Cambridge. Dr. Thusyanthan was awarded the Institution of Civil Engineers' 'Roscoe Prize for Soil Mechanics' in 2001. His research interests include offshore pipeline behaviour, pipe–soil interaction, tsunami wave loading, seismic behaviour of landfills and liquefaction. He has extensive experience in centrifuge testing for different geotechnical problems. He has worked in various geotechnical projects for Mott MacDonald and WS Atkins.

Martin S. Williams is a professor in the Department of Engineering Science at Oxford University, and a fellow of New College, Oxford. He has led numerous research projects in earthquake engineering and structural dynamics, including the development of the real-time hybrid test method, analysis and testing of passive energy dissipation devices, modelling of grandstand vibrations and investigation of dynamic human–structure interaction. He is a fellow of ICE and IStructE, and has held visiting academic posts at the University of British Columbia, UNAM Mexico City and the University of Queensland.

Chapter 1

Introduction

Seismic design and Eurocode 8

Philippe Bisch

CONTENTS

1.1 THE EUROCODES

The European directive 'Construction Products' issued in 1989 comprises requirements relating to the strength, stability and fire resistance of construction. In this context, the structural Eurocodes are technical rules, unified at the European level, which aim at ensuring the fulfilment of these requirements. They are a set of 58 standards compiled into 10 Eurocodes and provide the basis for the analysis and design of structures and the constitutive materials. Complying with Eurocodes makes it possible to declare the conformity of structures and construction products and to apply the Conformité Européenne (CE) marking to them (a requirement for many products, including most construction products, marketed within the European Union). Thus, Eurocodes constitute a set of standards of structural design, consistent in principle, which facilitates free distribution of products and services in the construction sector within the European Union.

Beyond the political goals pursued by the Union, the development of Eurocodes has also given rise to considerable technical progress, by taking into account the most recent knowledge in structural design, and producing technical standardisation across the European construction sector. The Eurocodes have been finalised in the light of extensive feedback from practitioners, since codes should reflect recognised practices current at the time of issue, without, however, preventing the progress of knowledge.

The methodology used to demonstrate the reliability (in particular safety assessment) of structures is the approach referred to as 'semi-probabilistic', which makes use of partial coefficients applied to actions, materials properties and covering the imperfections of analysis models and construction. The verification consists of analysing the failure modes of the structure, associated with limit states, in design situations with associated combinations of actions, which can reasonably be expected to occur simultaneously.

Inevitably, the Eurocodes took many years to complete, since to reach general consensus, it was necessary to reconcile differing national experiences and requirements coming from both researchers and practising engineers.

1.2 STANDARDISATION OF SEISMIC DESIGN

The first concepts for structural design in seismic areas, the subject of Eurocode 8 (EC8), was developed from experience gained in catastrophes such as those due to the San Francisco earthquake in 1906 and the Messina earthquake in 1908.

At the very beginning, in the absence of experimental data, the method used was to design structures to withstand uniform horizontal accelerations of the order of 0.1 g. After the Long Beach earthquake in 1933, the experimental data showed that the ground accelerations could be much higher, for instance 0.5 g. Consequently, the resistance of certain structures could be explained only by the energy dissipation that occurred during the movement of the structure caused by the earthquake. The second generation of codes took into account, on the one hand, the amplification due to the dynamic behaviour of the structures, and, on the other hand, the energy dissipation. However, the way to incorporate this dissipation remained very elementary and did not allow correct differentiation between the behaviour of the various materials and types of lateral resisting systems.

The current third generation of codes makes it possible, on the one hand, to specify the way to take the energy dissipation into account, according to the type of lateral resistance and the type of structural material used, and, on the other hand, to widen the scope of the codes, for instance by dealing with geotechnical aspects. Moreover, these new rules take into account the semi-probabilistic approach for verification of safety, as defined in EN 1990.

The appearance of displacement-based analysis methods makes it possible to foresee an evolution towards a fourth generation of seismic design codes, where the various components of the seismic behaviour will be better controlled, in particular those that relate to energy dissipation. From this point of view, EC8 is at the junction between the third generation codes, of which it still forms part of its present configuration, and of fourth generation codes.

1.3 IMPLEMENTATION OF EUROCODE 8 IN MEMBER STATES

The clauses of Eurocodes are divided into two types: (i) Principles, which are mandatory, and (ii) Application Rules, which are acceptable procedures to demonstrate compliance with the Principles. However, unless explicitly specified in the Eurocode, the use of alternative Application Rules to those given does not allow the design to be made in conformity with the code. Also, in a given Member State, the basic Eurocode text is accompanied for each of its parts by a National Annex specifying the values of certain parameters (Nationally determined parameters [NDPs]) to be used in this country, as well as the choice of methods when the Eurocode part allows such a choice. NDPs are ones which relate to the levels of safety to be achieved, and include for example partial factors for material properties. In the absence of a National Annex, the recommended values given in the relevant Eurocode can be adopted for a specific project, unless the project documentation specifies otherwise.

For the structures and in the zones concerned, the application of EC8 involves that of other Eurocodes. EC8 only brings additional rules to those given in other Eurocodes, to which it refers. Guides or handbooks can also supplement EC8 as application documents for certain types of structural elements.

To allow the application of EC8 in a given territory, it is necessary to have a seismic zoning map and associated data-defining peak ground accelerations and spectral shapes. This set of data, which constitutes an essential basis for analysis, can be directly introduced into the National Annex. However, in certain countries, seismic design codes are regulated by statute, and where this applies, zoning maps and associated data are defined separately by the national authorities.

1.4 CONTENTS OF EUROCODE 8

EC8 comprises six parts relating to different types of structures (Table 1.1). Parts 1 and 5 form the basis for the seismic design of new buildings and its foundations; their rules are aimed both at protecting human life and also limiting economic loss. It is interesting to note that EC8 Part 1 also provides design rules for base isolated structures.

Particularly because of its overlap with other Eurocodes and the cross-referencing that this implies, EC8 presents some difficulties at first reading. Although these can be easily overcome by a good comprehension of the underlying principles, they point to the need for application manuals to assist the engineer in design of the most common types of structure.

1.5 FUTURE DEVELOPMENTS

Although all the Eurocodes have undergone a rigorous process of development, drafting and revision before publication, it is only through their application by practicing engineers on real projects that areas for improvement become apparent, whether it is the need for clarification or omissions that should be addressed. Against this background, there is a need to collate comments from code users and compile these into a coherent list of improvements and developments to be implemented in future updates of the codes.

The current revision cycle for the Structural Eurocodes, including EC8, has already started in 2015/2016, but is likely to take at least five years before it is approved and made available to practicing engineers. Two revision mechanisms are involved, namely a 'systematic review' and an 'evolution process'. The 'systematic review' takes the form of a line by line review of the code, focusing on clauses that require editorial or technical correction or improvement in clarity, on identifying aspects that need to be extended or shortened, or on clauses whose application may lead to excessive design effort or uneconomic construction. The 'systematic review' informs the 'evolution process' that incorporates the drafting of revisions including, where necessary, additional codified rules. The evolution process is nonetheless constrained by the need to maintain the 'stability' of the code, by avoiding radical changes, and by the desire to maintain 'ease of use' and to reduce the number of NDPs that are set separately by Member States.

Although the main emphasis in this book is on the existing provisions in EC8, the scope of several chapters extends beyond this in order to address specific aspects that are currently being addressed within the 'evolution process' of EC8. These aspects are clearly highlighted, where relevant, within the individual chapters.

Table 1.1 Parts of Eurocode 8

Title	Reference
Part 1: General Rules, Seismic Actions and Rules for Buildings	EN 1998-1:2004
Part 2: Bridges	EN 1998-2:2005
Part 3: Assessment and Retrofitting of Buildings	EN 1998-3:2005
Part 4: Silos, Tanks and Pipelines	EN 1998-4:2006
Part 5: Foundations, Retaining Structures and Geotechnical Aspects	EN 1998-5:2004
Part 6: Towers, Masts and Chimneys	EN 1998-6:2005

1.6 OVERVIEW OF THIS BOOK

Seismic design of structures aims at ensuring, in the event of occurrence of a reference earthquake, the protection of human lives, the limitation of damage to the structures and operational continuity of constructions important for civil safety. These goals are linked to seismic actions. Chapter 2 provides a detailed review of methods used in determining seismic hazards and earthquake actions. It covers seismicity and ground motion models, with specific reference to the stipulations of EC8.

To design economically a structure subjected to severe seismic actions, post-elastic behaviour is allowed. The default method of analysis uses linear procedures, and post-elastic behaviour is accounted for by simplified methods. More detailed analysis methods are normally only utilised in important or irregular structures. These aspects are addressed in Chapter 3, which presents a review of basic dynamics, including the response of single- and multi-degree of freedom systems and the use of earthquake response spectra, leading to the seismic analysis methods used in EC8. This chapter also introduces an example building, which is used throughout most of this book to illustrate the use of EC8 in practical building design. The structure was specifically selected to enable the presentation and examination of various provisions in EC8.

The design of buildings benefits from respecting certain general principles conducive to good seismic performance, and in particular to principles regarding structural regularity. The provisions relating to general considerations for the design of buildings are dealt with in Chapter 4. These relate to the shape and regularity of structures, the proper arrangement of the lateral resisting elements and a suitable foundation system. Chapter 4 also introduces the commonly adopted approach of design and dimensioning referred to as 'capacity design', which is used to control the yielding mechanisms of the structure and to organise the hierarchy of failure modes. The selected building introduced in Chapter 3 is then used to provide examples for the use of EC8 for siting as well as for assessing structural regularity.

Chapter 5 focuses on the design of reinforced concrete structures to EC8. It starts by describing the design concepts related to structural types, behaviour factors, ductility provisions and other conceptual considerations. The procedures associated with the design for various ductility classes are discussed, with particular emphasis on the design of frames and walls for the intermediate (medium) ductility class. In order to illustrate the design of both frames and walls to EC8, the design of a dual frame/wall lateral resisting system is presented and discussed.

The design of steel structures is discussed in Chapter 6. This chapter starts by outlining the provisions related to structural types, behaviour factors, ductility classes and cross-sections. This is followed by a discussion of the design procedures for moment and braced frames. Requirements related to material properties, as well as the control of design and construction, are also summarised. The example building is then utilised in order to demonstrate the application of EC8 procedures for the design of moment and braced lateral resisting steel systems.

Due to the similarity of various design approaches and procedures used for steel and composite steel/concrete structures in EC8, Chapter 7 focuses primarily on discussing additional requirements, which are imposed when composite dissipative elements are adopted. Important design aspects are also highlighted by considering the design of the example building used in previous chapters.

There has been a renewed interest in recent years for the use of timber as a main structural material largely due to a combination of the rising environmental concerns and the wider availability of newly developed high-performance timber-materials. Chapter 8 of this book starts by reviewing fundamental concepts of the cyclic behaviour of timber structures with particular emphasis on current building practices, followed by a summary of current EC8 provisions, which are still rather limited in scope. Finally the same example building used in previous chapters is used for illustrating the design of a laminated-timber shear-wall system.

Chapter 9 deals primarily with the seismic design of unreinforced masonry structures, although reinforced and confined masonry, for which limited guidance is provided in EC8, are also discussed. General structural behaviour and design concepts for the specific building typologies of reinforced, confined and unreinforced masonry are presented. The primary analysis methods available to design masonry structures are then discussed, followed by a design example that highlights the benefits and limitations of these analysis options. The eight-storey building considered in previous chapters could not be realistically used and, instead, an alternative three-storey unreinforced masonry building is considered in this chapter.

Seismic isolation involves the introduction of low lateral stiffness bearings to detune the building from the predominant frequencies of an earthquake. On the other hand, supplemental damping involves the addition of damping elements to the structure. EC8 covers seismically isolated structures in general, as well as specific rules for base isolation of buildings. In Chapter 10, background concepts on seismic isolation, types of devices available, numerical results and design criteria from EC8 are discussed and an illustrative example using the same case study building used in most chapters of this book is adopted. Moreover, although EC8 does not cover passive energy dissipation systems (supplemental damping) that are distributed over several storeys or levels, Chapter 10 includes a discussion of this topic together with a summary of available devices, as well as an outline of recommended design criteria for viscous and viscoelastic damping systems.

It is clearly necessary to ensure the stability of soils and adequate performance of foundations under earthquake loading. This is addressed in Chapters 11 and 12 for shallow and deep foundations, respectively. Chapter 11 provides background information on the behaviour of soils and on seismic loading conditions, and covers issues related to liquefaction and settlement. Focus is given to the behaviour and design of shallow foundations. The design of a raft foundation for the example building according to the provisions of EC8 is also illustrated. On the other hand, Chapter 12 focuses on the design of deep foundations. It covers the assessment of capacity of piled foundations and pile buckling in liquefied soils as well as comparison of static and dynamic performance requirements. These aspects of design are illustrated through numerical applications for the example building.

In the illustrative design examples presented in Chapters 3 through 12, reference is made to the relevant rules and clauses in EC8, such that the discussions and calculations can be considered in conjunction with the code procedures. To this end, it is important to note that this publication is not intended as a complete description of the code requirements or as a replacement for any of its provisions. The purpose of this book is mainly to provide background information on seismic design in general, and to offer discussions and comments on the use of EC8 in the design of buildings and their foundations.

Chapter 2

Seismic hazard and earthquake actions

Julian J. Bommer and Peter J. Stafford

CONTENTS

2.1 INTRODUCTION

Earthquake-resistant design can be considered as the art of balancing the seismic capacity of structures with the expected seismic demand to which they may be subjected. In this sense, earthquake-resistant design is the mitigation of seismic risk, which may be defined as the possibility of losses (human, social or economic) due to the effects of future earthquakes. Seismic risk is often considered as the convolution of seismic hazard, exposure and vulnerability. Exposure refers to the people, buildings, infrastructure, commercial and industrial facilities located in an area where earthquake effects may be felt; exposure is usually determined by planners and investors, although in some cases avoidance of major geo-hazards may lead to relocation of new infrastructure. Vulnerability is the susceptibility of structures to earthquake effects and is generally defined by the expected degree of damage that would result under different levels of seismic demand; this is the component of the risk equation that can be controlled by engineering design. Seismic hazards are the potentially damaging effects of earthquakes at a particular location, which may include surface rupture, tsunami runup, liquefaction and landslides, although the most important cause of damage on a global scale is earthquake-induced ground shaking (Bird and Bommer, 2004). The focus in this chapter is exclusively on this particular hazard and the definition of seismic actions in terms of strong ground motions. In the context of probabilistic seismic hazard analysis (PSHA), seismic hazard actually refers to the probability of exceeding a specific level of ground shaking within a given window of time.

If resources were unlimited, seismic protection would be achieved by simply providing as much earthquake resistance as possible to structures. In practice, it is not feasible to reduce seismic vulnerability to an absolute minimum because the costs would be prohibitive and certainly not justified since they would be for protection against a loading case that is unlikely to occur during the useful life of the structure. Seismic design therefore seeks to balance the investment in provision of seismic resistance against the level of damage, loss or disruption that earthquake loading could impose. For this reason, quantitative assessment and characterisation of the expected levels of ground shaking constitute an indispensable first step of seismic design, and it is this process of seismic hazard analysis that is introduced in this chapter.

The assessment of ground-shaking hazard due to future earthquakes invariably involves three steps: the development of a seismic source model that quantifies the rates of occurrence of future earthquake scenarios in the region; the development, or application, of a ground-motion model that defines the distribution of levels of shaking at a given site as a result of each of these earthquake scenarios; and, the integration of these two components into a model for the expected rates of exceedance of levels of shaking at the site of interest (Figure 2.1).

The first three sections of this chapter deal with the three steps illustrated in Figure 2.1, that is, seismic source models (Section 2.2), ground-motion models (Section 2.3) and seismic hazard analysis (Section 2.4). The remaining two sections then explore in more detail specific representations of the ground motion for engineering analysis and design, namely response spectra (Section 2.5) and acceleration time-histories (Section 2.6), both with specific reference to the stipulations of EC8. This chapter closes with brief conclusions and recommendations regarding both the use of EC8 as the basis for defining seismic design loads and possible improvements to the code that could be made in future revisions.

2.2 EARTHQUAKE PARAMETERS AND SEISMIC SOURCE MODELS

An entire book, let alone a chapter, could be dedicated to the issue of developing seismic source models. Herein, however, a very brief overview, with key references, is presented, with the aim of introducing definitions for the key parameters and the main concepts behind seismic source models.

With the exception of some classes of volcanic seismicity and very deep events, earthquakes are generally produced by sudden rupture of geological faults, resulting in the release of elastic strain energy stored in the surrounding crust, which then radiates from the fault rupture in the form of seismic waves. The location of the earthquake is specified by the location of the focus or hypocentre, which is the point on the fault where the rupture initiates and from where the first seismic waves are generated. This point is specified by the geographical coordinates of the epicentre, which is the projection of the hypocentre on the earth's surface, and the focal depth, which is the distance of the hypocentre below the earth's surface, measured in kilometres. Although for the purposes of observational seismology, using recordings obtained on sensitive instruments at distances of hundreds or thousands of kilometres from the earthquake, the source can be approximated as a point, it is

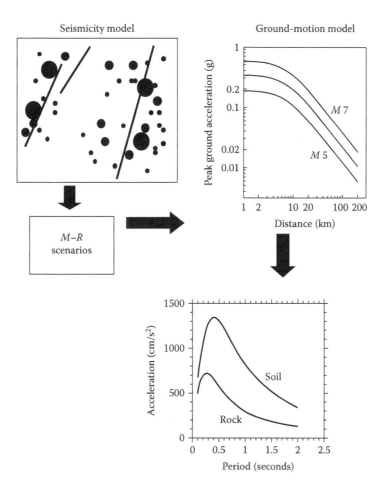

Figure 2.1 Schematic overview of seismic hazard analysis. The seismicity model defines scenarios of earth-quakes of magnitude, *M*, at a distance, *R*, from the site of interest, and the ground-motion model predicts the shaking parameter of interest for this *M–R* combination. The results in this case are expressed in terms of acceleration response spectra (see Chapter 3 for definition and detailed explanation of response spectra).

important to emphasise that in reality the earthquake source can be very large. The source is ultimately the part of the crust that experiences relaxation as a result of the fault slip; the dimensions of the earthquake source are controlled by the length of the fault rupture and, to a lesser extent, the amount of slip on the fault during the earthquake. The rupture and slip lengths both grow exponentially with the magnitude of the earthquake, as shown in Figure 2.2. Two good texts on the geological origin of earthquakes and the nature of fault-ing are Yeats et al. (1997) and Scholz (2002).

The magnitude of an earthquake is in effect a measure of the total amount of energy released in the form of seismic waves. There are several different magnitude scales, each of which is measured from the amplitude of different waves at different periods. The first magnitude scale proposed was the Richter scale, generally denoted by M_L, where the sub-script stands for local. Global earthquake catalogues generally report event size in terms of body-wave magnitude, m_b, or surface-wave magnitude, M_s, which will often give dif-ferent values for the same earthquake. All of the scales mentioned so far share a common

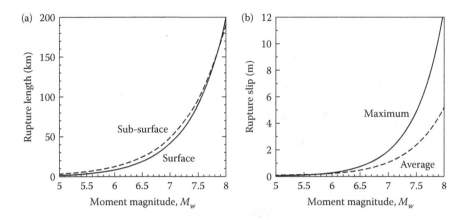

Figure 2.2 Median predicted values of rupture length (a) and slip (b) from the empirical equations of Wells and Coppersmith (1994).

deficiency in that they saturate at a certain size and are therefore unable to distinguish the sizes of the very largest earthquakes. This shortcoming does not apply to moment magnitude, designated as M_w or $\mathbf{M}$, which is determined from the very long-period part of the seismic radiation. This scale is based on the parameter seismic moment, which is the product of the area of the fault rupture, the average slip on the fault plane, and the rigidity of the crust.

A seismicity model needs to specify the expected location and frequency of future earthquakes of different magnitudes. A wide range of data can be used to build up seismic source models, generally starting with regional earthquake catalogues. Instrumental recordings of earthquakes are only available since the end of the nineteenth century and even then the sparse nature of early networks and low sensitivity of the instruments means that catalogues are generally incomplete for smaller magnitudes prior to the 1960s. The catalogue for a region can be extended through the study of historical accounts of earthquakes and the inference, through empirical relationships derived from twentieth century earthquakes, of magnitudes. For some parts of the world, historical seismicity can extend the catalogue from one hundred years to several centuries. The record can be extended even further through palaeo-seismological studies (McCalpin, 1996), which essentially means the field study of geological faults to assess the date and amplitude of previous co-seismic ruptures. Additional constraint on the seismicity model can be obtained from the tectonic framework and more specifically from the field study of potentially active structures and their signature in the landscape. Measurements of current crustal deformation, using traditional geodesy or satellite-based techniques, also provide useful input to estimating the total moment budget (e.g. Jackson, 2001).

The seismicity model needs to first specify the spatial distribution of future earthquake events, which is achieved by the definition of seismic sources. Where active geological faults are identified and their degree of activity can be characterised, the seismic sources will be lines or planes that reflect the location of these structures. Since in many cases active faults will not have been identified and also because it is generally not possible to unambiguously assign all events in a catalogue to known faults, source zones will often be defined. These are general areas in which it is assumed that seismicity is uniform in terms of mechanism and type of earthquake, and that events are equally likely to occur at any location within the source. Even where fault sources are specified, these will generally lie within areal sources that capture the seismicity that is not associated with the fault.

Once the boundaries of the source zones are defined, which fixes the spatial distribution of the seismicity model, the next step is to produce a model for the temporal distribution of seismicity. These models are generally referred to as recurrence models as they define the average rates of occurrence of earthquakes of magnitude greater than or equal to a particular value. The most widely used model is known as the Gutenberg–Richter (G–R) relationship, which defines a simple power law relationship between the number of earthquakes per unit time and magnitude. The relationship is defined by two parameters: the activity (i.e. the annual rate of occurrence of earthquakes of magnitude greater than or equal to zero or some other threshold level) and the b-value, which is the slope of the recurrence relation and defines the relative proportions of small and large earthquakes; b-values for large areas in much of the world are very often close to unity. The relationship must be truncated at an upper limit, M_{max}, which is the largest earthquake that the seismic source zone is considered capable of producing; this may be inferred from the dimensions of capable geological structures and empirical relations such as that shown in Figure 2.2 or simply by adding a small increment to the largest historical event in the earthquake catalogue. The typical form of the G–R relationship is illustrated in Figure 2.3a and d.

For major faults, it is believed that the G–R recurrence relationship may not hold and that large magnitude earthquakes occur quasi-periodically with relatively little activity at moderate magnitudes. This leads to alternative models, also illustrated in Figure 2.3b,c,e and f: if only large earthquakes occur, then the maximum magnitude model is adopted, whereas if there is also some activity in the smaller magnitude ranges, then a model is adopted, which combines a G–R relationship for lower magnitudes with the occurrence of larger characteristic earthquakes at higher rates than would be predicted by the extrapolation of

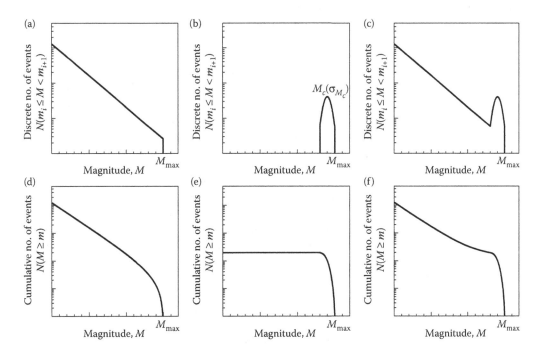

Figure 2.3 Typical forms of earthquakes recurrence relationships, shown in non-cumulative (a–c) and cumulative (d–f) formats. (a,d) G–R model, (b,e) maximum magnitude model and (c,f) characteristic earthquake model.

the G–R relationship. The recurrence rate of characteristic events will generally be inferred from palaeo-seismological studies rather than from the earthquake catalogue, since such earthquakes are generally too infrequent to have multiple occurrences in catalogues. Highly recommended references on recurrence relationships include Reiter (1990), Utsu (1999) and McGuire (2004).

2.3 GROUND-MOTION CHARACTERISATION AND PREDICTION

The crux of specifying earthquake actions for seismic design lies in estimating the ground motions caused by earthquakes. The inertial loads that are ultimately induced in structures are directly related to the motion of the ground upon which the structure is built. The present section is concerned with introducing the tools developed, and used, by engineering seismologists for the purpose of relating what occurs at the source of an earthquake to the ground motions that can be expected at any given site.

2.3.1 Accelerograms: Recording and processing

Most of the developments in the field of engineering seismology have spawned from the acquisition of high-quality recordings of strong ground-motions using accelerographs. The first of these was not obtained until March 1933 during the Long Beach, California, earthquake but since that time thousands of strong-motion records have been acquired through various seismic networks across the globe. Prior to the acquisition of the first accelerograms, recordings of earthquake ground-motions had been made using seismographs but the relatively high sensitivity of these instruments precluded truly strong ground motions from being recorded. It was not until the fine balance between creating a robust yet sensitive instrument was achieved, through the invention of the accelerograph, that the field of engineering seismology was born.

Accelerographs currently come in two main forms: analogue and digital. The first instruments were analogue and, while modern instruments are now almost exclusively digital, many analogue instruments remain in operation and continue to provide important recordings of strong ground motions. The records obtained from both types of instrument must be processed before being used for most applications. Accelerographs simultaneously record accelerations with respect to time in three orthogonal directions (usually two in the horizontal plane and one vertical) yet, despite this configuration, it is never possible to fully capture the true three-dimensional motion of the ground as the instruments do not 'see' all of the ground motion. The acceleration time-series that are recorded may be viewed in the frequency domain following a discrete Fourier transform. Upon performing this operation and comparing the recorded Fourier amplitude spectrum with the spectrum associated with the background noise recorded by the instrument, one finds that all accelerographs have a finite bandwidth over which the signal-to-noise ratio is sufficiently high that one can be confident that the recorded motions are genuinely associated with earthquake-induced ground shaking. Beyond the lower and upper limits of this bandwidth, and even at the peripheries if proper filtering is not performed, the record may become contaminated by noise. Boore and Bommer (2005) provide extensive guidance on how one should process accelerograms in order to ensure that the records are not contaminated. Boore and Bommer (2005) highlight the fundamental importance of applying an appropriate low-cut filter, particularly when using an accelerogram to obtain spectral ordinates of displacement. However, the key issue is to identify the maximum period up to which the filtered data can be reliably used.

Figure 2.4 Acceleration, velocity and displacement from analogue (a) and digital (b) recordings. Grey traces were obtained from the originally recorded records by removing the overall mean and the pre-event mean for the analogue and digital records, respectively. The black traces show the velocities and displacements derived from acceleration time series filtered as indicated. The displacement axis labels for the unfiltered motions (grey) are given on the right side of the graphs. (Adapted from Boore, D.M and Bommer, J.J. 2005, *Soil Dynamics and Earthquake Engineering*, 25, 93–115).

Akkar and Bommer (2006) explored the usable period ranges for processed analogue and digital accelerograms and concluded that for rock, stiff and soft soil sites, analogue recordings can be used for determining the elastic response at periods up to 0.65, 0.65 and 0.7 of the long-period filter cut-off respectively, whereas for digital recordings these limits increase to 0.8, 0.9 and 0.97. This issue is of great relevance as displacement-based design methods (Priestley et al., 2007), which rely upon the specification of long-period displacement spectral ordinates, become more widely adopted. An example of the influence of proper record processing is shown in Figure 2.4 in which both an analogue and a digital record are shown before and after processing – this example clearly shows how sensitive the displacement is to the presence of noise.

2.3.2 Ground-motion parameters

Once an accelerogram has been recorded and properly processed, many quantitative parameters of the ground motion may be calculated (for a description of many of these, see Kramer, 1996). Each of these parameters provides information about a different characteristic of the recorded ground motion. As far as engineering design is concerned, very few of these parameters are actually considered or used during the specification of design loads. Of those that may be calculated, peak ground acceleration (PGA) and ordinates of 5% damped elastic acceleration response spectra, $S_a(T, \xi = 5\%)$, have been used by far the most frequently.

Figure 2.5 shows many of the possible ground-motion parameters that may be calculated for an individual earthquake record. Each one of these descriptive parameters provides some

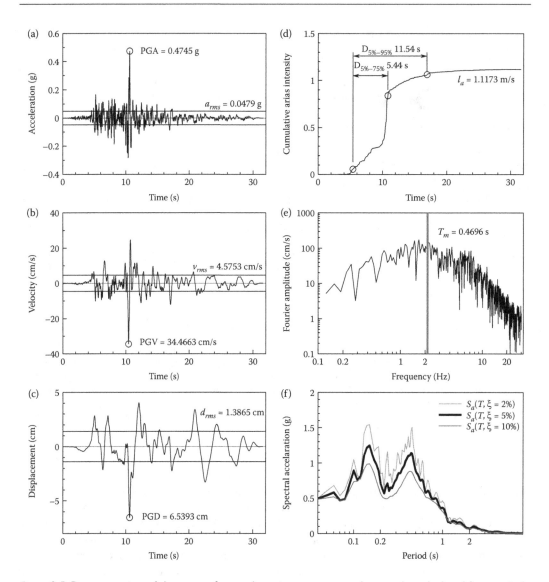

Figure 2.5 Demonstration of the types of ground-motion parameters that may be calculated from a single record. The record in this case is the 020° component recorded during the 1994 Northridge earthquake at the Saturn St. station in Los Angeles. The three panels (a–c) show the acceleration, velocity and displacement time series, respectively, as well as the peak and root-mean-square (*rms*) values. The panels (d–f) show, a Husid plot of the build up of Arias intensity as well as significant durations between 5%–75% and 5%–95% of the total Arias intensity, (e) the Fourier amplitude spectrum along with the mean period and (f) the acceleration response spectrum at damping levels of 2%, 5% and 10% of critical, respectively.

degree of information that may be used to help understand the demands imposed upon a structure. Although methodological frameworks are in place to simultaneously specify more than one ground-motion parameter (Bazzurro and Cornell, 2002) and to carry these parameters through to a structural analysis (Shome and Cornell, 2006) the additional complexity that is required for their implementation is excessively prohibitive without justifiable benefit in many cases. However, it is inevitable that earthquake engineers will seek to account for more characteristics of ground motions in the future.

2.3.3 Empirical ground-motion prediction equations

We have seen the numerous options that are available for describing the characteristics of ground motions in the previous section. Now, given a large number of records, one can calculate values for any of these parameters and develop a mathematical model to relate these values with any other parameter relevant to this suite of records, such as the magnitude of the earthquake from which they came. This type of reasoning is the basis for the development of empirical predictive equations for strong ground motions. Usually, a relationship is sought between a suite of observed ground-motion parameters and an associated set of independent variables, including a measure of the size of the earthquake, a measure of distance from the source to the site, some classification of the style-of-faulting involved and some description of the geological and geotechnical conditions at the recording site. An empirical ground-motion prediction equation is simply a mathematical function of these independent variables that provides an estimate of the expected value of the ground-motion parameter in consideration as well as some measure of the distribution of values about this expected value.

Thus far the development of empirical ground-motion prediction equations has been almost exclusively focussed upon the prediction of peak ground motions, particularly PGA and, to a far lesser extent, PGV, and ordinates of 5% damped elastic acceleration response spectra (Douglas, 2003; Bommer and Alarcón, 2006). Predictive equations have also been developed for most of the other parameters of the previous section, as well as others not mentioned, but as seismic design actions have historically been derived from PGA or $S_a(T)$ the demand for such equations is relatively weak. However, the performance of PGA (Wald et al., 1999) and, to a lesser extent, even $S_a(T)$ (Priestley, 2003; Akkar and Özen, 2005) for the purposes of predicting structural damage has been questioned. Improvements in the collaboration between engineering seismologists and structural earthquake engineers has prompted the emergence of research into what really are the key descriptors (such as inelastic spectral ordinates and elastic spectral ordinates for damping ratios other than 5%) of the ground motion that are of importance to structural response and to the assessment of damage in structures (Bozorgnia et al., 2006; Tothong and Cornell, 2006), although their uptake remains limited in practice.

Regardless of the ground-motion measure in consideration, a ground-motion prediction equation can be represented as a generic function of predictor variables, $\mu(M,R,\theta)$, and a variance term, $\varepsilon\sigma_T$, as in Equation 2.1, where M represents magnitude, R is a measure of distance and θ denotes a vector of any other predictor variables used by the prediction equation.

$$\log y = \mu(M, R, \theta) + \varepsilon\sigma_T \tag{2.1}$$

Many developers of ground-motion prediction equations attempt to assign physical significance to the terms in the empirically derived function $\mu(M,R,\theta)$. In some cases, it is possible to derive theoretical equations that may be used as the basis for selecting appropriate functional forms (e.g. Douglas, 2002). Although these theoretical considerations enable us to select appropriate functional forms, once the regression analysis has been conducted the actual values of regression coefficients should not be interpreted as having physical meaning as correlations of varying degrees always exist between the coefficients for different terms of the model.

For most ground-motion measures, the values will increase with increasing magnitude and decrease with increasing distance. These two scaling effects form the backbone of prediction equations and many functional forms have been proposed to capture the variation of motions with respect to these two predictors (Douglas, 2003). For modern relationships distinctions are also made between ground motions that come from earthquakes having different styles of faulting, with reverse faulting earthquakes tending to generate larger ground motions than either strike-slip or normal faulting events (Bommer et al., 2003). Historically, account was also taken for site conditions by adding modifying terms similar to those used for the style-of-faulting effects – stiff soil sites have larger motions than rock and soft soil sites have larger motions still. In Europe, this use of dummy variables for generic site classes (e.g. Ambraseys et al., 2005; Akkar and Bommer, 2007a,b) remained the adopted approach up until the latest generation of prediction equations were released (Douglas et al., 2014). However, these new ground-motion models are not yet routinely being used to define hazard maps in European national annexes. The new European models now follow the lead of researchers in the US and site response is now modelled using the average shear-wave velocity over the upper 30 m, as first introduced by Boore et al. (1997). Furthermore, the influence of non-linear site response, whereby weaker motions tend to be amplified more so than stronger motions due to the increased damping and reduced strength associated with the latter, is also taken into account (Abrahamson and Silva, 1997; Choi and Stewart, 2005). Figure 2.6 demonstrates the form of the non-linear site amplification functions adopted in two prediction equations developed as part of the next generation of attenuation relations (NGA-West 1) project in the United States. The difference in site amplification relative to rock for sites with differing shear-wave velocities and varying input rock ground motion is striking, with both models predicting de-amplification at strong levels of input rock motion.

In addition to the basic scaling of ground motions with magnitude, distance, site conditions, etc., there are additional seismological circumstances that may result in ground motions that differ from the scaling implied by basic source, path and site scaling that are commonly either omitted from developed equations or are later applied as correction factors to the base models. The most common examples include accounting for differences between sites located on the hanging or foot wall of dip-slip fault sources (Abrahamson and Somerville, 1996; Chang et al., 2004), accounting for rupture directivity effects

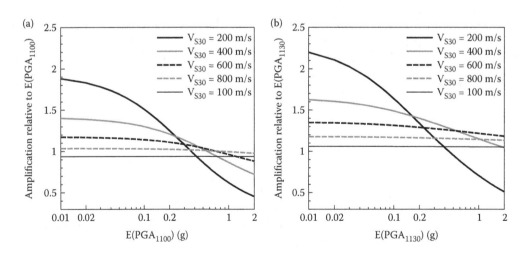

Figure 2.6 Comparison of two non-linear site response models for peak ground acceleration. Both models are from the NGA project with Abrahamson and Silva (2007) and Chiou and Youngs (2006) on parts (a) and (b), respectively. The Abrahamson and Silva (2007) model shows amplification with respect to the expected value of PGA at a site with $V_{S30} = 1100$ m/s while the Chiou and Youngs (2006) model shows the amplification with respect to expected motions on a site with $V_{S30} = 1130$ m/s.

(Somerville et al., 1997; Abrahamson, 2000), including models for the velocity pulse associated with directivity effects (Bray and Rodriguez-Marek, 2004), basin effects (Choi et al., 2005) and topographic modifiers (Toshinawa et al., 2004). The most recent predictor variable to be included in prediction equations for peak ground motions and spectral ordinates is the depth to the top of the rupture (Kagawa et al., 2004; Somerville and Pitarka, 2006). Currently, none of these effects are incorporated into any predictive equations for ground motions in Europe. Again, this is primarily a result of the lack of well-recorded strong earthquakes in the region. As the European strong-motion recording network continues to develop, our ability to capture some of these additional characteristics of ground motions improves.

2.3.4 Ground-motion variability

For any particular ground-motion record, the total variance term given in Equation 2.1 may be partitioned into two components as in Equation 2.2.

$$\log y_{ij} = \mu(m_i, r_{ij}, \theta_{ij}) + \delta_{e,i} + \delta_{a,ij} \tag{2.2}$$

The terms $\delta_{e,i}$ and $\delta_{a,ij}$ represent the inter-event and intra-event residuals respectively and quantify how far away from the mean estimate of $\log y_{ij}$ the motions from the ith event and the jth recording from the ith event are respectively (Abrahamson and Youngs, 1992). Alternatively, these terms may be expressed in terms of standard normal variates ($z_{e,i}$ and $z_{a,ij}$) and the standard deviations of the inter-event (τ) and intra-event (σ) components, that is, $\delta_{e,i} = z_{e,i}\tau$ and $\delta_{a,ij} = z_{a,ij}\sigma$. The total standard deviation for a predictive equation is obtained from the square root of the sum of the inter-event and intra-event variances, $\sigma_T^2 = \tau^2 + \sigma^2$. Later, in Section 2.4 regarding PSHA, mention will be made of epsilon, ε, representing the number of total standard deviations from the median predicted ground motion. Often ground motion modellers represent the terms $\delta_{e,i}$ and $\delta_{a,ij}$ by η_i and ε_{ij}, respectively. Under this convention, care must be taken to not confuse the epsilon, ε, with the intra-event residual, ε_{ij}, term – the two are related via the expression $\varepsilon = (\eta_i + \varepsilon_{ij})/\sigma_T$, that is, $\varepsilon = (\delta_{e,i} + \delta_{a,ij})/\sigma_T$ using our notation. There are also far more elaborate frameworks and naming conventions that have been proposed to deal with more sophisticated site-specific applications, for example, Al Atik et al. (2010), but a detailed discussion of these is far beyond the scope of this overview chapter.

Each of these components of variability may be modelled as functions of other parameters such as the magnitude of the earthquake (Youngs et al., 1995), the shear-wave velocity at the site (Abrahamson and Silva, 2007) or the amplitude of the ground motion (Campbell, 1997). Exactly how these components are calculated depends upon the regression methodology that is used to derive the equations. However, the most common approach is to adopt random effects procedures where the correlation between ground motions observed within any particular event is assumed to be the same across events and is equal to $\rho = \tau^2/(\tau^2 + \sigma^2)$. This concept is shown schematically in Figure 2.7.

Many people think of ground-motion variability as a measure of the lack of fit of a particular predictive equation. However, in most cases, it is better to think of a predictive equation as providing an estimate of the distribution of ground motions given a set of predictor variables such as magnitude and distance. From this perspective, the real misfit of the model is related to how well the model's distribution represents the true distribution of ground motions rather than how large are the variance components. People tend not to like large variability, reasoning that this implies that we cannot predict this measure of ground

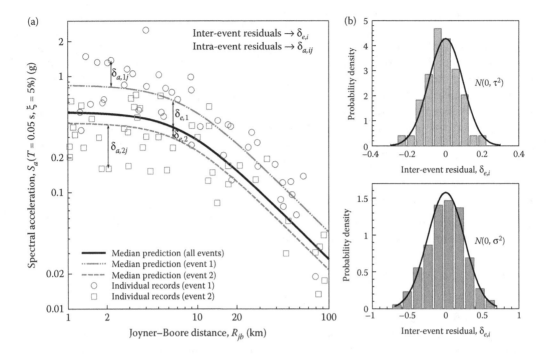

Figure 2.7 Explanation of the variance components specified in ground-motion prediction equations. The part (a) shows how the median prediction for an individual event may be higher or lower than the median prediction for all events – the inter-event residuals, $\delta_{e,i}$. About this median prediction for each event are random variations in ground motion – the intra-event residuals, $\delta_{a,ij}$. The histograms on part (b) show how both the inter- and intra-event residuals are normally distributed with zero means and variances of τ^2 and σ^2, respectively. The median predictions are generated for an M_w 6.5 earthquake with an R_{JB} distance of 10 km for strike-slip faulting and rock conditions using the equations of Akkar and Bommer (2007b). (Based on a concept from Youngs et al. 1995, *Bulletin of the Seismological Society of America*, 85, 1161–1176.)

motion with much certainty. However, this perspective is closely related to the paradigm that ground motions are ultimately predictable and that it is only through a result of inadequate modelling and incomplete knowledge that the apparent variability arises. If, on the other hand, one views ground motions as being inherently unpredictable (beyond a certain resolution), then one must view the variability not as a measure of the misfit, but rather as an additional part of the model that describes the range of observable ground motions given an event. Under this latter paradigm, there is no reason to like or dislike a particular ground-motion measure simply because predictive equations for this measure have a broad distribution. The only rational basis for judging the importance of a ground-motion measure is to assess the ability of this measure to accurately predict structural response. That said, in most cases, less variability in the ground motion estimate will translate into less variability in the response.

2.4 SEISMIC HAZARD ANALYSIS

The primary objective of engineering seismology is to enable seismic hazard analyses to be conducted. The two previous sections have provided most of the essential background required to understand seismic hazard analysis at its most basic level. As will soon be demonstrated, the mechanics of hazard analysis are relatively straightforward. However, a thorough understanding of the concepts laid out in the sections thus far, as well as many others, is a prerequisite for conducting a high-quality hazard analysis. Unfortunately, in current practice, this prerequisite is all too often not met.

2.4.1 Probabilistic versus deterministic approaches

Bommer (2002) presents a comprehensive discussion of the differences and similarities between probabilistic and deterministic approaches to seismic hazard analysis. While the proponents of deterministic methods would like to perpetuate the conception that there is ongoing academic debate regarding which is the superior method, the truth of the matter is that deterministic seismic hazard analysis (DSHA) is simply a special case of probabilistic seismic hazard analysis (PSHA) in which only a small number of earthquake scenarios (combinations of magnitude, distance and epsilon) are considered. In contrast, in PSHA all possible scenarios that are deemed to be of engineering interest are considered (Abrahamson, 2006; Bommer and Abrahamson, 2006). Much of the discussion regarding PSHA and DSHA has focussed on apparent issues that really stem from misunderstandings of the terminology that is often loosely used in PSHA (Bommer and Abrahamson, 2007). Bommer (2003) highlights some of the most common misunderstandings, particularly in relation to the treatment of uncertainty and urges the proponents of DSHA to try to develop a consistent set of terminology for their approaches.

2.4.2 Basics of PSHA, hazard curves and return periods

It is perhaps unfortunate that the mathematical formulation of PSHA is somewhat intimidating for some, as the mechanics behind the framework are actually very simple. For example, imagine one wanted to know how often a particular level of some ground-motion measure is exceeded at a site. Now, suppose that there is a seismic source near this site that regularly generates earthquakes of a particular magnitude and further suppose that the rate at which these earthquakes occur may be quantified. Once this rate is obtained, it may be combined with an estimate of how often the ground-motion level at the site is exceeded when this earthquake scenario occurs. For example, an event of magnitude M may occur once every six months and each time it does there is a 50% chance of exceeding a target ground-motion – this target level is then exceeded by this scenario, on average, once every year. If one then considered another earthquake scenario, and repeated the above procedure, one would determine how often the ground-motion level in consideration was exceeded for this alternative scenario. If the first scenario resulted in an exceedance of the ground-motion level λ_1 times per year and the second λ_2 times per year then for these two scenarios the ground-motion level is exceeded $\lambda_1 + \lambda_2$ times per year. This is how a PSHA is conducted: all one has to do to complete the process is to repeat the above steps for all of the possible earthquake scenarios that may affect the site, calculate the rates at which these scenarios result in ground motions above the target level, and then add them all up. Of course, it is not always straightforward to ascertain how often different earthquake scenarios occur, nor is it always obvious how to most appropriately determine the rate at which the ground motions are exceeded given these scenarios. However, none of these issues change the simplicity of the underlying framework that constitutes PSHA (Cornell, 1968; Cornell, 1971). With this simple explanation firmly in mind, it is now timely to relate this to what is more commonly seen in the literature on this subject.

Formally, basic PSHA may be represented as in Equation 2.3 (Bazzurro and Cornell, 1999):

$$\lambda_{GM}(gm^*) = \sum_i \left\{ \iiint I[GM > gm^* \mid m, r, \varepsilon] v_i f_{M,R,E}(m, r, \varepsilon)_i \, dm \, dr \, d\varepsilon \right\} \tag{2.3}$$

where the capital letters represent random variables (GM = a chosen ground-motion parameter, M = magnitude, R = distance, and E = epsilon) while their lower-case counterparts represent realisations of these random variables. The total rate at which earthquakes occur having a magnitude greater than the minimum considered for source i is denoted by v_i (as this term is a constant for each source it may be taken outside of the triple integral, as is commonly done in many representations of this equation). The joint probability density function of magnitude, distance and epsilon is given by $f_{M,R,E}(m,r,\varepsilon)_i$ and $I[GM > gm^* \mid m,r,\varepsilon]$ is an indicator function equal to one if $GM > gm^*$ and zero otherwise. Finally, and most importantly, $\lambda_{GM}(gm^*)$ is the total annual rate at which the target ground-motion value, gm^*, is exceeded. This is often the way that PSHA is presented in the literature; however, the nature of the joint probability density function in magnitude, distance and epsilon may be intractable for the non-cognoscenti and it is consequently worth spending some time to describe this key term of Equation 2.3. Using some basic concepts of probability theory, we may decompose the joint probability density function (pdf) into more tractable parts as in Equation 2.4.

$$\underbrace{v_i f_{M,R,E}(m, r, \varepsilon)_i}_{\substack{\text{How many times per year do} \\ \text{all possible levels of ground} \\ \text{motion occur from source } i?}} = \underbrace{v_i f_M(m \mid \mathbf{x}_{hyp}) f_{\mathbf{X}_{hyp}}(\mathbf{x}_{hyp})}_{\substack{\text{How many times per year does an} \\ \text{earthquake of } M = m \text{ occur in source} \\ i \text{ with a hypocentre at } \mathbf{x}_{hyp}?}} \underbrace{f_R(r \mid m, \mathbf{x}_{hyp}, \theta_i)}_{\substack{\text{When this event occurs,} \\ \text{what sort of rupture does} \\ \text{it produce?}}} \underbrace{f_E(\varepsilon)}_{\substack{\text{How likely are} \\ \text{the possible } GM \\ \text{values for this} \\ \text{scenario?}}} \tag{2.4}$$

Each of these components of the joint pdf, while already annotated, deserves some additional comment and explanation:

$f_{\mathbf{X}_{hyp}}(\mathbf{x}_{hyp})$ – the pdf for an event having a hypocentre equal to $\mathbf{x}_{hyp}$, where $\mathbf{x}_{hyp}$ = (longitude, latitude, depth) is any position within source i. A common assumption that is made, and that was made in Cornell's original presentation of PSHA, is that hypocentres are equally likely to occur anywhere within a seismic source. This assumption requires the least amount of information regarding the nature of activity for the seismic source.

$f_M(m \mid \mathbf{x}_{hyp})$ – the conditional pdf of magnitude given the hypocentral position. In many hazard analyses, this term is not implicitly considered; instead analysts simply take the previous assumption that earthquakes may occur with equal probability anywhere within a seismic source and also assume that these events may have the full range of magnitudes deemed possible for the source. In this case this term is not conditioned upon the hypocentre position and one simply recovers $f_M(m)$, the pdf of magnitude. However, some analysts may wish to address this problem more thoroughly and make alternative assumptions using analyses such as those of Somerville et al. (1999) and Mai et al. (2005). For example, it may be assumed that large earthquakes tend to have relatively deep hypocentres and the pdf may be modified accordingly. The pdf of magnitude is often assumed to follow a doubly bounded exponential distribution for areal sources (Cornell and Vanmarcke, 1969); a modified form of the famous G–R equation (Gutenberg and Richter, 1944), and a characteristic distribution for fault sources (Schwartz and Coppersmith, 1984) as mentioned in Section 2.2. However, any distribution that relates the relative rates of occurrence of earthquakes of different sizes is permissible.

$f_R(r|m,\mathbf{x}_{hyp},\theta_i)$ – the conditional pdf of the distance measure used in the ground-motion prediction equation given the rupture surface of the earthquake. The rupture surface depends upon the hypocentre, the size of the event and various other parameters encapsulated in θ_i including the strike and dip of the fault plane (for fault sources), the depth boundaries of the seismogenic zone, the segment of the fault on which the rupture starts, etc. This term is important as it translates the assumptions regarding the potential locations of earthquakes into measures of distance that are appropriate for use in empirical prediction equations. Note that this term is necessarily different for each distance measure that is considered.

$f_E(\varepsilon)$ – the pdf of epsilon. It is important to note that this term is always simply the pdf of the standard normal distribution. For this reason, it is not necessary to make this a conditional pdf with respect to anything else. Although standard deviations from ground-motion predictive equations may be dependent upon predictor variables such as magnitude, the variable epsilon remains statistically independent of these other variables (Bazzurro and Cornell, 1999).

Given this more complete representation of Equation 2.4, one must now also modify the integral to be expressed in terms of the relevant variables in Equation 2.3. In reality, this is not at all cumbersome as the integrals are not evaluated analytically anyway and all that is required is to discretise the range of possible parameter values and to determine the contribution to the hazard from each permissible permutation of each parameter. The general process alluded to in the introductory example and elaborated upon in the above is further represented schematically in Figure 2.8. In this figure, the method via which the probability that the ground motion exceeds the target level is represented in two ways: (1) in

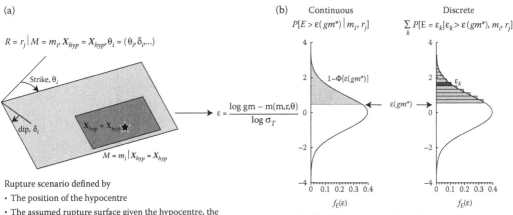

Figure 2.8 Schematic representation of the PSHA process. (a) A portion (dark grey) of a fault source (light grey) ruptures about the hypocentral position given by the star. The geometry of this rupture surface depends upon various characteristics of the source as well as the magnitude of the earthquake. (b) The probability of the target ground-motion (gm^*) being exceeded given this scenario is shown using two equivalent approaches.

a continuous manner through the use of the cumulative distribution function of the standard normal distribution, and (2) in a discrete manner whereby the range of epsilon values is discretised and the contribution to the total hazard is determined for each increment. Both of these approaches will give very similar answers but the latter approach offers advantages in terms of later representing the total hazard and also for the selection of acceleration time histories to be used in seismic design (McGuire, 1995; Bazzurro and Cornell, 1999; Baker and Cornell, 2006).

Thus far, we have only been concerned with calculating the rate at which a single target ground motion is exceeded. If we now select a series of target ground-motion levels and calculate the total rate at which each level is exceeded, we may obtain a hazard curve, which is the standard output of a PSHA, that is, a plot of λ_{GM} (gm^*) against gm^*. Examples of the form of typical hazard curves are given in Figure 2.9 where the ground-motion measure in this case is PGA.

The curves shown in Figure 2.9 demonstrate the strong influence that the aleatory variability in the ground-motion prediction equation has on the results of a seismic hazard analysis. Bommer and Abrahamson (2006) discussed this issue in detail, reviewing the historical development of PSHA as well as bringing to light the reason why modern hazard analyses often lead to higher hazard estimates. The answer to this question often lies in the inappropriate treatment, in early studies, of the aleatory variability in ground-motion prediction equations, with the worst practice being to simply ignore this component of PSHA in a manner akin to most deterministic hazard analyses.

Once a hazard curve has been developed, the process of obtaining a design ground motion is straightforward. The hazard curve represents values of the average annual rate of exceedance for any given ground-motion value. Under the assumption that ground motions may be described by a Poisson distribution over time, the average rate corresponding to the

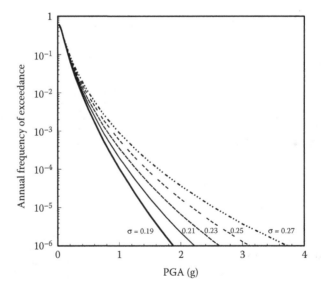

Figure 2.9 Example hazard curves for a fictitious site. Each hazard curve is calculated using a different value for the total standard deviation for the ground-motion prediction equation; the values presented on the figure correspond to typical values for prediction equations using base 10 logarithms. (Courtesy of Bommer, J.J. and Abrahamson, N.A. 2006, *Bulletin of the Seismological Society of America*, **96**, 1967–1977.)

probability of at least one exceedance within a given time period may be determined using Equation 2.5.

$$\lambda = \frac{-\ln(1-P)}{T} \tag{2.5}$$

For example, the ubiquitous, yet arbitrary (Bommer, 2006a), 475-year return period used in most seismic design codes throughout the world comes from specifying ground motions having a 10% chance of being exceeded at least once in any 50-year period. Inserting $P = 0.1$ and $T = 50$ years into Equation 2.5 yields the average annual rate corresponding to this condition, the reciprocal of which is the return period, that in this case is equal to 475 years. Note that because λ is a function of both P and T there are infinitely many combinations of P and T that result in a 475-year return period. Once this design criterion is specified, one simply finds the level of ground motion that corresponds to this rate on the hazard curve in order to obtain the design ground motion.

2.4.3 Uncertainty and logic trees

The PSHA methodology laid out thus far is capable of accounting for all of the aleatory variability that exists within the process. However, there is another important component of uncertainty that must also be accounted for – the uncertainty associated with not knowing the applicability of available models. This type of uncertainty is known as epistemic uncertainty within the context of PSHA. Aleatory variability and epistemic uncertainty can further be partitioned into modelling and parametric components as is described in Table 2.1 (here the focus is on ground-motion modelling, but the concepts hold for any other component of the PSHA process). These distinctions are not just semantics, each aspect of the overall uncertainty must be treated prudently and each must be approached in a different manner. As implied in Table 2.1, the logic-tree is the mechanism via which the epistemic uncertainty is usually accounted for in PSHA. As with any conceptual framework, practical application often reveals nuances that require further investigation and many such issues have been brought to light as a result high-level projects in practice, the PEGASOS project (Abrahamson et al., 2002) serving as a prime example. Aspects such as model selection,

Table 2.1 Proper partitioning of the total uncertainty associated with ground-motion modelling into distinct modelling and parametric components of both aleatory variability and epistemic uncertainty

	Aleatory variability	Epistemic uncertainty
Modelling	Variability based on the misfit between model predictions and observed ground-motions (unexplained randomness) σ_m	Uncertainty that the model is correct. Relative weights given to alternative credible models. (Alternative estimates of median ground-motions and σ_m)
Parametric	Variability based on propagating the aleatory variability of additional source parameters through a model (understood randomness) σ_p	Uncertainty that the distribution of the additional source parameters is correct. Relative weights given to alternative models of the parameter distributions. (Alternative estimates of σ_p for each model)
Total	$\sqrt{\sigma_m^2 + \sigma_p^2}$ (the modelling and parametric variabilities are uncorrelated)	Logic-trees for both components (the modelling and parametric logic-trees will be correlated)

Source: Adapted from Bommer, J.J. and Abrahamson, N.A. 2007, *Bulletin of the Seismological Society of America*, 97: 2208–2211.

Figure 2.10 Example of a suite of PGA hazard curves obtained from a logic tree for a fictitious site.

model compatibility and the overall sensitivity of PSHA to logic-tree branches for ground-motion models have all been addressed (Scherbaum et al., 2004a,b, 2005; Bommer et al., 2005; Sabetta et al., 2005; Beyer and Bommer, 2006; Cotton et al., 2006) as have issues associated with how the outputs (suites of hazard curves) of the logic-tree are harvested (Abrahamson and Bommer, 2005; McGuire et al., 2005; Musson, 2005).

Figure 2.10 shows a suite of hazard curves, including the mean, the median and four other fractiles, obtained from a hypothetical PSHA conducted using a logic tree. This figure highlights two important aspects associated with the outputs of logic trees: (1) the range of ground-motion values corresponding to a given hazard level may vary considerably across fractiles and (2) as one moves to longer return periods the difference between the mean and median hazard curves may become very large. The first aspect reinforces the importance of taking into account different interpretations of the regional seismotectonics as well as different models or approaches to estimating ground motions (see Table 2.1), while the second aspect demonstrates that one must be clear about how the design ground motion is to be specified as the results corresponding to the mean hazard and various fractiles may differ considerably.

2.4.4 Hazard maps and zonations

For the purpose of representing seismic hazard over a broad spatial region, separate hazard analyses are conducted at a sufficiently large number of points throughout the region such that contours of ground-motion parameters may be plotted. Such maps could be used directly for the specification of seismic design loads, but what is more common is to take these maps and to identify zones over which the level of hazard is roughly consistent. If the hazard map is produced with a high enough spatial resolution, then changes in hazard over small distances are always relatively subtle. However, for zonation maps, there will often be locations where small differences in position will mean the difference between being in one zone or another with the associated possibility of non-trivial changes in ground-motions.

Under such a circumstance, regulatory authorities must be take care in defining the boundaries of the relevant sources; the common practice is to adjust the zone limits to coincide with political boundaries in order to prevent ambiguity.

Prior to the introduction of EC8, a comparative analysis of the state of national hazard maps within 16 European countries was undertaken (García-Mayordomo et al., 2004). The study highlights the numerous methodological differences that existed between hazard maps developed for various countries across Europe. Many of the differences that exist do so as a result of the differing degrees of seismicity that exist throughout the region, but some of these differences were exacerbated as a result of parochialism despite geological processes not being concerned with human-made or political boundaries. There are, however, other examples of efforts that have been made to develop consistent seismic hazard maps over extended regions. The best example of a concerted effort to harmonise the characterisation of seismic hazard throughout Europe is provided by the SHARE project (http://www.share-eu.org). It is interesting to compare the outcomes of this recent project to two older examples of similar efforts: the GSHAP (Giardini et al., 1999) and SESAME (Jiménez et al., 2001) projects. An example of one of these older regional hazard maps may be viewed at http://www.seismo.ethz.ch/GSHAP/.

For truly robust hazard maps to be developed, the best of both approaches must be drawn upon. For example, ground-motion prediction equations developed from large regional datasets, such as those of Ambraseys et al. (2005) or Akkar and Bommer (2007a,b), are likely to be more robust when applied within individual countries than those developed from a more limited national dataset (Bommer, 2006b). Furthermore, ground-motion modellers working in low-seismicity regions, such as in most parts of Europe, often make inferences regarding the scaling of ground motions with magnitude on the basis of the small magnitude data that is available to them. In doing so, researchers find apparent regional differences that exist when making comparisons between their data and the predictions of regional ground-motion models derived from predominantly from recordings of larger magnitude earthquakes (i.e. Marin et al., 2004). Recent work has shown that such inferences may be unfounded and that particular care must be taken when extrapolating empirical ground-motion models beyond the range of magnitudes from which they were derived (Bommer et al., 2007). On the other hand, the detailed assessments of seismogenic sources that are often included for national hazard map and zonation purposes are often not fully incorporated into regional studies where the spatial resolution is relatively poor.

2.5 ELASTIC DESIGN RESPONSE SPECTRA

Most seismic design is based on representing the earthquake actions in the form of an equivalent static force applied to the structure. These forces are determined from the maximum acceleration response of the structure under the expected earthquake-induced ground shaking, which is represented by the acceleration response spectrum. The starting point is an elastic response spectrum, which is subsequently reduced by factors that account for the capacity of the structure to dissipate the seismic energy through inelastic deformations. The definition of the elastic response spectrum and its conversion to an inelastic spectrum are presented in Chapter 3; this section focuses on how the elastic design response spectra are presented in seismic design codes, with particular reference to EC8.

The purpose of representing earthquake actions in a seismic design code such as EC8 is to circumvent the necessity of carrying out a site-specific seismic hazard analysis for every engineering project in seismically active regions. For non-critical structures, it is generally considered sufficient to provide a zonation map indicating the levels of expected ground motions throughout the region of applicability of the code and then to use the parameters represented in these zonations, together with a classification of the near-surface geology, in order to construct the elastic design response spectrum at any given site.

2.5.1 Uniform hazard spectra and code spectra

The primary output from a PSHA is a suite of hazard curves for response spectral ordinates for different response periods. A design return period is then selected – often rather arbitrarily as noted previously (e.g. Bommer, 2006a) – and then the response parameter at this return period is determined at each response period and used to construct the elastic response spectrum. A spectrum produced in this way, for which it is known that the return period associated with several response periods is the same, is known as a uniform hazard spectrum (UHS) and it is considered an appropriate probabilistic representation of the basic earthquake actions at a particular location. The UHS will often be an envelope of the spectra associated with different sources of seismicity, with short-period ordinates controlled by nearby moderate-magnitude earthquakes and the longer-period part of the spectrum dominated by larger and more distant events. As a consequence, the motion represented by the UHS may not be particularly realistic if interpreted as a scenario spectrum and this becomes an issue when the motions need to be represented in the form of acceleration time histories, as discussed in Section 2.6. If the only parameter of interest to the engineer is the maximum acceleration that the structure will experience in its fundamental mode of vibration, regardless of the origin of this motion or any other of its features (such as duration), then the UHS is a perfectly acceptable format for the representation of the earthquake actions. In the following discussion, it is assumed that the UHS is a desirable objective.

Until the late 1980s, seismic design codes invariably presented a single zonation map, usually for a return period of 475 years, showing values of a parameter that in essence was the PGA. This value was used to anchor a spectral shape specified for the type of site, usually defined by the nature of the surface geology, and thus obtain the elastic design spectrum. In many codes, the ordinates could also be multiplied by an importance factor, which would increase the spectral ordinates (and thereby the effective return period) for the design of structures required to perform to a higher level under the expected earthquake actions, either because of the consequences of damage (e.g. large occupancy or toxic materials) or because the facility would need to remain operational in a post-earthquake situation (e.g. fire station or hospital).

A code spectrum constructed in this way would almost never be a UHS. Even at zero period, where the spectral acceleration is equal to PGA, the associated return period would often not be the target value of 475 years since the hazard contours were simplified into zones with a single representative PGA value over the entire area. More importantly, this spectral construction technique did not allow the specification of seismic loads to account for the fact that the shape of response spectrum varies with earthquake magnitude as well as with site classification (Figure 2.11), with the result that even if the PGA anchor value was associated with the exact design return period, it is very unlikely indeed that the spectral ordinates at different periods would have the same return period (McGuire, 1977). Consequently, the objective of a UHS is usually not met by anchoring spectral shapes to the zero-period acceleration.

Various different approaches have been introduced in order to achieve a better approximation to the UHS in design codes, generally by using more than one parameter to construct the spectrum. The 1984 Colombian and 1985 Canadian codes both introduced a second zonation map for peak ground velocity (PGV) and in effect used PGA to anchor the short-period part of the spectrum and PGV for the intermediate spectral ordinates. Since

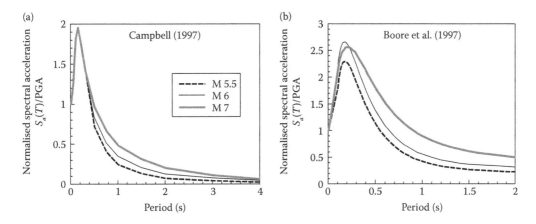

Figure 2.11 Median predicted response spectra, normalised to PGA, for a rock site at 10 km from earth-quakes of different magnitudes from the Californian equations of (a) Campbell (1997) and (b) Boore et al. (1997).

the zonation maps for the two parameters were different, the shape of the resulting elastic design spectrum varied from place to place, reflecting the influence of earthquakes of different magnitude in controlling the hazard. The 1997 edition of the Uniform Building Code (UBC) used two parameters, C_a and C_v, for the short- and intermediate-period portions of the spectra (with the subscripts indicating relations with acceleration and velocity) but curiously the ratio of the two parameters was the same in each zone with the result that the shape of the spectrum did not vary except with site classification.

In the Luso-Iberian peninsula, seismic hazard is the result of moderate-magnitude local earthquakes and large-magnitude earthquakes offshore in the Atlantic. The Spanish seismic code handles their relative influence by anchoring the response spectrum to PGA but then introducing a second set of contours, of a factor called the 'contribution coefficient' K that controls the relative amplitude of the longer-period spectral ordinates; high values of K occur to the west, reflecting the stronger influence of the large offshore events. The Portuguese seismic code goes one step further and simply presents separate response spectra, with different shapes, for local and distant events. The Portuguese code is an interesting case because it effectively abandons the UHS concept, although it is noteworthy that the return period of the individual spectra is 975 years, in effect twice the value of 475 years associated with the response spectra in most European seismic design codes (Figure 2.12).

Within the drafting committee for EC8, there were extensive discussions about how the elastic design spectra should be constructed, with the final decision being an inelegant and almost anachronistic compromise to remain with spectral shapes anchored only to PGA. In order to reduce the divergence from the target UHS, however, the code introduced two different sets of spectral shapes (for different site classes), one for the higher seismicity areas of southern Europe (Type 1) and the other for adoption in the less active areas of northern Europe (Type 2). The Type 1 spectrum is in effect anchored to earthquakes of magnitude close to $M_s \sim 7$ whereas the Type 2 spectrum is appropriate to events of M_s 5.5 (e.g. Rey et al., 2002). At any location where the dominant earthquake event underlying the hazard is different from one or other of these magnitudes, the spectrum will tend to diverge from the target 475-year UHS, especially at longer periods.

The importance of the vertical component of shaking in terms of the demand on structures is a subject of some debate (e.g. Papazoglou and Elnashai, 1996) but there are certain types of structures and structural elements, such as cantilever beams, for which the vertical

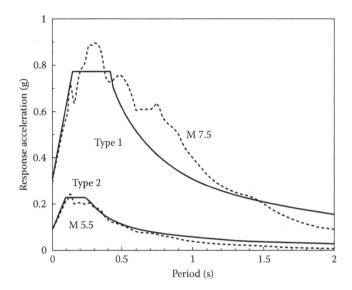

Figure 2.12 Median predicted spectral ordinates from the European ground-motion prediction equations of Ambraseys et al. (1996) for rock sites at 10 km from small and large magnitude events, compared with the EC8 Type 1 and 2 rock spectra anchored to the median predicted PGA.

loading could be important. Many seismic codes do not provide a vertical spectrum at all and those that do generally specify it as simply the horizontal spectrum with the ordinates reduced by one third. Near-source recordings have shown that the short-period motions in the vertical direction can actually exceed the horizontal motion, and it has also been clearly established that the shape of the vertical response spectrum is very different from the horizontal components of motion (e.g. Bozorgnia and Campbell, 2004). In this respect, EC8 has some merit in specifying the vertical response spectrum separately rather than through scaling of the horizontal spectrum; this approach was based on the work of Elnashai and Papazoglou (1997). As a result, at least for a site close to the source of an earthquake, the EC8 vertical spectrum provides a more realistic estimation of the vertical motion than is achieved in many seismic design codes (Figure 2.13).

2.5.2 The influence of near-surface geology on response spectra

The fact that locations underlain by soil deposits generally experience stronger shaking than rock sites during earthquakes has been recognised for many years, both from field studies of earthquake effects and from recordings of ground motions. The influence of surface geology on ground motions is now routinely included in predictive equations. The nature of the near-surface deposits is characterised either by broad site classes, usually defined by ranges of shear-wave velocities (V_s), or else by the explicit value of the V_s over the uppermost 30 m at the site. Figure 2.14 shows the influence of different soil classes on the predicted spectral ordinates from European attenuation equations and for two different magnitudes.

Code specifications of spectral shapes for different site classes generally reflect the amplifying effect of softer soil layers, resulting in increased spectral ordinates for such sites, and the effect on the frequency content, which leads to a wider constant acceleration plateau and higher ordinates at intermediate and long response periods. The EC8 Type 1 spectra for different site classes are illustrated in Figure 2.15.

Figure 2.13 The implied vertical-to-horizontal (V/H) ratio of the Type I spectra for soil sites in Eurocode 8 compared with the median ratios predicted by Bozorgnia and Campbell (2004) for soil sites at different distances from the earthquake source.

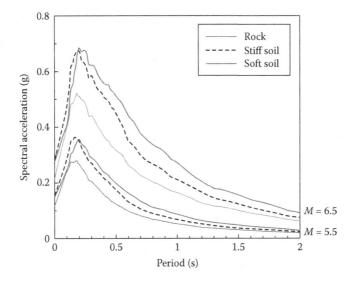

Figure 2.14 Median predicted spectral ordinates from the equations of Bommer et al. (2003) for different site classes at 10 km from strike-slip earthquakes of M_s 5.5 and M_s 6.5 as indicated.

As previously mentioned in Section 2.3.3, the response of a soil layer to motions propagating upwards from an underlying rock layer depends on the strength of the incoming rock motions as a result of the non-linear response of soil (see Figure 2.6). The greater the shear strain in the soil, the higher the damping and the lower the shear modulus of the soil, whence weak input motion tends to be amplified far more than stronger shaking. As a rule-of-thumb, non-linear soil response can be expected to be invoked by rock accelerations beyond 0.1–0.2 g (Beresnev and Wen, 1996). In recent years, ground-motion prediction equations developed for California have included the influence of soil non-linearity with greater ratios of soft soil to rock motions for magnitude–distance combinations resulting in weak rock motions

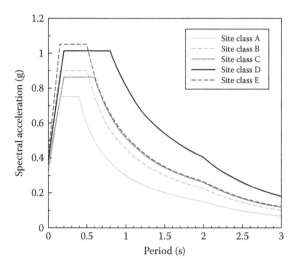

Figure 2.15 Type 1 spectra from Eurocode 8 for different site classes, anchored for a PGA in rock of 0.3 g.

than those for which strong shaking would be expected (Figure 2.6). Improvements in the numbers of accelerograms recorded throughout Europe coupled with improvements in the characterisation of the recording stations have recently enabled such effects to also be incorporated within European ground-motion models (see Douglas et al., 2014), although not all models have constrained this behaviour. To adequately capture these effects, and to enable more models to incorporate these effects through empirical means, good-quality data on site characteristics and more recordings of genuinely strong motion in the European area are needed. Earlier studies, such as that of Bommer et al. (2007) that looked specifically for evidence of non-linear response, were unable to find this within the European data at the time. Some design codes, most notably the 1997 edition of UBC, have included the effects of soil non-linearity in the specification of amplification factors for spectral loads. The implied amplification factors for rock motions from a few attenuation equations and design regulations for intermediate-period spectral response ordinates are compared in Figure 2.16.

A number of interesting observations can be made regarding the curves in Figure 2.16, the first being the wide range of proposed amplification factors for different sites, especially those overlain by soft soil layers. The second observation that can be made is that amplification factors assigned to broad site classes will often be rather crude approximations to those obtained for specific sites where the V_s profile is known. The UBC spectra for Zone 1 (low hazard) and Zone 4 (high hazard) have quite different amplification factors, with non-linear soil response leading to much lower soil amplification in the high hazard zone. A similar feature seems to be captured by the Type 1 and Type 2 spectra from EC8.

2.5.3 Displacement response spectra

In relatively recent years, exclusively force-based approaches to seismic design have been questioned, both because of the poor correlation between transient accelerations and structural damage, and also because for post-yield response the forces effectively remain constant and damage control requires limitation of the ensuing displacements. Most of the recently introduced performance-based design methodologies can be classified as being based either on displacement modification techniques or else equivalent linearisation. FEMA-440 (ATC, 2005) presents both approaches, allowing the designer to select the one felt to be

Figure 2.16 Amplification factors for 1.0-second spectral acceleration for different site shear-wave velocity values relative to rock motions; for Boore et al. (1997), rock has been assigned a shear-wave velocity of 800 m/s.

more appropriate, acknowledging, in effect, that opinions are currently divided as to which is the preferred approach. EC8 also envisages the potential application of these two general approaches to the computation of displacement demand, and provides guidelines on the appropriate seismic actions in informative annexes A and B.

The equivalent linearisation approach to displacement-based seismic design requires the characterisation of the design motions in the form of elastic displacement response spectra. The inelastic deformation of the structure is reflected in the longer effective period of vibration, which requires the spectral ordinates to be specified for a wider range of periods than has normally been the case in design codes. The dissipation of energy through hysteresis is modelled through an increased equivalent damping. Based on a proposal by Bommer et al. (2000), the EC8 acceleration spectrum can be transformed to a displacement spectrum multiplying the ordinates by $T^2/4\pi^2$, where T is the natural period of vibration. The critical question is at which period should the constant displacement plateau begin, which, as can be discerned in Figure 2.17, was set at 2 seconds for the Type 1 spectrum in EC8. This value has since been recognised to be excessively small; the corner period of the spectrum increases with earthquake magnitude, and for the larger events expected in Europe ($M \sim 7$), the period could be expected to be on the order of 10 seconds (e.g. Bommer and Pinho, 2006). The inadequacy of the corner period, T_D, being set at 2 seconds has been demonstrated by European equations developed after the release of EC8 for the prediction of response spectral ordinates up to 4 seconds (Akkar and Bommer, 2007b). Figure 2.17 compares the displacement spectra from EC8 with those from Akkar and Bommer (2007b) for stiff soil sites at 10 km from earthquakes of different magnitudes. In each case, the EC8 spectra has been anchored to the predicted median PGA from the equation of Akkar and Bommer (2007b). A number of interesting observations can be made, the first being that the fixed spectral shape of EC8 is unable to capture the influence of varying magnitude, with the result that the short-period spectral ordinates are severely over-predicted for the smaller magnitudes. The second observation is that the fixed corner period of 2 seconds is clearly inadequate and the dependence of this parameter on magnitude is very clear; for

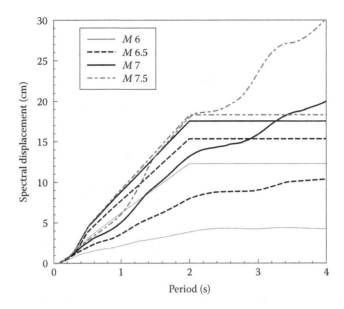

Figure 2.17 Comparison of 5%-damped displacement response spectra for a stiff soil site at 10 km from earthquakes of different magnitudes from Akkar and Bommer (2007b) with the EC8 Type I spectra for the same conditions, anchored to the PGA value predicted by the equation presented in the same study.

earthquakes of greater than magnitude 6, the corner period is longer than 2 seconds, and for the larger events greater than 4 seconds. The most recent European ground-motion models also corroborate these findings and continue to indicate a significant problem with the location of the start of this constant displacement plateau.

The spectral ordinates with damping ratios higher than the nominal 5% of critical are obtained by multiplying the spectral ordinates at intermediate periods by a factor, derived by Bommer et al. (2000), that is a function only of the target damping level. These factors replaced those in an early draft of EC8, and many other factors have since been proposed in the literature and in other seismic design codes. Bommer and Mendis (2005) explored the differences amongst the various factors and found that the amount of reduction of the 5%-damped ordinates required to match the ordinates at higher damping levels increases with the duration of the ground motion. Since the Type 2 spectrum in EC8 corresponds to relatively small magnitude earthquakes, which will generate motions of short duration, it was proposed that the existing scaling equation in EC8 to obtain spectral displacements (*SD*) and different damping values, ξ:

$$SD(\xi) = SD(5\%)\sqrt{\frac{10}{5+\xi}} \qquad (2.6)$$

should be retained for the Type 1 spectrum, whereas for the Type 2 spectrum this should be replaced by the following expression derived by Mendis and Bommer (2006):

$$SD(\xi) = SD(5\%)\sqrt{\frac{35}{30+\xi}} \qquad (2.7)$$

2.6 ACCELERATION TIME-HISTORIES

Although seismic design invariably begins with methods of analysis in which the earthquake actions are represented in the form of response spectra, some situations require fully dynamic analyses to be performed and in these cases the earthquake actions must be represented in the form of acceleration time-histories. Such situations include the design of safety-critical structures, highly irregular buildings, base-isolated structures, and structures designed for a high degree of ductility. For such projects, the simulation of structural response using a scaled elastic response spectrum is not considered appropriate and suites of accelerograms are required for the dynamic analyses. The guidance given in the majority of seismic design codes on the selection and scaling of suites of acceleration time-histories for such purposes is either very inadequate or else so prescriptive as to make it practically impossible to identify realistic accelerograms that meet the specified criteria (Bommer and Ruggeri, 2002). A point that cannot be emphasised too strongly is that time-histories should never be matched to a uniform hazard spectrum, but rather to a spectrum corresponding to a particular earthquake scenario that accounts for spectral shape implied by the hazard results, that is, a conditional (mean) spectrum (Baker, 2011). In the case of codes, this may be difficult since the code generally provides an approximation, albeit a crude one, to the UHS and offers no possibility to generate a disaggregated event-specific spectrum. However, there is no reason why this information could be provided through National Annexes.

There are a number of options for obtaining suites of acceleration time-histories for dynamic analysis of structures, including the generation of spectrum-compatible accelerograms from white noise, a method which is now widely regarded as inappropriate because the resulting signals are so unlike earthquake ground motions. The most popular option is to use real accelerograms, which can be selected either on the basis of having response spectra similar, at least in shape, to the elastic design spectrum, or else matching an earthquake scenario in terms of magnitude, source-to-site distance and possibly also site geology (Bommer and Acevedo, 2004). The latter approach, however, is generally not feasible in the context of seismic design code applications, because information regarding the underlying earthquake scenarios is usually not available to the user. Selecting records from earthquakes of appropriate magnitude is primarily an issue if the duration of the shaking (or some other metric related to energy content) is considered an important parameter in determining the degree of seismic demand that the records impose. This issue has received a significant amount of attention in the technical literature (e.g. Hancock and Bommer, 2006; Iervolino et al., 2006), and while some general consensus views have emerged, the issues are not resolved in all cases (Chandramohan et al., 2016).

Once a suite of records has been selected, ideally using a conditional (mean) spectrum or at least an earthquake scenario (rather than the approximate UHS suggested through EC8), the next question for the design engineer to address is how many records are needed. Most of the seismic design codes that address this issue, including EC8, specify that a minimum of three records should be used, and that if less than seven records are used then the maximum structural response must be used as the basis for design whereas if seven or more time-histories are employed then the average structural response can be used. The use of the maximum inelastic response obtained from dynamic analyses should result in estimates of the demand that are biased high with respect to the expected level of demand. However, it is not possible to appreciate how large this bias is likely to be without making some estimate of what the distribution of demands should be for this design scenario.

The key question then becomes how many records are required to obtain a stable estimate of the mean inelastic response, which will depend on how the records are adjusted so

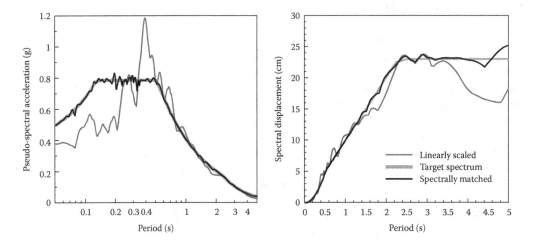

Figure 2.18 Comparison of the difference between scaled and matched spectra. (Modified from Hancock, J. et al. 2006, *Journal of Earthquake Engineering*, 10(special issue 1), 67–89.)

that their spectral ordinates approximate to those of the elastic design spectrum: the more closely the adjusted records match the target elastic design spectrum, the fewer analyses will be needed. Options include scaling the records to match the design spectrum at the natural period of the structure or scaling to match or exceed, the average ordinates over a period range around this value, the extended range accounting for both the contributions to the response from higher modes and also for the elongation of response period due to inelastic deformations. Scaling the records in amplitude is legitimate given that the whilst the amplitude of the motion is highly dependent on distance – especially within a few tens of kilometres from the source – the shape of the response spectrum is actually rather insensitive to distance over the range of distances of normal engineering interest (Bommer and Acevedo, 2004). Although scaling limits of a factor of 2 were proposed at one time, and became embedded in the 'folklore' of engineering practice, much larger scaling factors can be applied (Watson-Lamprey and Abrahamson, 2006). Adjusting records by scaling the time axis, however, is to be avoided.

An alternative to linear scaling of the records is to make adjustments, using fast Fourier transform or wavelet transformations, to achieve a spectral shape that approximates to that of the target design spectrum (Bommer and Acevedo, 2004). The most elegant way to achieve this is using the wavelet transformation, which minimises the alteration of the original accelerogram but at the same time can achieve a very good spectral match (Hancock et al., 2006). An example of the difference between linearly scaling a record and matching spectra through wavelet transformations is given in Figure 2.18. However, one must appreciate that while such adjustments are permitted by EC8 they do not appropriately reflect the inter-period variability that natural earthquake records possess (Stafford and Bommer, 2010).

2.7 CONCLUSIONS AND RECOMMENDATIONS

For most engineering projects in seismic zones, the earthquake loading can be represented by an acceleration response spectrum, modified to account for inelastic deformation of the structure. The elastic design spectrum will most frequently be obtained through probabilistic seismic hazard analysis, which provides the most rational framework for handling the large uncertainties associated with the models for seismicity and ground-motion prediction. Most seismic design codes present zonation maps and response spectra derived probabilistically, even though these design loads are often associated with a return period whose origin is a fairly arbitrary selection, and the resulting response spectrum is generally a poor approximation to the concept of a uniform hazard spectrum.

The main advantage that seismic codes offer in terms of earthquake loading is allowing the engineer to bypass the very considerable effort, expense and time required for a full site-specific hazard assessment. This should not, however, be interpreted to mean that the engineer should not be aware of the assumptions underlying the derivation and presentation of the earthquake actions, as well as their limitations.

EC8 is unique amongst seismic design codes in that it is actually a template for a code rather than a complete set of definitions of earthquake actions for engineering design. Each member state of the European Union has to produce its own National Application Document, including a seismic hazard map showing PGA values for the 475-year return period, select either the Type 1 or Type 2 spectrum and, if considered appropriate, adapt details of the specification of site classes and spectral parameters. Interestingly, although the stated purpose of EC8 is harmonisation of seismic design across Europe, there could well be jumps in the level of seismic design loads across national borders as currently there is no official project for a community-wide hazard zonation map (although the SHARE project clearly focussed upon achieving this objective; http://www.share-eu.org).

Although there are a number of innovative features in EC8 with regards to the specification of design earthquake actions, such as the separate definition of the vertical response spectrum and the provision of input for displacement-based design approaches, the basic mechanism for defining the horizontal elastic design spectrum is outdated and significantly behind innovations in recent codes from other parts of the world, most notably the United States. It is to be hoped that the first major revision of EC8, will modify the spectral construction technique, incorporating at least one more anchoring parameter in addition to PGA. Several other modifications are also desirable, including to the long-period portion of the displacement spectrum and the adjustment for damping levels higher than 5% of critical.

Although seismic codes provide useful guidance for the earthquake-resistant design of many structures, there are cases where the code specifications will not be sufficient. Examples may include the following:

- Projects located in proximity to active faults for which near-source directivity effects associated with the fault rupture need to be considered in the design (such effects are considered in the 1997 edition of UBC but not in EC8).
- Projects in areas where active faults are known or suspected to be present, and for which surface displacements would be a critical consideration for the performance of the structure.
- Projects on sites with deep and/or very soft soil deposits, for which the effects of the near-surface geology on the ground motions are unlikely to be well captured by the simplified site classes and corresponding spectral shapes in the code.
- Projects for which return periods significantly longer than the nominal 475 years are considered appropriate.
- Any project for which fully dynamic analysis is required (since the EC8 guidelines on preparing time-history input for such analyses is lacking in many respects).

If it is judged that a site-specific seismic hazard assessment is required, then this needs to be planned carefully and in good time – it should be considered as an integral part of the site investigation, and scheduled and budgeted accordingly. If investigations of active geological faults are to be part of the assessment, then the time and budget requirements are likely to increase very significantly.

Seismic hazard analysis is a highly specialised discipline that is constantly evolving and advancing, and in which a great deal of expert judgement is required. Nowadays it is fairly straightforward to obtain geological maps, satellite imagery, earthquake catalogues, published ground-motion prediction equations and software for performing hazard calculations, in many cases from the Internet and free of charge. The art of seismic hazard analysis, however, lies not primarily in accessing and analysing these resources but rather in judging their completeness and quality, and assessing the uncertainties associated with the data and the applicability of models to the specific region and site under consideration.

REFERENCES

Abrahamson, N.A. 2000, Effects of rupture directivity on probabilistic seismic hazard analysis, *Proceedings of the 6th International Conference on Seismic Zonation*, Palm Springs, CA.

Abrahamson, N.A. 2006, Seismic hazard assessment: Problems with current practice and future developments, *First European Conference on Earthquake Engineering and Seismology*, Geneva, Switzerland.

Abrahamson, N.A., Birkhauser, P., Koller, M., Mayer-Rosa, D., Smit, P., Sprecher, C., Tinic, S. and Graf, R. 2002, PEGASOS: A comprehensive probabilistic seismic hazard assessment for nuclear power plants in Switzerland, *12th European Conference on Earthquake Engineering*, London, U.K.

Abrahamson, N.A. and Bommer, J.J. 2005, Probability and uncertainty in seismic hazard analysis, *Earthquake Spectra*, 21:603–607.

Abrahamson, N.A. and Silva, W.J. 1997, Empirical response spectral attenuation relations for shallow crustal earthquakes, *Seismological Research Letters*, 68:94–127.

Abrahamson, N.A. and Silva, W.J. 2007, Abrahamson & Silva NGA ground motion relations for the geometric mean horizontal component of peak and spectral ground motion parameters, *Interim Reports of Next Generation Attenuation (NGA) Models*, Pacific Earthquake Engineering Research Center, Richmond, CA.

Abrahamson, N.A. and Somerville, P.G. 1996, Effects of the hanging wall and footwall on ground motions recorded during the Northridge earthquake, *Bulletin of the Seismological Society of America*, 86:S93–S99.

Abrahamson, N.A. and Youngs, R.R. 1992, A stable algorithm for regression analyses using the random effects model, *Bulletin of the Seismological Society of America*, 82:505–510.

Akkar, S. and Bommer, J.J. 2006, Influence of long-period filter cut-off on elastic spectral displacements, *Earthquake Engineering and Structural Dynamics*, 35:1145–1165.

Akkar, S. and Bommer, J.J. 2007a, New empirical prediction equations for peak ground velocity derived from strong-motion records from Europe and the Middle East, *Bulletin of the Seismological Society of America*, 97:511–530.

Akkar, S. and Bommer, J.J. 2007b, Prediction of elastic displacement response spectra in Europe and the Middle East, *Earthquake Engineering and Structural Dynamics*, 36:1275–1301.

Akkar, S. and Özen, O. 2005, Effect of peak ground velocity on deformation demands for SDOF systems, *Earthquake Engineering and Structural Dynamics*, 34:1551–1571.

Al Atik, L., Abrahamson, N., Bommer, J.J., Scherbaum, F., Cotton, F. and Kuehn, N. 2010, The variability of ground-motion prediction models and its components, *Seismological Research Letters*, 71:794–801.

Ambraseys, N.N., Douglas, J., Sarma, S.K. and Smit, P.M. 2005, Equations for the estimation of strong ground motions from shallow crustal earthquakes using data from Europe and the Middle East: Horizontal peak ground acceleration and spectral acceleration, *Bulletin of Earthquake Engineering*, 3:1–53.

Ambraseys, N.N., Simpson, K.A. and Bommer, J.J. 1996, The prediction of horizontal response spectra in Europe, *Earthquake Engineering and Structural Dynamics*, 25:371–400.

ATC 2005, *Improvement of Nonlinear Static Seismic Analysis Procedures, FEMA-440*, Redwood, California: Applied Technology Council.

Baker, J.W. 2011, Conditional mean spectrum: Tool for ground motion selection, *Journal of Structural Engineering – ASCE*, 137:322–331.

Baker, J.W. and Cornell, C.A. 2006, Spectral shape, epsilon and record selection, *Earthquake Engineering and Structural Dynamics*, 35:1077–1095.

Bazzurro, P. and Cornell, C.A. 1999, Disaggregation of seismic hazard, *Bulletin of the Seismological Society of America*, 89:501–520.

Bazzurro, P. and Cornell, C.A. 2002, Vector-valued probabilistic seismic hazard analysis (VPSHA), *Proceedings of the Seventh U.S. National Conference on Earthquake Engineering (7NCEE)*, Boston, Massachusetts, USA.

Beresnev, I.A. and Wen, K.L. 1996, Nonlinear soil response – A reality?, *Bulletin of the Seismological Society of America*, 86:1964–1978.

Beyer, K. and Bommer, J.J. 2006, Relationships between median values and between aleatory variabilities for different definitions of the horizontal component of motion, *Bulletin of the Seismological Society of America*, 96:1512–1522.

Bird, J.F. and Bommer, J.J. 2004, Earthquake losses due to ground failure, *Engineering Geology*, 75:147–179.

Bommer, J.J. 2002, Deterministic vs. probabilistic seismic hazard assessment: An exaggerated and obstructive dichotomy, *Journal of Earthquake Engineering*, 6:43–73.

Bommer, J.J. 2003, Uncertainty about the uncertainty in seismic hazard analysis, *Engineering Geology*, 70:165–168.

Bommer, J.J. 2006a, Re-thinking seismic hazard mapping and design return periods, *First European Conference on Earthquake Engineering and Seismology*, Geneva, Switzerland.

Bommer, J.J. 2006b, Empirical estimation of ground motion: Advances and issues, *Third International Symposium on the Effects of Surface Geology on Seismic Motion*, Grenoble, France.

Bommer, J.J. and Abrahamson, N.A. 2006, Why do modern probabilistic seismic-hazard analyses often lead to increased hazard estimates?, *Bulletin of the Seismological Society of America*, 96:1967–1977.

Bommer, J.J. and Abrahamson, N.A. 2007, Reply to "Comment on 'Why do modern probabilistic seismic hazard analyses often lead to increased hazard estimates?' by Julian J. Bommer and Norman A. Abrahamson or 'How not to treat uncertainties in PSHA'" by J.-U. Klügel, *Bulletin of the Seismological Society of America*, 97: 2208–2211.

Bommer, J.J. and Acevedo, A.B. 2004, The use of real earthquake accelerograms as input to dynamic analyses, *Journal of Earthquake Engineering*, 8(special issue 1):43–91.

Bommer, J.J. and Alarcón, J.E. 2006, The prediction and use of peak ground velocity, *Journal of Earthquake Engineering*, 10:1–31.

Bommer, J.J., Douglas, J. and Strasser, F.O. 2003, Style-of-faulting in ground-motion prediction equations, *Bulletin of Earthquake Engineering*, 1:171–203.

Bommer, J.J., Elnashai, A.S. and Weir, A.G. 2000, Compatible acceleration and displacement spectra for seismic design codes. *Proceedings of 12th World Conference on Earthquake Engineering*, Auckland, New Zealand, paper no. 207.

Bommer, J.J. and Mendis, R. 2005, Scaling of spectral displacement ordinates with damping ratios, *Earthquake Engineering and Structural Dynamics*, 33:145–165.

Bommer, J.J. and Pinho, R. 2006, Adapting earthquake actions in Eurocode 8 for performance-based seismic design, *Earthquake Engineering and Structural Dynamics*, 35:39–55.

Bommer, J.J. and Ruggeri, C. 2002, The specification of acceleration time-histories in seismic design codes, *European Earthquake Engineering*, 16:3–18.

Bommer, J.J., Scherbaum, F., Bungum, H., Cotton, F., Sabetta, F. and Abrahamson, N.A. 2005, On the use of logic trees for ground-motion prediction equations in seismic-hazard analysis, *Bulletin of the Seismological Society of America*, 95:377–389.

Bommer, J.J., Stafford, P.J., Alarcón, J.E. and Akkar, S. 2007, The influence of magnitude range on empirical ground-motion prediction, *Bulletin of the Seismological Society of America*, 97(6): 2152–2170. doi: 10.1785/0120070081.

Boore, D.M. and Bommer, J.J. 2005, Processing of strong-motion accelerograms: Needs, options and consequences, *Soil Dynamics and Earthquake Engineering*, 25:93–115.

Boore, D.M., Joyner, W.B. and Fumal, T.E. 1997, Equations for estimating horizontal response spectra and peak acceleration from Western North American earthquakes: A summary of recent work, *Seismological Research Letters*, 68:128–153.

Bozorgnia, Y. and Campbell, K.W. 2004, The vertical-to-horizontal spectra ratio and tentative procedures for developing simplified V/H and vertical design spectra, *Journal of Earthquake Engineering*, 8:175–207.

Bozorgnia, Y., Hachem, M.M. and Campbell, K.W. 2006, Attenuation of inelastic and damage spectra, *8th U.S. National Conference on Earthquake Engineering*, San Francisco, CA.

Bray, J.D. and Rodriguez-Marek, A. 2004, Characterization of forward-directivity ground motions in the near-fault region, *Soil Dynamics and Earthquake Engineering*, 24:815–828.

Campbell, K.W. 1997, Empirical near-source attenuation relationships for horizontal and vertical components for peak ground acceleration, peak ground velocity, and psuedo-absolute acceleration response spectra, *Seismological Research Letters*, 68:154–179.

Chandramohan, R., Baker, J.W. and Deierlein, G.G. 2016, Quantifying the influence of ground motion duration on structural collapse capacity using spectrally equivalent records, *Earthquake Spectra*, 32(2):927–950.

Chang, T.-Y., Cotton, F., Tsai, Y.-B. and Angelier, J. 2004, Quantification of hanging-wall effects on ground motion: Some insights from the 1999 Chi-Chi earthquake, *Bulletin of the Seismological Society of America*, 94:2186–2197.

Chiou, B.S.-J. and Youngs, R.R. 2006, Chiou and Youngs PEER-NGA empirical ground motion model for the average horizontal component of peak acceleration and psuedo-spectral acceleration for spectral periods of 0.01 to 10 seconds, interim report for USGS review, June 14, 2006 (revised editorially July 10, 2006), *Interim Reports of Next Generation Attenuation (NGA) Models*, Richmond, California: Pacific Earthquake Engineering Research Center.

Choi, Y. and Stewart, J.P. 2005, Nonlinear site amplification as function of 30 m shear wave velocity, *Earthquake Spectra*, 21:1–30.

Choi, Y.J., Stewart, J.P. and Graves, R.W. 2005, Empirical model for basin effects accounts for basin depth and source location, *Bulletin of the Seismological Society of America*, 95:1412–1427.

Cornell, C.A. 1968, Engineering seismic risk analysis, *Bulletin of the Seismological Society of America*, 58:1583–1606.

Cornell, C.A. 1971, Probabilistic analysis of damage to structures under seismic load, in Howells, D.A., Haigh, I.P. and Taylor, C. (eds) *Dynamic Waves in Civil Engineering*, London, England: John Wiley & Sons Ltd.

Cornell, C.A. and Vanmarcke, E.H. 1969, The major influences on seismic risk, *Proceedings of the Fourth World Conference on Earthquake Engineering*, Santiago, Chile.

Cotton, F., Scherbaum, F., Bommer, J.J. and Bungum, H. 2006, Criteria for selecting and adjusting ground-motion models for specific target regions: Application to central Europe and rock sites, *Journal of Seismology*, 10:137–156.

Douglas, J. 2002, Note on scaling of peak ground acceleration and peak ground velocity with magnitude, *Geophysical Journal International*, 148:336–339.

Douglas, J. 2003, Earthquake ground motion estimation using strong-motion records: A review of equations for the estimation of peak ground acceleration and response spectral ordinates, *Earth-Science Reviews*, 61:43–104.

Douglas, J., Akkar, S., Ameri, G., Bard, P.-Y., Bindi, D., Bommer, J.J., Bora, S.S. et al. 2014, Comparisons among the five ground-motion models developed using RESORCE for the prediction of response spectral accelerations due to earthquakes in Europe and the Middle East, *Bulletin of Earthquake Engineering*, 12:341–358.

Elnashai, A.S. and Papazoglou, A.J. 1997, Procedure and spectra for analysis of RC structures subjected to strong vertical earthquake loads, *Journal of Earthquake Engineering*, 1:121–155.

García-Mayordomo, J., Faccioli, E. and Paolucci, R. 2004, Comparative study of the seismic hazard assessments in European national seismic codes, *Bulletin of Earthquake Engineering*, 2:51–73.

Giardini, D., Grünthal, G., Shedlock, K.M. and Zhang, P. 1999, The GSHAP global seismic hazard map, *Annali Di Geofisica*, 42:1225–1230.

Gutenberg, B. and Richter, C.F. 1944, Frequency of earthquakes in California, *Bulletin of the Seismological Society of America*, 34:185–188.

Hancock, J. and Bommer, J.J. 2006, A state-of-knowledge review of the influence of strong-motion duration on structural damage, *Earthquake Spectra*, 22:827–845.

Hancock, J., Watson-Lamprey, J., Abrahamson, N.A., Bommer, J.J., Markatis, A., McCoy, E. and Mendis, R. 2006, An improved method of matching response spectra of recorded earthquake ground motion using wavelets, *Journal of Earthquake Engineering*, 10(special issue 1):67–89.

Iervolino, I., Manfredi, G. and Cosenza, E. 2006, Ground motion duration effects on nonlinear seismic response, *Earthquake Engineering and Structural Dynamics*, 35:21–38.

Jackson, J.A. 2001, Living with earthquakes: Know your faults, *Journal of Earthquake Engineering*, 5(special issue 1):5–123.

Jiménez, M.J., Giardini, D., Grünthal, G. and the SESAME Working Group 2001, Unified seismic hazard modelling throughout the Mediterranean region, *Bollettino di Geofisica Teorica ed Applicata*, 42:3–18.

Kagawa, T., Irikura, K. and Somerville, P.G. 2004, Differences in ground motion and fault rupture process between the surface and buried rupture earthquakes, *Earth Planets and Space*, 56:3–14.

Kramer, S.L. 1996, *Geotechnical Earthquake Engineering*, Upper Saddle River, New Jersey: Prentice-Hall.

Mai, P.M., Spudich, P. and Boatwright, J. 2005, Hypocenter locations in finite-source rupture models, *Bulletin of the Seismological Society of America*, 95:965–980.

Marin, S., Avouac, J.P., Nicolas, M. and Schlupp, A. 2004, A probabilistic approach to seismic hazard in metropolitan France, *Bulletin of the Seismological Society of America*, 94:2137–2163.

McCalpin, J.P. 1996, *Paleoseismology*, San Diego: Academic Press.

McGuire, R.K. 1977, Seismic design and mapping procedures using hazard analysis based directly on oscillator response, *Earthquake Engineering and Structural Dynamics*, 5:211–234.

McGuire, R.K. 1995, Probabilistic seismic hazard analysis and design earthquakes: Closing the loop, *Bulletin of the Seismological Society of America*, 85:1275–1284.

McGuire, R.K. 2004, *Seismic Hazard and Risk Analysis*, Oakland, California: Earthquake Engineering Research Institute.

McGuire, R.K., Cornell, C.A. and Toro, G.R. 2005, The case for using mean seismic hazard, *Earthquake Spectra*, 21:879–886.

Mendis, R. and Bommer, J.J. 2006, Modification of the Eurocode 8 damping reduction factors for displacement spectra, *Proceedings of the First European Conference on Earthquake Engineering and Seismology*, Geneva, Switzerland, paper no. 1203.

Musson, R.M.W. 2005, Against fractiles, *Earthquake Spectra*, 21:887–891.

Papazoglou, A.J. and Elnashai, A.S. 1996, Analytical and field evidence of the damaging effect of vertical earthquake ground motion, *Earthquake Engineering and Structural Dynamics*, 25:1109–1137.

Priestley, M.J.N. 2003, Myths and fallacies in earthquake engineering – Revisited, *The Mallet Milne Lecture*, Pavia, Italy: IUSS Press.

Priestley, M.J.N., Calvi, G.M. and Kowalsky, M.J. 2007, *Displacement-based Seismic Design of Structures*, Pavia, Italy: IUSS Press.

Reiter, L. 1990, *Earthquake Hazard Analysis: Issues and Insights*, Columbia, USA: Columbia University Press.

Rey, J., Faccioli, E. and Bommer, J.J. 2002, Derivation of design soil coefficients (S) and response spectral shapes for Eurocode 8 using the European Strong-Motion Database, *Journal of Seismology*, 6:547–555.

Sabetta, F., Lucantoni, A., Bungum, H. and Bommer, J.J. 2005, Sensitivity of PSHA results to ground motion prediction relations and logic-tree weights, *Soil Dynamics and Earthquake Engineering*, 25:317–329.

Scherbaum, F., Bommer, J.J., Bungum, H., Cotton, F. and Abrahamson, N.A. 2005, Composite ground-motion models and logic trees: Methodology, sensitivities, and uncertainties, *Bulletin of the Seismological Society of America*, 95:1575–1593.

Scherbaum, F., Cotton, F. and Smit, P. 2004a, On the use of response spectral-reference data for the selection and ranking of ground-motion models for seismic-hazard analysis in regions of moderate seismicity: The case of rock motion, *Bulletin of the Seismological Society of America*, 94:2164–2185.

Scherbaum, F., Schmedes, J. and Cotton, F. 2004b, On the conversion of source-to-site distance measures for extended earthquake source models, *Bulletin of the Seismological Society of America*, 94:1053–1069.

Scholz, C.H. 2002, *The Mechanics of Earthquakes and Faulting*, Cambridge, UK: Cambridge University Press.

Schwartz, D.P. and Coppersmith, K.J. 1984, Fault behavior and characteristic earthquakes – Examples from the Wasatch and San-Andreas fault zones, *Journal of Geophysical Research*, 89:5681–5698.

Shome, N. and Cornell, C.A. 2006, Probabilistic seismic demand analysis for a vector of parameters, *8th U.S. National Conference on Earthquake Engineering*, San Francisco, CA.

Somerville, P., Irikura, K., Graves, R., Sawada, S., Wald, D., Abrahamson, N., Iwasaki, Y., Kagawa, T., Smith, N. and Kowada, A. 1999, Characterizing earthquake slip models for the prediction of strong ground motion, *Seismological Research Letters*, 70:59–80.

Somerville, P. and Pitarka, A. 2006, Differences in earthquake source and ground motion characteristics between surface and buried faulting earthquakes, *8th U.S. National Conference on Earthquake Engineering*, San Francisco, CA.

Somerville, P.G., Smith, N.F., Graves, R.W. and Abrahamson, N.A. 1997, Modification of empirical strong ground motion attenuation relations to include the amplitude and duration effects of rupture directivity, *Seismological Research Letters*, 68:199–222.

Stafford, P.J. and Bommer, J.J. 2010, Theoretical consistency of common record selection strategies in performance-based earthquake engineering, in: Fardis, M. (ed) *Advances in Performance-based Earthquake Engineering*, Springer, the Netherlands, pp. 49–58. doi: 10.1007/978-90-481-8746-1_5.

Toshinawa, T., Hisada, Y., Konno, K., Shibayama, A., Honkawa, Y. and Ono, H. 2004, Topographic site response at a quaternary terrace in Hachioji, Japan, observed in strong motions and microtremors, *13th World Conference on Earthquake Engineering*, Vancouver, B.C., Canada.

Tothong, P. and Cornell, C.A. 2006, An empirical ground-motion attenuation relation for inelastic spectral displacement, *Bulletin of the Seismological Society of America*, 96:2146–2164.

Utsu, T. 1999, Representation and analysis of the earthquake size distribution: A historical review and some new approaches, *Pure and Applied Geophysics*, 155:509–535.

Wald, D.J., Quitoriano, V., Heaton, T.H. and Kanamori, H. 1999, Relationships between peak ground acceleration, peak ground velocity, and modified Mercalli intensity in California, *Earthquake Spectra*, 15:557–564.

Watson-Lamprey, J. and Abrahamson, N. 2006, Selection of ground motion time series and limits on scaling, *Soil Dynamics and Earthquake Engineering*, 26:477–482.

Wells, D.L. and Coppersmith, K.J. 1994, New empirical relationships among magnitude, rupture length, rupture width, rupture area, and surface displacement, *Bulletin of the Seismological Society of America*, 84:974–1002.

Yeats, R.S., Sieh, K. and Allen, C.R. 1997, *The Geology of Earthquakes*, Oxford: Oxford University Press.

Youngs, R.R., Abrahamson, N., Makdisi, F.I. and Sadigh, K. 1995, Magnitude-dependent variance of peak ground acceleration, *Bulletin of the Seismological Society of America*, 85:1161–1176.

Chapter 3

Structural analysis

Martin S. Williams

CONTENTS

3.1 INTRODUCTION

This chapter presents a brief account of the basics of dynamic behaviour of structures, the representation of earthquake ground motion by response spectra and the principal methods of seismic structural analysis.

Dynamic analysis is normally a two-stage process: we first estimate the dynamic properties of the structure (natural frequencies and mode shapes) by analysing it in the absence of external loads and then use these properties in the determination of earthquake response.

Earthquakes often induce non-linear response in structures. However, the most practical seismic design continues to be based on linear analysis. The effect of non-linearity is generally to reduce the seismic demands on the structure, and this is normally accounted for by a simple modification to the linear analysis procedure.

A fuller account of this basic theory can be found in Clough and Penzien (1993) or Craig (1981).

3.2 BASIC DYNAMICS

This section outlines the key properties of structures that govern their dynamic response, and introduces the main concepts of dynamic behaviour with reference to single-degree-of-freedom (SDOF) systems.

3.2.1 Dynamic properties of structures

For linear dynamic analysis, a structure can be defined by three key properties: its stiffness, mass and damping. For non-linear analysis, estimates of the yield load and the post-yield behaviour are also required. This section will concentrate on the linear properties, with non-linearity introduced later on.

First, consider how mass and stiffness combine to give oscillatory behaviour. The mass m of a structure, measured in kg, should not be confused with its weight, mg, which is a force measured in N. Stiffness (k) is the constant of proportionality between force and displacement, measured in N/m. If a structure is displaced from its equilibrium position, then a restoring force is generated equal to stiffness × displacement. This force accelerates the structure back towards its equilibrium position. As it accelerates, the structure acquires momentum (equal to mass × velocity), which causes it to overshoot. The restoring force then reverses sign and the process is repeated in the opposite direction, so that the structure oscillates about its equilibrium position. The behaviour can also be considered in terms of energy – vibrations involve repeated transfer of strain energy into kinetic energy as the structure oscillates around its unstrained position.

In addition to the above, all structures gradually dissipate energy as they move, through a variety of internal mechanisms that are normally grouped together and known as damping. Without damping, a structure, once set in motion, would continue to vibrate indefinitely. There are many different mechanisms of damping in structures. However, analysis methods are based on the assumption of linear viscous damping, in which a viscous dashpot generates a retarding force proportional to the velocity difference across it. The damping coefficient (c) is the constant of proportionality between force and velocity, measured in Ns/m. Whereas it should be possible to calculate values of m and k with some confidence, c is a rather nebulous quantity, which is difficult to estimate. It is far more convenient to convert it to a dimensionless parameter ξ, called the damping ratio:

$$\xi = \frac{c}{2\sqrt{km}} \tag{3.1}$$

ξ can be estimated based on experience of similar structures. In civil engineering, it generally takes a value in the range 0.01 to 0.1, and an assumed value of 0.05 is widely used in earthquake engineering.

In reality, all structures have distributed mass, stiffness and damping. However, in most cases, it is possible to obtain reasonably accurate estimates of the dynamic behaviour using lumped parameter models, in which the structure is modelled as a number of discrete masses connected by light spring elements representing the structural stiffness and dashpots representing damping.

Each possible displacement of the structure is known as a degree of freedom. Obviously a real structure with distributed mass and stiffness has an infinite number of degrees of freedom, but in lumped-parameter idealisations, we are concerned only with the possible displacements of the lumped masses. For a complex structure, the finite element method may be used to create a model with many degrees of freedom, giving a very accurate representation of the mass and stiffness distributions. However, the damping is still represented by the approximate global parameter ξ.

3.2.2 Equation of motion of a linear SDOF system

An SDOF system is one whose deformation can be completely defined by a single displacement. Obviously most real structures have many degrees of freedom, but a surprisingly large number can be modelled approximately as SDOF systems.

Figure 3.1 shows a SDOF system subjected to a time-varying external force $F(t)$, which causes a displacement x. The movement of the mass generates restoring forces in the spring and damper as shown in the free body diagram in Figure 3.1.

By Newton's second law, resultant force = mass × acceleration:

$$F(t) - kx - c\dot{x} = m\ddot{x} \quad \text{or} \quad m\ddot{x} + c\dot{x} + kx = F(t) \tag{3.2}$$

where each dot represents one differentiation with respect to time, so that $\dot{x}$ is the velocity and $\ddot{x}$ is the acceleration. This is known as the equation of motion of the system. An

Figure 3.1 Dynamic forces on a mass-spring-damper system.

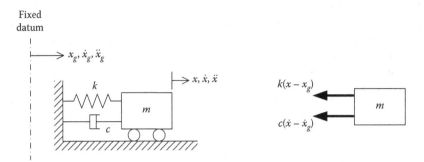

Figure 3.2 Mass-spring-damper system subjected to base motion.

alternative way of coming to the same result is to treat the term $m\ddot{x}$ as an additional internal force, the inertia force, acting on the mass in the opposite direction to the acceleration. The equation of motion is then an expression of dynamic equilibrium between the internal and external forces:

Inertia force + damping force + stiffness force = external force

In an earthquake, there is no force applied directly to the structure. Instead, the ground beneath it is subjected to a (predominantly horizontal) time-varying motion as shown in Figure 3.2.

In the absence of any external forces, Newton's second law now gives

$$-k(x - x_g) - c(\dot{x} - \dot{x}_g) = m\ddot{x} \quad \text{or} \quad m\ddot{x} + c(\dot{x} - \dot{x}_g) + k(x - x_g) = 0 \tag{3.3}$$

Note that the stiffness and damping forces are proportional to the relative motion between the mass and the ground, while the inertia force is proportional to the absolute acceleration experienced by the mass. Let the relative displacement between the mass and the ground be $y = x - x_g$, with similar expressions for velocity and acceleration. The equation of motion can then be written as

$$m\ddot{y} + c\dot{y} + ky = -m\ddot{x}_g \tag{3.4}$$

So a seismic ground motion results in a similar equation of motion to an applied force, but in terms of motion relative to the ground and with the forcing function proportional to the ground acceleration.

Before looking at solutions to this equation, we will look at the free vibration case (i.e. no external excitation). This will provide us with the essential building blocks for solution of the case when the right hand side of Equation 3.2 or 3.4 is non-zero.

3.2.3 Free vibrations of SDOF systems

Consider first the theoretical case of a simple mass-spring system with no damping and no external force. The equation of motion is simply

$$m\ddot{x} + kx = 0 \tag{3.5}$$

If the mass is set in motion by giving it a small initial displacement x_0 from its equilibrium position, then it undergoes free vibrations at a rate known as the natural frequency. The solution to Equation 3.5 is

$$x(t) = x_0 \cos \omega_n t \quad \text{where} \quad \omega_n = \sqrt{\frac{k}{m}} \tag{3.6}$$

where ω_n is called the circular natural frequency (measured in rad/s). It can be thought of as the angular speed of an equivalent circular motion, such that one complete revolution of the equivalent motion takes the same time as one complete vibration cycle. More easily visualized parameters are the natural frequency f_n (measured in cycles per second, or Hz) and the natural period T_n (the time taken for one complete cycle, measured in s). These are related to ω_n by

$$f_n = \frac{\omega_n}{2\pi} = \frac{1}{2\pi} \sqrt{\frac{k}{m}} \tag{3.7}$$

$$T_n = \frac{1}{f_n} = 2\pi \sqrt{\frac{m}{k}} \tag{3.8}$$

Next consider the vibration of an SDOF system with damping included but still with no external force, again set in motion by applying an initial displacement x_0. The equation of motion is

$$m\ddot{x} + c\dot{x} + kx = 0 \tag{3.9}$$

The behaviour of this system depends on the relative magnitudes of c, k and m. If $c = 2\sqrt{km}$, the system is said to be critically damped and will return to its equilibrium position without oscillating. In general, c is much smaller than this, giving an underdamped system. Critical damping is useful mainly as a reference case against which others can be scaled to give the damping ratio defined earlier in Equation 3.1:

$$\xi = \frac{c}{2\sqrt{km}}$$

For an underdamped system, the displacement is given by

$$x = x_0 e^{-\xi \omega_n t} \cos \sqrt{1 - \xi^2}\, \omega_n t \tag{3.10}$$

An example is given in Figure 3.3, which shows the response of SDOF systems with natural period 1 s and different damping ratios, when released from an initial unit displacement. These damped responses differ from the undamped case in two ways: first the oscillations are multiplied by an exponential decay term, so that they die away quite quickly; second, the natural frequency has been altered by the factor $\sqrt{1 - \xi^2}$. However, for practical values of damping, this factor is very close to unity. It is therefore acceptable to neglect damping when calculating natural frequencies.

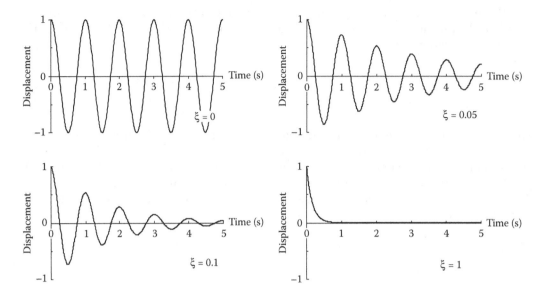

Figure 3.3 Effect of damping on free vibrations.

Using the relationships between ω_n, ξ, m, c and k, the Equation 3.4 can conveniently be written as

$$\ddot{y} + 2\xi\omega_n\dot{y} + \omega_n^2 y = -\ddot{x}_g \qquad (3.11)$$

3.2.4 Response to a sinusoidal base motion

Suppose first of all that the ground motion varies sinusoidally with time at a circular frequency ω, with corresponding period $T = 2\pi/\omega$:

$$x_g = X_g \sin\omega t \qquad (3.12)$$

Of course, a real earthquake ground motion is more complex, but this simplification serves to illustrate the main characteristics of the response.

Equation 3.11 can be solved by standard techniques and the response computed. Figure 3.4 shows the variation of structural acceleration $\ddot{x}$ with time for a structure with a natural period of 0.5 s and 5% damping, for a variety of frequencies of ground shaking. Three regimes of structural response can be seen:

a. The ground shaking is at a much slower rate than the structure's natural oscillations, so that the behaviour is quasi-static; the structure simply moves with the ground, with minimal internal deformation and its absolute displacement amplitude is approximately equal to the ground displacement amplitude.
b. When the ground motion period and natural period are similar, resonance occurs and there is a large dynamic amplification of the motion. In this region, the stiffness and inertia forces at any time are approximately equal and opposite, so that the main resistance to motion is provided by the damping of the system.

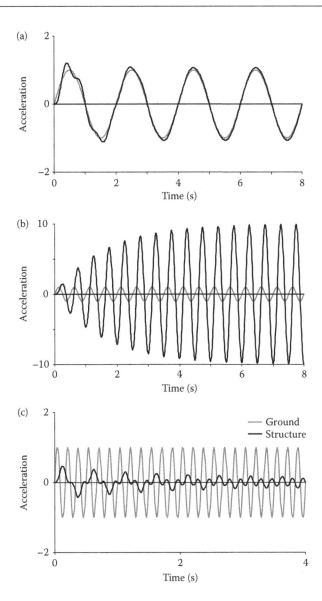

Figure 3.4 Acceleration (in arbitrary units) of a 0.5 s natural period SDOF structure subject to ground shaking at a period of: (a) 2 s, (b) 0.5 s, (c) 0.167 s.

c. If the ground motion is much faster than the natural oscillations of the structure, then the mass undergoes less motion than the ground, with the spring and damper acting as vibration absorbers.

The effect of the loading rate on the response of a SDOF structure is summarised in Figure 3.5, for different damping levels. Here the peak absolute displacement of the structure X (normalised by the peak ground displacement X_g) is plotted against the ratio of the natural period T_n to the period of the sinusoidal loading T.

The same three response regimes are evident in this figure, with the structural motion equal to the ground motion at the left-hand end of the graph, then large resonant amplifications at

Figure 3.5 Displacement amplification curves for an SDOF structure subject to sinusoidal ground shaking.

around $T_n/T = 1$, and finally very low displacements when T_n/T is large. At pure resonance ($T_n/T = 1$) the ratio X/X_g roughly equals $1/(2\xi)$. The peak displacement at resonance is thus very sensitive to damping and is infinite for the theoretical case of zero damping. For a more realistic damping ratio of 0.05, the displacement of the structure is around 10 times the ground displacement.

This illustrates the key principles of dynamic response, but it is worth noting here that the dynamic amplifications observed under real earthquake loading are rather lower than those discussed above, both because an earthquake time history is not a simple sinusoid, and because it has a finite (usually quite short) duration.

3.3 RESPONSE SPECTRA AND THEIR APPLICATION TO LINEAR STRUCTURAL SYSTEMS

We now go on to consider the linear response of structures to realistic earthquake time histories. An earthquake can be measured and represented as the variation of ground acceleration with time in three orthogonal directions (N-S, E-W and vertical). An example, recorded during the 1940 El Centro earthquake in California, is shown in Figure 3.6. Obviously, the exact nature of an earthquake time history is unknown in advance, will be different for every earthquake, and indeed will vary over the affected region due to factors such as local ground conditions, epicentral distance, etc.

3.3.1 Earthquake response

The time-domain response to an earthquake ground motion by can be determined by a variety of techniques, all of which are quite mathematically complex. For example, in the Duhamel's integral approach, the earthquake record is treated as a sequence of short impulses, and the time-varying responses to each impulse are summed to give the total response.

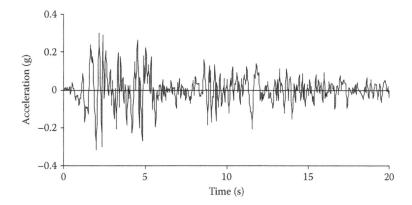

Figure 3.6 Accelerogram for 1940 El Centro earthquake (N-S component).

Although the method of evaluation is rather complex, the behaviour under a general dynamic load can be quite easily understood by comparison with the single-frequency, sinusoidal load case discussed in Section 3.2.4. In that case, we saw that large dynamic amplifications occur if the loading period is close to the natural period of the structure. Irregular dynamic loading can be thought of as having many different components at different periods. Often the structure's natural period will lie within the band of periods contained in the loading. The structure will tend to pick up and amplify those components close to its own natural period just as it would with a simple sinusoid. The response will therefore be dominated by vibration at or close to the natural period of the structure. However, because the loading does not have constant amplitude and is likely to have only finite duration, the amplifications achieved are likely to be much smaller than for the sinusoidal case. An example is shown in Figure 3.7, where a 0.5 s period structure is subjected to the El Centro earthquake record plotted in Figure 3.6. The earthquake contains a wide band of frequency components, but it can be seen that the 0.5 s component undergoes a large amplification and dominates the response.

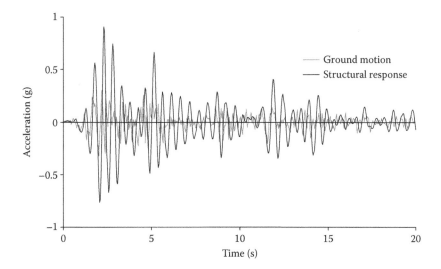

Figure 3.7 Acceleration of 0.5 s period SDOF structure subject to the El Centro (N-S) earthquake record.

3.3.2 Response spectrum

The response of a wide range of structures to a particular earthquake can be summarised using a response spectrum. The time-domain response of numerous SDOF systems having different natural periods is computed, and the maximum absolute displacement (or acceleration, or velocity) achieved is plotted as a function of the SDOF system period. If desired, a range of curves can be plotted for SDOF systems having different damping ratios.

So the response spectrum shows the peak response of a SDOF structure to a particular earthquake, as a function of the natural period and damping ratio of the structure. For example, Figure 3.8 shows the response spectrum for the El Centro (N-S) accelerogram in Figure 3.6, for SDOF structures with 5% damping.

A key advantage of the response spectrum approach is that earthquakes that look quite different when represented in the time domain may actually contain similar frequency contents, and so result in broadly similar response spectra. This makes the response spectrum a useful design tool for dealing with a future earthquake whose precise nature is unknown. To create a design spectrum, it is normal to compute spectra for several different earthquakes, then envelope and smooth them, resulting in a single curve, which encapsulates the dynamic characteristics of a large number of possible earthquake accelerograms.

Figure 3.9 shows the elastic response spectra defined by EC8. EC8 (2004) specifies two categories of spectra: type 1 for areas of high seismicity (defined as $M_s > 5.5$), and type 2 for areas of moderate seismicity ($M_s \leq 5.5$). Within each category, spectra are given for five different soil types: A – rock; B – very dense sand or gravel, or very stiff clay; C – dense sand or gravel, or stiff clay; D – loose-to-medium cohesionless soil, or soft-to-firm cohesive soil; E – soil profiles with a surface layer of alluvium of thickness 5 to 20 m. The vertical axis is the peak, or spectral acceleration of the elastic structure, denoted by S_e, normalised by a_g, the design peak ground acceleration on type A ground. The spectra are plotted for an assumed structural damping ratio of 5%. See EC8 Cl. 3.2.2.2 for mathematical definitions of these curves and Table 3.1 of EC8 for fuller descriptions of ground types A–E.

As with the harmonic load case, there are three regimes of response. Very stiff, short period structures simply move with the ground. At intermediate periods, there is dynamic amplification of the ground motion, though only by a factor of 2.5–3, and at long periods the structure moves less than the ground beneath it. In the region of the spectra between

Figure 3.8 5% damped response spectrum for 1940 El Centro earthquake (N-S component).

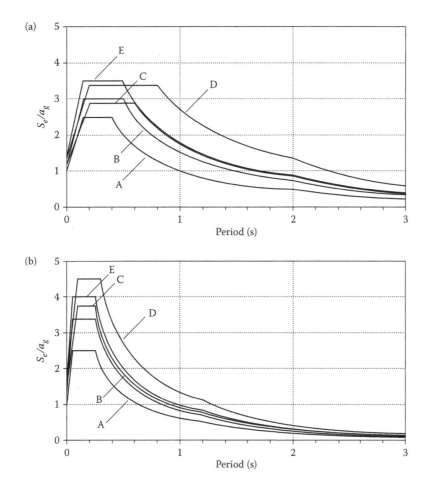

Figure 3.9 EC8 5% damped, elastic spectra, (a) Type 1, (b) Type 2.

T_B and T_C the spectra acceleration is constant with period. The region between T_C and T_D represents constant velocity and beyond T_D is the constant displacement region.

It can be seen that in the high seismicity events (type 1 spectra), the spectral amplifications tend to occur at longer periods, and over a wider period range, than in the moderate seismicity events. It is also noticeable that the different soil types give rise to varying levels of amplification of the bedrock motions, and affect the period range over which amplification occurs. The EC8 values for T_D have caused some controversy – it has been argued that the constant velocity region of the spectra should continue to higher periods, which would result in a more onerous spectral acceleration for long-period (e.g. very tall) structures.

3.3.3 Application of response spectra to elastic SDOF systems

In a response spectrum analysis of a SDOF system, we generally wish to determine the force to which the structure is subjected, and its maximum displacement. We start by estimating the natural period T_n and damping ratio ξ. The peak (spectral) acceleration S_e experienced by the mass can then be read directly from the response spectrum. Now the maximum acceleration in a vibrating system occurs when it is at its point of extreme displacement, at which instant the velocity (and therefore the damping force) is zero. The peak force is then just equal to the inertia force experienced by the mass:

$$F = mS_e \qquad (3.13)$$

This must be in dynamic equilibrium with the stiffness force developed within the structure. If we define the spectral displacement S_D as the peak absolute displacement corresponding to the spectral acceleration S_e then we must have $kS_D = mS_e$ which, using the relationships between mass, stiffness and natural period given in Equation 3.8, leads to

$$S_D = \frac{F}{k} = mS_e \cdot \frac{T_n^2}{4\pi^2 m} = \frac{S_e T_n^2}{4\pi^2} \qquad (3.14)$$

Note that, while the force experienced depends on the mass, the spectral acceleration and displacement do not – they are functions only of the natural period and damping ratio.

It should be remembered that the spectral acceleration is absolute (i.e. it is the acceleration of the mass relative to the ground plus the ground acceleration, hence proportional to the inertia force experienced by the mass), but the spectral displacement is the displacement of the mass relative to the ground (and hence proportional to the spring force).

While elastic spectra are useful tools for design and assessment, they do not account for the inelasticity which will occur during severe earthquakes. In practice, energy absorption and plastic redistribution can be used to reduce the design forces significantly. This is dealt with in EC8 by the modification of the elastic spectra to give design spectra S_d, as described in Section 3.4.2.

3.3.4 Analysis of linear MDOF systems

Not all structures can be realistically modelled as SDOF systems. Structures with distributed mass and stiffness may undergo significant deformations in several modes of vibration and therefore need to be analysed as multi-degree-of-freedom (MDOF) systems. These are not generally amenable to hand solution and so computer methods are widely used – see, for example, Hitchings (1992) or Petyt (1998) for details.

For a system with N degrees of freedom, it is possible to write a set of equations of motion in matrix form, exactly analogous to Equation 3.4:

$$\mathbf{m}\ddot{\mathbf{y}} + \mathbf{c}\dot{\mathbf{y}} + \mathbf{k}\mathbf{y} = \mathbf{m}\iota\ddot{x}_g \qquad (3.15)$$

where $\mathbf{m}$, $\mathbf{c}$ and $\mathbf{k}$ are the mass, damping and stiffness matrices (dimensions $N \times N$), $\mathbf{y}$ is the relative displacement vector and ι is an $N \times 1$ influence vector containing ones corresponding to the DOFs in the direction of the earthquake load, and zeroes elsewhere. $\mathbf{k}$ is derived in the same way as for a static analysis and is a banded matrix.

$\mathbf{m}$ is most simply derived by dividing the mass of each element between its nodes. This results in a lumped mass matrix, which contains only diagonal terms. To get a sufficiently detailed description of how the mass is distributed, it may be necessary to divide the structure into smaller elements than would be required for a static analysis. Alternatively, many finite element programs give the option of using a consistent mass matrix, which allows a more accurate representation of the mass distribution without the need for substantial mesh refinement. A consistent mass matrix includes off-diagonal terms.

In practice, **c** is very difficult to define accurately and is not usually formulated explicitly. Instead, damping is incorporated in a simplified form. We shall see how this is done later.

3.3.5 Free vibration analysis

As with SDOF systems, before attempting to solve Equation 3.15, it is helpful to consider the free vibration problem. Because it has little effect on free vibrations, we also omit the damping term, leaving

$$\mathbf{m\ddot{y} + ky} = 0 \tag{3.16}$$

The solution to this equation has the form

$$\mathbf{y} = \phi \sin \omega t \tag{3.17}$$

where ϕ is the mode shape, which is a function solely of position within the structure. Differentiating and substituting into Equation 3.16 gives

$$(\mathbf{k} - \omega^2 \mathbf{m})\phi = 0 \tag{3.18}$$

This can be solved to give N circular natural frequencies ω_1, $\omega_2 \ldots \omega_i \ldots \omega_N$, each associated with a mode shape ϕ_i. Thus an N-DOF system is able to vibrate in N different modes, each having a distinct deformed shape and each occurring at a particular natural frequency (or period). The modes of vibration are system properties, independent of the external loading. Figure 3.10 shows the sway modes of vibration of a four-storey shear-type building (i.e. one with relatively stiff floors, so that lateral deformations are dominated by shearing deformation between floors), with the modes numbered in order of ascending natural frequency (or descending period).

Often approximate formulae are used for estimating the fundamental natural period of multi-storey buildings. EC8 recommends the following formulae. For multi-storey frame buildings:

$$T_1 = C_t H^{0.75} \tag{3.19}$$

where T_1 is measured in seconds, the building height H is measured in metres and the constant C_t equals 0.085 for steel moment-resisting frames (MRFs), 0.075 for concrete MRFs

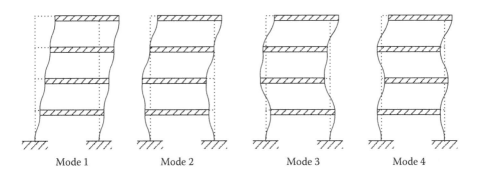

Mode 1   Mode 2   Mode 3   Mode 4

Figure 3.10 Mode shapes of a four-storey building.

or steel eccentrically braced frames, and 0.05 for other types of frame. For shear-wall type buildings:

$$C_t = \frac{0.075}{\sqrt{A_c}} \qquad (3.20)$$

where A_c is the total effective area of shear walls in the bottom storey, in m^2.

3.3.6 Multi-modal response spectrum analysis

Having determined the natural frequencies and mode shapes of our system, we can go on to analyse the response to an applied load. Equation 3.15 is a set of N coupled equations in terms of the N degrees of freedom. This can be most easily solved using the principle of modal superposition, which states that any set of displacements can be expressed as a linear combination of the mode shapes:

$$y = Y_1\phi_1 + Y_2\phi_2 + Y_3\phi_3 + \cdots + Y_N\phi_N = \sum_i Y_i\phi_i \qquad (3.21)$$

The coefficients Y_i are known as the generalized or modal displacements. The modal displacements are functions only of time, while the mode shapes are functions only of position. Equation 3.21 allows us to transform the equations of motion into a set of equations in terms of the modal displacements rather than the original degrees of freedom:

$$\mathbf{M}\ddot{\mathbf{Y}} + \mathbf{C}\dot{\mathbf{Y}} + \mathbf{K}\mathbf{Y} = \phi^T \mathbf{m}\iota\,\ddot{x}_g \qquad (3.22)$$

where $\mathbf{Y}$ is the vector of modal displacements, and $\mathbf{M}$, $\mathbf{C}$ and $\mathbf{K}$ are the modal mass, stiffness and damping matrices. Because of the orthogonality properties of the modes, it turns out that $\mathbf{M}$, $\mathbf{C}$ and $\mathbf{K}$ are all diagonal matrices, so that the N equations in (3.22) are uncoupled, that is, each mode acts as a SDOF system and is independent of the responses in all other modes. Each line of Equation 3.22 has the form:

$$M_i\ddot{Y}_i + C_i\dot{Y}_i + K_iY_i = L_i\ddot{x}_g \qquad (3.23)$$

or, by analogy with Equation 3.11 for an SDOF system:

$$\ddot{Y}_i + 2\xi\omega_i\dot{Y}_i + \omega_i^2 Y = \frac{L_i}{M_i}\ddot{x}_g \qquad (3.24)$$

where

$$L_i = \sum_j m_j\phi_{ij} \qquad (3.25)$$

$$M_i = \sum_j m_j\phi_{ij}^2 \qquad (3.26)$$

Here the subscript i refers to the mode shape and j to the degrees of freedom in the structure. So ϕ_{ij} is the value of mode shape i at DOF j. L_i is an earthquake excitation factor, representing the extent to which the earthquake tends to excite response in mode i. M_i is called the modal mass. The dimensionless factor L_i/M_i is the ratio of the response of an MDOF structure in a particular mode to that of an SDOF system with the same mass and period.

Note that Equation 3.24 allows us to define the damping in each mode simply by specifying a damping ratio ξ, without having to define the original damping matrix c.

While Equation 3.24 could be solved explicitly to give Y_i as a function of time for each mode, it is more normal to use the response spectrum approach. For each mode, we can read off the spectral acceleration S_{ei} corresponding to that mode's natural period and damping – this is the peak response of an SDOF system with period T_i to the ground acceleration $\ddot{x}_g$. For our MDOF system, the way we have broken it down into separate modes has resulted in the ground acceleration being scaled by the factor L_i/M_i. Since the system is linear, the structural response will be scaled by the same amount. So the acceleration amplitude in mode i is $(L_i/M_i).S_{ei}$ and the maximum acceleration of DOF j in mode i is

$$\ddot{x}_{ij}(\text{max}) = \frac{L_i}{M_i} S_{ei}\phi_{ij} \tag{3.27}$$

Similarly for displacements, by analogy with Equation 3.14:

$$y_{ij}(\text{max}) = \frac{L_i}{M_i} S_{ei}\phi_{ij}.\frac{T_i^2}{4\pi^2} \tag{3.28}$$

To find the horizontal force on mass j in mode i, we simply multiply the acceleration by the mass:

$$F_{ij}(\text{max}) = \frac{L_i}{M_i} S_{ei}\phi_{ij}m_j \tag{3.29}$$

and the total horizontal force on the structure (usually called the base shear) in mode i is found by summing all the storey forces to give

$$F_{bi}(\text{max}) = \frac{L_i^2}{M_i} S_{ei} \tag{3.30}$$

The ratio L_i^2/M_i is known as the effective modal mass. It can be thought of as the amount of mass participating in the structural response in a particular mode. If we sum this quantity for all modes of vibration, the result is equal to the total mass of the structure.

To obtain the overall response of the structure, in theory we need to apply Equations 3.27 through 3.30 to each mode of vibration and then combine the results. Since there are as many modes as there are degrees of freedom, this could be an extremely long-winded process. In practice, however, the scaling factors L_i/M_i and L_i^2/M_i are small for the higher modes of vibration. It is therefore normally sufficient to consider only a subset of the modes. EC8 offers a variety of ways of assessing how many modes need to be included in the response analysis. The normal approach is either to include sufficient modes that the sum of their effective modal masses is at least 90% of the total structural mass, or to include all modes with an effective modal mass greater than 5% of the total mass. If these conditions are difficult to satisfy, a permissible alternative is that the number of modes should be at least $3\sqrt{n}$ where n is the number of storeys and should include all modes with periods below 0.2 s.

Another potential problem is the combination of modal responses. Equations 3.27 through 3.30 give only the peak values in each mode, and it is unlikely that these peaks will all occur at the same point in time. Simple combination rules are used to give an estimate of the total response. Two methods are permitted by EC8. If the difference in natural period between any two modes is at least 10% of the longer period, then the modes can be regarded as independent. In this case, the simple SRSS method can be used, in which the peak overall response is taken as the square root of the sum of the squares of the peak modal responses. If the independence condition is not met, then the SRSS approach may be non-conservative and a more sophisticated combination rule should be used. The most widely accepted alternative is the complete quadratic combination (CQC) method (Wilson et al., 1981), which is based on calculating a correlation coefficient between two modes. Although it is more mathematically complex, the additional effort associated with using this more general and reliable method is likely to be minimal, since it is built into many dynamic analysis computer programs.

In conclusion, the main steps of the mode superposition procedure can be summarised as follows:

1. Perform free vibration analysis to find natural periods T_i and corresponding mode shapes ϕ_i. Estimate damping ratio ξ.
2. Decide how many modes need to be included in the analysis.
3. For each mode
   Compute the modal properties L_i and M_i from Equations 3.25 and 3.26
   Read the spectral acceleration from the design spectrum
   Compute the desired response parameters using Equations 3.27 through 3.30
4. Combine modal contributions to give estimates of total response.

3.3.7 Equivalent static analysis of MDOF systems

A logical extension of the process of including only a subset of the vibrational modes in the response calculation is that, in some cases, it may be possible to approximate the dynamic behaviour by considering only a single mode. It can be seen from Equation 3.29 that, for a single mode of vibration, the force at level j is proportional to the product of mass and mode shape at level j, the other terms being modal parameters that do not vary with position. If the structure can reasonably be assumed to be dominated by a single (normally the fundamental) mode, then a simple static analysis procedure can be used, which involves only minimal consideration of the dynamic behaviour. For many years, this approach has been the mainstay of earthquake design codes. In EC8, the procedure is as follows.

Estimate the period of the fundamental mode T_1 – usually by some simplified approximate method rather than a detailed dynamic analysis (e.g. Equation 3.19). It is then possible to check whether equivalent static analysis is permitted – this requires that $T_1 < 4T_C$ where T_C is the period at the end of the constant-acceleration part of the design response spectrum. The building must also satisfy the EC8 regularity criteria. If these two conditions are not met, the multi-modal response spectrum method outlined above must be used.

For the calculated structural period, the spectral acceleration S_e can be obtained from the design response spectrum. The base shear is then calculated as

$$F_b = \lambda m S_e \tag{3.31}$$

where m is the total mass. This is analogous to Equation 3.30, with the ratio L_i^2/M_i replaced by λm. λ takes the value 0.85 for buildings of more than two storeys with $T_1 < 2T_C$, and is 1.0 otherwise. The total horizontal load is then distributed over the height of the building in proportion to (mass × mode shape). Normally this is done by making some simple assumption about the mode shape. For instance, for simple, regular buildings EC8 permits the assumption that the first mode shape is a straight line (i.e. displacement is directly proportional to height). This leads to a storey force at level k given by

$$F_k = F_b \frac{z_k m_k}{\sum_j z_j m_j} \tag{3.32}$$

where z represents storey height. Finally, the member forces and deformations can be calculated by static analysis.

3.4 PRACTICAL SEISMIC ANALYSIS TO EC8

3.4.1 Ductility and behaviour factor

Designing structures to remain elastic in large earthquakes is likely to be uneconomic in most cases, as the force demands will be very large. A more economical design can be achieved by accepting some level of damage short of complete collapse, and making use of the ductility of the structure to reduce the force demands to acceptable levels.

Ductility is defined as the ability of a structure or member to withstand large deformations beyond its yield point (often over many cycles) without fracture. In earthquake engineering, ductility is expressed in terms of demand and supply. The ductility demand is the maximum ductility that the structure experiences during an earthquake, which is a function of both the structure and the earthquake. The ductility supply is the maximum ductility the structure can sustain without fracture. This is purely a structural property.

Of course, if one calculates design forces on the basis of a ductile response, it is then essential to ensure that the structure does indeed fail by a ductile mode well before brittle failure modes develop, that is, that ductility supply exceeds the maximum likely demand – a principle known as capacity design. Examples of designing for ductility include

- Ensuring plastic hinges form in beams before columns
- Providing adequate confinement to concrete using closely spaced steel hoops
- Ensuring that steel members fail away from connections
- Avoiding large irregularities in structural form
- Ensuring flexural strengths are significantly lower than shear strengths

Probably the easiest way of defining ductility is in terms of displacement. Suppose we have a SDOF system with a clear yield point – the displacement ductility is defined as the maximum displacement divided by the displacement at first yield.

$$\mu = \frac{x_{\max}}{x_y} \tag{3.33}$$

Yielding of a structure also has the effect of limiting the peak force that it must sustain. In EC8, this force reduction is quantified by the behaviour factor, q

$$q = \frac{F_{el}}{F_y} \tag{3.34}$$

where F_{el} is the peak force that would be developed in a SDOF system if it responded to the earthquake elastically, and F_y is the yield load of the system.

A well-known empirical observation is that, at long periods ($>T_C$), yielding and elastic structures undergo roughly the same peak displacement. It follows that, for these structures, the force reduction is simply equal to the ductility (see Figure 3.11). At shorter periods, the amount of force reduction achieved for a given ductility reduces. EC8 therefore uses the following expressions:

$$\begin{aligned}\mu &= q & \text{for } T \geq T_C \\ \mu &= 1 + (q-1)\frac{T_C}{T} & \text{for } T < T_C\end{aligned} \tag{3.35}$$

When designing structures taking account of non-linear seismic response, a variety of analysis options are available. The simplest and most widely used approach is to use the linear analysis methods set out above, but with the design forces reduced on the basis of a single, global behaviour factor q. EC8 gives recommended values of q for common structural forms. This approach is most suitable for regular structures, where inelasticity can be expected to be reasonably uniformly distributed.

In more complex cases, the q-factor approach can become inaccurate and a more realistic description of the distribution of inelasticity through the structure may be required. In these cases, a fully non-linear analysis should be performed, using either the non-linear static (pushover) approach, or non-linear time-history analysis. Rather than using a single factor, these methods require representation of the non-linear load–deformation characteristics of each member within the structure.

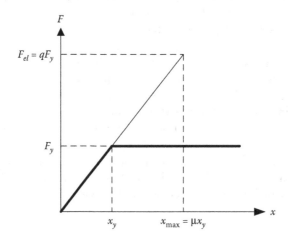

Figure 3.11 Equivalence of ductility and behaviour factor with equal elastic and inelastic displacements.

3.4.2 Ductility-modified response spectra

To make use of ductility requires the structure to respond non-linearly, meaning that the linear methods introduced above are not appropriate. However, for an SDOF system, an approximate analysis can be performed in a very similar way to above by using a ductility-modified response spectrum. In EC8, this is known as the design spectrum, S_d. Figure 3.12 shows EC8 design spectra based on the type 1 spectrum and soil type C, for a range of behaviour factors. Over most of the period range (for $T \geq T_B$), the spectral accelerations S_d (and hence the design forces) are a factor of q times lower than the values S_e for the equivalent elastic system. For a theoretical, infinitely stiff system (zero period), ductility does not imply any reduction in spectral acceleration, since an infinitely stiff structure will not undergo any deformation and will simply move with the ground beneath it. Therefore, the curves all converge to the same spectral acceleration at zero period. A linear interpolation is used between periods of zero and T_B.

When calculating displacements using the design spectrum, it must be noted that the relationship between peak displacement and acceleration in a ductile system is different from that derived in Equation 3.14 for an elastic system. The ductile value is given by

$$S_D(\text{ductile}) = \mu \frac{F_y}{k} = \mu \, mS_d \cdot \frac{T_n^2}{4\pi^2 m} = \mu \frac{S_d T_n^2}{4\pi^2} \tag{3.36}$$

Comparing with Equation 3.14, we see that the ratio between spectral displacement and acceleration is μ times larger for a ductile system than for an elastic one. Thus, the seismic analysis of a ductile system can be performed in exactly the same way as for an elastic system, but with spectral accelerations taken from the design spectrum rather than the elastic spectrum, and with the calculated displacements scaled up by the ductility factor μ.

For long period structures ($T > T_C$), the result of this approach will be that design forces are reduced by the factor q compared to an elastic design, and the displacement of the ductile system is the same as for an equivalent elastic system (since $q = \mu$ in this period range). For $T_B < T < T_C$, the same force reduction will be achieved but displacements will be slightly greater than the elastic case. For very stiff structures ($T < T_B$), the benefits of ductility are reduced, with smaller force reductions and large displacements compared to the elastic case.

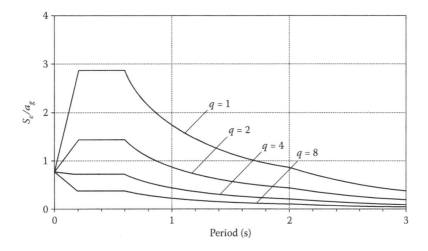

Figure 3.12 EC8 design response spectra (type 1 spectrum, soil type C).

Lastly, it should be noted that the use of ductility-modified spectra is reasonable for SDOF systems, but should be applied with caution to MDOF structures. For elastic systems, we have seen that an accurate dynamic analysis can be performed by considering the response of the structure in each of its vibration modes, then combining the modal responses. A similar approach is widely used for inelastic structures, that is, each mode is treated as an SDOF system and its ductility-modified response determined as above. The modal responses are then combined by a method such as SRSS. While this approach forms the basis of much practical design, it is important to realize that it has no theoretical justification. For linear systems, the method is based on the fact that any deformation can be treated as a linear combination of the mode shapes. Once the structure yields its properties change and these mode shapes no longer apply.

When yielding is evenly spread throughout the structure, the deformed shape of the plastic structure is likely to be similar to the elastic one, and the ductility-modified response spectrum analysis may give reasonable (though by no means precise) results. If, however, yielding is concentrated in certain parts of the structure, such as an soft storey, then this procedure is likely to be substantially in error and one of the non-linear analysis methods described below should be used.

3.4.3 Non-linear static analysis

In recent years, there has been a substantial growth of interest in the use of non-linear static, or pushover analysis (Lawson et al. 1994; Krawinkler and Seneviratna, 1998; Fajfar, 2002) as an alternative to the ductility-modified spectrum approach. In this approach, appropriate lateral load patterns are applied to a numerical model of the structure and their amplitude is increased in a stepwise fashion. A non-linear static analysis is performed at each step, until the building forms a collapse mechanism. A pushover curve (base shear against top displacement) can then be plotted. This is often referred to as the capacity curve since it describes the deformation capacity of the structure. To determine the demands imposed on the structure by the earthquake, it is necessary to equate this to the demand curve (i.e. the earthquake response spectrum) to obtain peak displacement under the design earthquake – termed the target displacement. The non-linear static analysis is then revisited to determine member forces and deformations at this point.

This method is considered a step forward from the use of linear analysis and ductility-modified response spectra, because it is based on a more accurate estimate of the distributed yielding within a structure, rather than an assumed, uniform ductility. The generation of the pushover curve also provides the engineer with a good feel for the non-linear behaviour of the structure under lateral load. However, it is important to remember that pushover methods have no rigorous theoretical basis, and may be inaccurate if the assumed load distribution is incorrect. For example, the use of a load pattern based on the fundamental mode shape may be inaccurate if higher modes are significant, and the use of a fixed load pattern may be unrealistic if yielding is not uniformly distributed, so that the stiffness profile changes as the structure yields.

The main differences between the various pushover analysis procedures that have been proposed are (i) the choices of load patterns to be applied and (ii) the method of simplifying the pushover curve for design use. The EC8 method is summarised below.

First, two pushover analyses are performed, using two different lateral load distributions. The most unfavourable results from these two force patterns should be adopted for design purposes. In the first, the acceleration distribution is assumed proportional to the fundamental mode shape. The inertia force F_k on mass k is then

$$F_k = \frac{m_k \phi_k}{\sum_j m_j \phi_j} F_b \tag{3.37}$$

where F_b is the base shear (which is increased steadily from zero until failure), m_k the kth storey mass and ϕ_k the mode shape coefficient for the kth floor. If the fundamental mode shape is assumed to be linear, then ϕ_k is proportional to storey height z_k and Equation 3.36 then becomes identical to Equation 3.32, presented earlier for equivalent static analysis. In the second case, the acceleration is assumed constant with height. The inertia forces are then given by

$$F_k = \frac{m_k}{\sum_j m_j} F_b \tag{3.38}$$

The output from each analysis can be summarised by the variation of base shear F_b with top displacement d, with maximum displacement d_m. This can be transformed to an equivalent SDOF characteristic (F^* vs d^*) using

$$F^* = \frac{F_b}{\Gamma}, \quad d^* = \frac{d}{\Gamma} \tag{3.39}$$

where

$$\Gamma = \frac{\sum_j m_j \phi_j}{\sum_j m_j \phi_j^2} \tag{3.40}$$

The SDOF pushover curve is likely to be piecewise linear due to the formation of successive plastic hinges as the lateral load intensity is increased, until a collapse mechanism forms. For determination of the seismic demand from a response spectrum, it is necessary to simplify this to an equivalent elastic-perfectly plastic curve as shown in Figure 3.13. The yield load F_y^* is taken as the load required to cause formation of a collapse mechanism, and the yield displacement d_y^* is chosen so as to give equal areas under the actual and idealised curves. The initial elastic period of this idealised system is then estimated as

$$T^* = 2\pi \sqrt{\frac{m^* d_y^*}{F_y^*}} \tag{3.41}$$

The target displacement of the SDOF system under the design earthquake is then calculated from

$$d_t^* = S_e \left(\frac{T^*}{2\pi}\right)^2 \qquad\qquad T^* \geq T_C$$

$$d_t^* = S_e \left(\frac{T^*}{2\pi}\right)^2 \frac{1}{q_u}\left[1 + (q_u - 1)\frac{T_C}{T^*}\right] \quad T^* < T_C \tag{3.42}$$

where $q_u = (S_e/(F_y^*/m^*))$ and $m^* = \Sigma_{j=1}^n m_j \phi_j$.

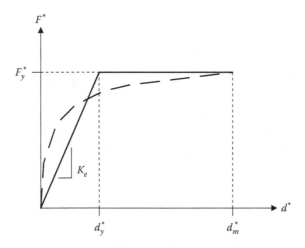

Figure 3.13 Idealisation of pushover curve in EC8.

Equation 3.41 is illustrated schematically in Figure 3.14, in which the design response spectrum has been plotted in acceleration versus displacement format rather than the more normal acceleration versus period. This enables both the spectrum (i.e. the demand curve) and the capacity curve to be plotted on the same axes, with a constant period represented by a radial line from the origin. For $T^* \geq T_C$, the target displacement is based on the equal displacement rule for elastic and inelastic systems. For shorter period structures, a correction is applied to account for the more complex interaction between behaviour factor and ductility (see Equation 3.35).

Having found the target displacement for the idealised SDOF system, this can be transformed back to that of the original MDOF system using Equation 3.38, and the forces and deformations in the structure can be checked by considering the point in the pushover analysis corresponding to this displacement value.

The EC8 procedure is simple and unambiguous, but can be rather conservative. Some other guidelines (mainly ones aimed at assessing existing structures rather than new construction) recommend rather more complex procedures, which may give more accurate results. For example, ASCE 41-13 (2014) allows the use of adaptive load patterns, which take account of load redistribution due to yielding, and simplifies the pushover curve to bilinear with a positive post-yield stiffness.

3.4.4 Non-linear time-history analysis

A final alternative, which remains comparatively rare, is the use of full non-linear dynamic analysis. In this approach, a non-linear model of the structure is analysed under a ground acceleration time history whose frequency content matches the design spectrum. The time history is specified as a series of data points at time intervals of the order of 0.01 s, and the analysis is performed using a stepwise procedure usually referred to as direct integration. This is a highly specialised topic, which will not be covered in detail here – see Clough and Penzien (1993) or Petyt (1998) for a presentation of several popular time integration methods and a discussion of their relative merits.

Since the design spectrum has been defined by enveloping and smoothing spectra corresponding to different earthquake time histories, it follows that there are many (in fact, an infinite

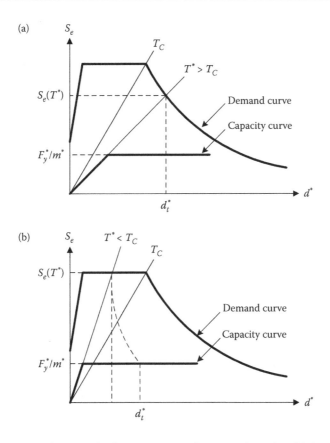

Figure 3.14 Determination of target displacement in pushover analysis for (a) long-period structure, (b) short-period structure.

number of) time histories that are compatible with the spectrum. These may be either recorded or artificially generated – specialised programs exist, such as SIMQKE, for generating suites of spectrum-compatible accelerograms. Different spectrum-compatible time histories may give rise to quite different structural responses, and so it is necessary to perform several analyses to be sure of achieving representative results. EC8 specifies that a minimum of three analyses under different accelerograms must be performed. If at least seven different analyses are performed, then mean results may be used, otherwise the most onerous result should be used.

Beyond being compatible with the design spectrum, it is important that earthquake time histories should be chosen whose time-domain characteristics (e.g. duration, number of cycles of strong motion) are appropriate to the regional seismicity and local ground conditions. Some guidance is given in Chapter 2, but this is a complex topic for which specialist seismological input is often needed.

3.5 CONCLUDING SUMMARY

A seismic analysis must take adequate account of dynamic amplification of earthquake ground motions due to resonance. The normal way of doing this is by using a response spectrum.

The analysis of the effects of an earthquake (or any other dynamic load case) has two stages:

1. Estimation of the dynamic properties of the structure – natural period(s), mode shape(s), damping ratio – these are structural properties, independent of the loading. The periods and mode shapes may be estimated analytically or using empirical formulae.
2. A response calculation for the particular load case under consideration. This calculation makes use of the dynamic properties calculated in (a), which influence the load the structure sustains under earthquake excitation.

Methods based on linear analysis (either multi-modal response analysis or equivalent static analysis based on a single mode of vibration) are widely used. In these cases, non-linearity is normally dealt with by using a ductility-modified response spectrum.

Alternative methods of dealing with non-linear behaviour (particularly static pushover methods) are growing in popularity and are permitted in EC8.

3.6 DESIGN EXAMPLE

3.6.1 Introduction

An example building structure has been chosen to illustrate the use of EC8 in practical building design. It is used to show the derivation of design seismic forces in the remaining part of this chapter, and the same building is used in subsequent chapters to illustrate checks for regularity, foundation design and alternative designs in steel and concrete. It is important to note that the illustrative examples presented herein and in subsequent chapters do not attempt to present complete design exercises. The main purpose is to illustrate the main calculations and design checks associated with seismic design to EC8 and to discussions of related approaches and procedures.

The example building represents a hotel, with a single-storey podium housing the public spaces of the hotel, surmounted by a seven-storey tower block, comprising a central corridor with bedrooms to either side. Figure 3.15 provides a schematic plan and section of the building, while Figure 3.16 gives an isometric view.

The building is later shown to be regular in plan and elevation (see Section 4.9). EC8 then allows the use of a planar structural model and the equivalent static analysis approach. There is no need to reduce q factors to account for irregularity. The calculation of seismic loads for equivalent static analysis can be broken down into the following tasks:

1. Estimate self-weight and seismic mass of building
2. Calculate seismic base shear in x-direction
3. Calculate distribution of lateral loads and seismic moment
4. Consider how frame type and spacing influence member forces

3.6.2 Weight and mass calculation

3.6.2.1 Dead load

For this preliminary load estimate, neglect weight of frame elements (resulting in same weight/mass for steel and concrete frame structures). Assume

- 150 mm concrete floor slabs throughout: $0.15 \times 24 = 3.6 \text{ kN/m}^2$
- Outer walls – brick/block cavity wall, each 100 mm thick, 12 mm plaster on inside face:
  - Brick: $0.1 \times 18 = 1.8$
  - Block: $0.1 \times 12 = 1.2$

Section

Plan

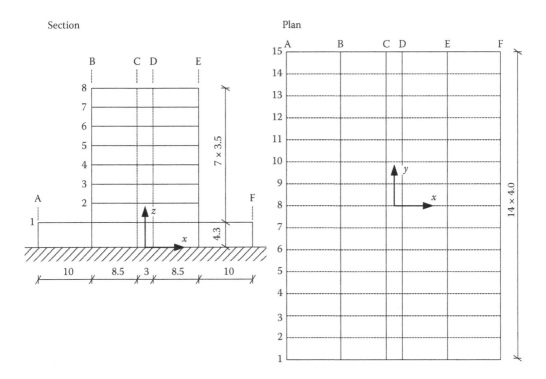

Figure 3.15 Schematic plan and section of example building.

Figure 3.16 Isometric view of example building.

Table 3.1 Dead load calculation

Level		Calculation	Load (kN)	Total (kN)
8	Slab	$(56 \times 20) \times 3.6$	4,032	
	Finishes	$(56 \times 20) \times 1.0$	1,120	5,152
2–7	Slab	$(56 \times 20) \times 3.6$	4,032	
	Finishes	$(56 \times 20) \times 1.0$	1,120	
	Outer walls	$(2 \times (56 + 20) \times 3.5) \times 3.25$	1,729	
	Internal walls (gl 2–14)	$(26 \times 8.5 \times 3.5) \times 1.7$	1,315	
	Internal walls (gl C, D)	$(2 \times 56 \times 3.5) \times 1.7$	666	8,862
1	Tower section (gl B-E)	As levels 2–7	8,862	
	Slab (gl A-B, E-F)	$(56 \times 20) \times 3.6$	4,032	
	Finishes (gl A-B, E-F)	$(56 \times 20) \times 1.0$	1,120	
	External glazing	$(2 \times (56 + 40) \times 4.3) \times 0.4$	330	14,344
Total dead load, G				**72,668**

- Plaster: $0.012 \times 21 = 0.25$
- Total = 3.25 kN/m²
- Internal walls – single leaf 100 mm blockwork, plastered both sides:
  - Block: $0.1 \times 12 = 1.2$
  - Plaster: $0.024 \times 21 = 0.5$
  - Total = 1.7 kN/m²
- Ground floor perimeter glazing: 0.4 kN/m²
- Floor finishes etc.: 1.0 kN/m²

The dead load calculations are set out in Table 3.1.

3.6.2.2 Imposed load

Imposed load calculations are set out in Table 3.2, assuming design values of 2.0 kN/m² for the bedrooms and roof, and 4.2 kN/m² elsewhere.

3.6.2.3 Seismic mass

Clause 3.2.4 states that the masses to be used in a seismic analysis should be those associated with the load combination:

$$G + \psi_{E,i}Q$$

Table 3.2 Imposed load calculation

Level		Calculation	Load (kN)	Total (kN)
8	Roof	$(56 \times 20) \times 2.0$	2,240	2,240
2–7	Corridors etc.	$((56 \times 3) + (8.5 \times 4) + (8.5 \times 8)) \times 4.0$	1,080	
	Bedrooms	$((56 \times 20)-270) \times 2.0$	1,700	2,780
1	Tower area	As levels 2–7	2,780	
	Roof terrace	$(56 \times 20) \times 4.0$	4,480	7,260
Total imposed load, Q				**26,180**

Table 3.3 Seismic mass calculation

Level	G (kN)	Q (kN)	$G + \psi_{E,i}Q$ (kN)	Mass (tonne)
8	5,152	2,240	5,824	593.7
2–7	8,862	2,780	9,696	988.4
I	14,344	7,260	16,522	1,684.2
			Total seismic mass	**8,208**

Take $\psi_{E,i}$ to be 0.3.

The seismic mass calculations are set out in Table 3.3.

The corresponding building weight is $8,208 \times 9.81 = 80,522$ kN.

3.6.3 Seismic base shear

First, define design response spectrum. Use Type 1 spectrum (for areas of high seismicity) soil type C. Spectral parameters are (from EC8 Table 3.2)

$$S = 1.15, \ T_B = 0.2 \text{ s}, \ T_C = 0.6 \text{ s}, T_D = 2.0 \text{ s}$$

The reference peak ground acceleration is $a_{gR} = 3.0$ m/s². The importance factor for the building is $\gamma_I = 1.0$, so the design ground acceleration $a_g = \gamma_I \, a_{gR} = 3.0$ m/s². The resulting design spectrum is shown in Figure 3.17 for $q = 1$ and $q = 4$, and design spectral accelerations can also be obtained from the equations in Cl. 3.2.2.5 of EC8.

The framing type has not yet been considered, so we will calculate base shear for three possible options:

- Steel moment-resisting frame (MRF)
- Concrete MRF
- Dual system (concrete core with either concrete or steel frame)

The procedure follows EC8 Cl. 4.3.3.2.2.

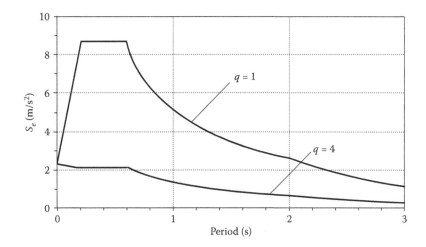

Figure 3.17 Design spectrum.

3.6.3.1 Steel MRF

Estimate natural period, EC8 Equation 4.6

$$T_1 = C_t \, H^{0.75}$$

For steel MRF $C_t = 0.085$, hence

$$T_1 = 0.085 \times 28.8^{0.75} = 1.06 \text{ s}$$

$T_C \leq T_1 \leq T_D$ so EC8 Equation 3.15 applies

$$S_d = a_g S \frac{2.5}{q} \frac{T_C}{T_1}$$

EC8 Table 6.2: Assuming ductility class medium (DCM), $q = 4$. Therefore

$$S_d = 3.0 \times 1.15 \times \frac{2.5}{4} \frac{0.6}{1.06} = 1.22 \text{ m/s}^2$$

EC8 Equation 4.5

$$F_b = \lambda m S_d$$

In this case $T_1 < 2T_C$ so $\lambda = 0.85$. Therefore

$$F_b = 0.85 \times 8{,}208 \times 1.22 = 8{,}515 \text{ kN}$$

Net horizontal force is $100 \times 8{,}515/80{,}522 = 10.6\%$ of total building weight.

3.6.3.2 Concrete MRF

Estimate natural period, EC8 Equation 4.6

$$T_1 = C_t \, H^{0.75}$$

For concrete MRF $C_t = 0.075$, hence

$$T_1 = 0.075 \times 28.8^{0.75} = 0.93 \text{ s}$$

$T_C \leq T_1 \leq T_D$ so EC8 Equation 3.15 applies

$$S_d = a_g S \frac{2.5}{q} \frac{T_C}{T_1}$$

EC8 Table 5.1: Assuming DCM, $q = 3.0\alpha_u/\alpha_1$, where α_u is the load factor to cause overall instability due to plastic hinge formation, and α_1 is the load factor at first yield in the structure.

Where these values have not been determined explicitly, for regular buildings, EC Cl. 5.2.2.2 allows default values of the ratio α_u/α_1 to be assumed. For our multi-storey, multi-bay frame, $\alpha_u/\alpha_1 = 1.3$, hence $q = 3 \times 1.3 = 3.9$.

Therefore

$$S_d = 3.0 \times 1.15 \times \frac{2.5}{3.9} \frac{0.6}{0.93} = 1.43 \, \text{m/s}^2$$

EC8 Equation 4.5

$$F_b = \lambda m S_d$$

In this case, $T_1 < 2\,T_C$ so $\lambda = 0.85$. Therefore

$$F_b = 0.85 \times 8,028 \times 1.43 = 9,954 \, \text{kN}$$

Net horizontal force is $100 \times 9,954/80,522 = 12.4\%$ of total building weight.

3.6.3.3 Dual system (concrete core with either concrete or steel frame)

Estimate natural period, EC8 Equation 4.6

$$T_1 = C_t \, H^{0.75}$$

For structures other than MRFs, EC8 gives $C_t = 0.05$, hence

$$T_1 = 0.05 \times 28.8^{0.75} = 0.62 \, \text{s}$$

(For buildings with shear walls, EC8 Equation 4.7 gives a permissible alternative method of evaluating C_t based on the area of shear walls in the lowest storey. This is likely to give a slightly shorter period than that calculated above. However, as the calculated value is very close to the constant-acceleration part of the response spectrum ($T_C = 0.6$ s), the lower period would result in very little increase in the spectral acceleration or the design base shear. This method has therefore not been pursued here.)

$T_C \leq T_1 \leq T_D$ so EC8 Equation 3.15 applies

$$S_d = a_g S \frac{2.5}{q} \frac{T_C}{T_1}$$

For dual systems, DCM, EC8 Table 5.1 gives $q = 3.0\alpha_u/\alpha_1$ and EC Cl. 5.2.2.2 gives a default value of the ratio $\alpha_u/\alpha_1 = 1.2$ for a wall-equivalent dual system. Hence $q = 3 \times 1.2 = 3.6$.
Therefore

$$S_d = 3.0 \times 1.15 \times \frac{2.5}{3.6} \frac{0.6}{0.62} = 2.32 \, \text{m/s}^2$$

EC8 Equation 4.5

$$F_b = \lambda m S_d$$

In this case, $T_1 < 2\,T_C$ so $\lambda = 0.85$. Therefore

$$F_b = 0.85 \times 8,208 \times 2.32 = 16,176 \, \text{kN}$$

Net horizontal force is $100 \times 16,176/80,522 = 20.1\%$ of total building weight.

3.6.4 Load distribution and moment calculation

The way the base shear is distributed over the height of the building is a function of the fundamental mode shape. For a regular building, EC8 Cl. 4.3.3.2.3 permits the assumption that the deflected shape is linear. With this assumption, the inertia force generated at a given storey is proportional to the product of the storey mass and its height from the base.

Since the assumed load distribution is independent of the form of framing chosen, and of the value of the base shear, we will calculate a single load distribution based on a base shear of 1,000 kN as shown in Table 3.4. This can then simply be scaled by the appropriate base shear value from Section 3.6.3.1, 3.6.3.2 or 3.6.3.3, as appropriate.

EC8 Equation 4.11 gives the force on storey k to be

$$F_k = F_b \frac{z_k m_k}{\sum_j z_j m_j}$$

The ratio of the total base moment to the base shear gives the effective height of the resultant lateral force:

$$h_{eff} = \frac{19,265}{1,000} = 19.3 \text{ m above the base}, \quad \text{and} \quad h_{eff}/h = 19.3/28.8 = 0.67.$$

3.6.5 Framing options

Although not strictly part of the loading and analysis task, it is helpful at this stage to consider the different possible ways of framing the structure.

3.6.5.1 Regularity and symmetry

The general structural form has already been shown to meet the EC8 regularity requirements in plan and elevation. A regular framing solution needs to be adopted to ensure that there is no large torsional eccentricity. Large reductions in section size with height should be avoided. If these requirements are satisfied, the total seismic loads calculated above can be assumed to be evenly divided between the transverse frames.

Table 3.4 Lateral load distribution using linear mode shape approximation

Level k	Height z_k (m)	Mass m_k (t)	$z_k m_k$ (m.t)	Force F_k (kN)	Moment $= F_k z_k$ (kNm)
8	28.8	593.7	17,098	139.6	4,020
7	25.3	988.4	25,006	204.1	5,165
6	21.8	988.4	21,547	175.9	3,835
5	18.3	988.4	18,087	147.7	2,702
4	14.8	988.4	14,628	119.4	1,767
3	11.3	988.4	11,169	91.2	1,030
2	7.8	988.4	7,709	62.9	491
1	4.3	1,684.2	7,242	59.2	254
Totals	–	**8,208**	**122,486**	**1,000.0**	**19,265**

3.6.5.2 Steel or concrete

Either material is suitable for a structure such as this, and the choice is likely to be made based on considerations other than seismic performance. The loads calculated above are based on a seismic mass, which has neglected the mass of the main frame elements. These will tend to be more significant for a concrete structure, which may therefore sustain somewhat higher loads than the initial estimates calculated here.

3.6.5.3 Frame type – moment-resisting, dual frame/shear wall system or braced frame

In the preceding calculations, both frame and dual frame/shear wall systems have been considered. In practice, it is likely to be advantageous to make use of the shear wall action of the service cores to provide additional lateral resistance. It can be seen that this reduces the natural period of the structure, shifting it closer to the peak of the response spectrum and thus increasing the seismic loads. However, the benefit in terms of the additional resistance would outweigh this disadvantage.

In general, MRFs provide the most economic solution for low-rise buildings, but for taller structures, they tend to sustain unacceptably large deflections and some form of bracing or shear wall action is then required. The height of this structure is intermediate in this respect, so that a variety of solutions are worth considering.

The load distributions for each of the frame types considered can be obtained by scaling the results from Section 3.6.4 by the base shears from Section 3.6.3 (Table 3.5).

Clearly the dual structure gives rise to significantly larger forces (because its lower period puts it closer to the peak of the response spectrum). However, it also provides a more efficient lateral load-resisting system, so it will not necessarily be uneconomic.

Steel braced frames have not been considered explicitly here. They would give rise to similar design forces to the dual system, since EC8 recommends the use of the same C_t value in the period calculation, and allows use of a slightly higher q factor (4 instead of 3.6).

3.6.5.4 Frame spacing

In the short plan (x) dimension, it is likely that columns would be provided at each of gridlines B, C, D and E, ensuring regularity and symmetry, and limiting beam spans to

Table 3.5 Total lateral forces for different frame types

Level	Total lateral forces (kN)		
	Steel MRF	Concrete MRF	Dual system
8	1,189	1,390	2,258
7	1,738	2,032	3,302
6	1,498	1,751	2,845
5	1,258	1,470	2,389
4	1,017	1,189	1,931
3	777	908	1,475
2	536	626	1,017
1	504	589	958
Base shear (kN)	**8,515**	**9,954**	**16,176**
Base moment (MN m)	**164.0**	**191.8**	**311.6**

reasonable levels. For vertical continuity, the framing of the tower should be continued down to ground level. It may then be desirable to pin the first floor roof terrace beams to the tower structure, so as to prevent them from picking up too much load.

In the long plan (y) dimension, the choice is between providing a frame at every gridline (i.e. at 4 m spacing) or at alternate gridlines (8 m spacing). With 8 m spacing, the seismic loads to be carried by a typical internal frame are simply those given above scaled by 8/56. With a 4 m spacing, these values would be halved.

3.6.5.5 Ductility class and its influence on q factor

All calculations so far have assumed DCM. If instead the structure is designed with high ductility (DCH), then higher q-factors may be used, further reducing the seismic loads. Since in all cases we are on the long-period part of the response spectrum, the spectral acceleration, and hence all seismic loads, are simply divided by q.

The design and detailing requirements to meet the specified ductility classes will be discussed in depth in the concrete design and steel design chapters. At this stage, it is worth noting that the EC8 DCH requirements for concrete are rather onerous and are unlikely to be achieved with the construction skills available. For steel, designing for DCH is likely to be more feasible.

Consider the effect of designing to DCH for the three frame types (refer to Tables 5.1 and 6.2 of EC8, and associated text).

For the steel MRF, EC8 Table 6.2 specifies that for DCH $q = 5\alpha_u/\alpha_1$. A default value of α_u/α_1 of 1.3 may be assumed, or a value of up to 1.6 may be used if justified by a static pushover analysis. Thus q may be taken as up to 6.5 by default, or up to 8.0 based on analysis. If we use a value of 6.5 (compared to 4.0 for DCM), then all seismic loads calculated above can be scaled by 4.0/6.5, that is, reduced by 38%.

For the concrete MRF, EC8 Table 5.1 specifies that for DCH $q = 4.5\alpha_u/\alpha_1$. A default value of α_u/α_1 of 1.3 may be assumed, or a value of up to 1.5 may be used if justified by a static pushover analysis. Thus q may be taken as up to 5.85 by default, or up to 6.75 based on analysis. If we use a value of 5.85 (compared to 3.9 for DCM) then the seismic loads calculated above can be scaled by 3.9/5.85, that is, reduced by 33%. A similar proportional reduction in loads can be achieved for the dual system.

If a steel concentrically braced frame were used, EC8 Table 6.2 specifies a maximum q value of 4.0 for both DCM and DCH, so changing to DCH would offer no benefit in terms of design loads.

REFERENCES

Clough R.W. and Penzien J. 1993. *Dynamics of Structures*, 2nd Ed, McGraw-Hill.

Craig R.R. 1981. *Structural Dynamics: An Introduction to Computer Methods*, Wiley.

EC8 2004. *Eurocode 8: Design of structures for earthquake resistance. General rules, seismic actions and rules for buildings.* EN1998-1:2004, European Committee for Standardisation, Brussels.

Fajfar P. 2002. Structural analysis in earthquake engineering – a breakthrough of simplified non-linear methods. *Proc. 12th European Conf. on Earthquake Engineering*, Elsevier, London, Paper 843.

ASCE. 2014. Seismic evaluation and retrofit of existing buildings. ASCE/SEI 41-13, American Society of Civil Engineers.

Hitchings D. 1992. *A Finite Element Dynamics Primer*, NAFEMS, London.

Krawinkler H. and Seneviratna G. 1998. Pros and cons of a pushover analysis for seismic performance evaluation. *Eng. Struct.*, 20, 452–464.

Lawson R.S., Vance V. and Krawinkler H. 1994. Nonlinear static pushover analysis – why, when and how? *Proc. 5th US Conference on Earthquake Engineering*, Chicago IL, Vol. 1, 283–292.

Petyt M. 1998. *Introduction to Finite Element Vibration Analysis*, Cambridge University Press.

Wilson E.L., der Kiureghian A. and Bayo E.R. 1981. A replacement for the SRSS method in seismic analysis. *Earthquake Engng Struct. Dyn.*, 9, 187–194.

Chapter 4

Basic seismic design principles for buildings

Edmund Booth and Zygmunt Lubkowski

CONTENTS

4.1 INTRODUCTION

Fundamental decisions taken at the initial stages of planning a building structure usually play a crucial role in determining how successfully the finished building achieves its performance objectives in an earthquake. This chapter describes how EC8 sets out to guide these decisions, with respect to siting considerations, foundation design and choice of superstructure.

4.2 FUNDAMENTAL PRINCIPLES

4.2.1 Introduction

In EC8 Part 1, the fundamental requirements for seismic performance are set out in Section 2. There are two main requirements. The first is to meet a 'no collapse' performance level, which requires that the structure retains its full vertical load-bearing capacity after an earthquake with a recommended return period of 475 years; longer return periods are given for special structures, for example casualty hospitals or high-risk petrochemical installations. After this earthquake, there should also be sufficient residual lateral strength and stiffness to protect life even during strong aftershocks. The second main requirement is to meet a 'damage limitation' performance level, which requires that the cost of damage and associated limitations of use should not be disproportionately high, in comparison with the total cost of the structure, after an earthquake with a recommended return period (for normal structures) of 95 years. Note that Section 2 of EC8 (and hence these basic requirements) applies to all types of structures, not just buildings.

EC8's rules for meeting the 'no collapse' performance level in buildings are given in Section 4 of Part 1 with respect to analysis procedures and in Sections 5 to 9 of Part 1 with respect to material-specific procedures to ensure sufficient strength and ductility in the structure. The rules for meeting the 'damage limitation' performance level in buildings are given in Section 4 of Part 1; they consist of simple restrictions on deflections to limit structural and non-structural damage, and some additional rules for protecting non-structural elements.

EC8 Part 1 Section 4.2.1 sets out some aspects of seismic design specifically for buildings, which should be considered at conceptual design stage, and which will assist in meeting the 'no collapse' and 'damage limitation' requirements. It is not mandatory that they should be satisfied, and indeed since they are qualitative in nature, it would be hard to enforce them, but they are sound principles which deserve study. Related, but quantified rules generally appear elsewhere in EC8; for example, the structural regularity rules in Section 4.2.3 supplement the uniformity and symmetry principles given in Section 4.2.1. Six guiding principles are given EC8 Part 1 Section 4.2.1 as follows, and these are now discussed in turn.

- Structural simplicity
- Uniformity, symmetry and redundancy
- Bidirectional resistance and stiffness
- Torsional resistance and stiffness
- Adequacy of diaphragms at each storey level
- Adequate foundations

4.2.2 Structural simplicity

This entails the provision of a clear and direct load path for transmission of seismic forces from the top of a building to its foundations. The load path must be clearly identified by the building's structural designer, who must ensure that all parts of the load path have adequate strength, stiffness and ductility.

Direct load paths will help to reduce uncertainty in assessing both strength and ductility, and also dynamic response. Complex load paths, for example involving transfer structures, tend to give rise to stress concentrations and make the assessment of strength, ductility and dynamic response more difficult. Satisfactory structures may still be possible with complex load paths but they are harder to achieve.

4.2.3 Uniformity, symmetry and redundancy

Numerous studies of earthquake damage have found that buildings with a uniform and symmetrical distribution of mass, strength and stiffness in plan and elevation generally perform much better than buildings lacking these characteristics.

Uniformity in plan improves dynamic performance by suppressing torsional response, as discussed further in Section 4.5. Irregular or asymmetrical plan shapes such as L or T configurations may be improved by dividing the building with joints to achieve compact, rectangular shapes (Figure 4.1), but this introduces a number of design issues that must be solved; these are avoiding 'buffeting' (impact) across the joint, and detailing the finishes, cladding and services which cross the joint to accommodate the associated seismic movements.

Uniformity of strength and stiffness in elevation helps avoid the formation of weak or soft storeys. Non-uniformity in elevation does not always lead to poor performance, however; for example, seismically isolated buildings are highly non-uniform in elevation but are found to perform very well in earthquakes.

Redundancy implies that more than one loadpath is available to transmit seismic loads, so that if a particular loadpath becomes degraded in strength or stiffness during an earthquake, another is available to provide a back-up. Redundancy should therefore increase reliability, since the joint probability of two parallel systems both having lower than expected capacity (or greater than expected demand) should be less than is the case for one system separately. Redundant systems, however, are inherently less 'simple' than determinate ones, which usually makes their assessment more complex.

4.2.4 Bi-directional resistance and stiffness

Unlike the situation that often applies to wind loads on buildings, seismic loads are generally similar along both principal horizontal axes of a building. Therefore, similar resistance in both directions is advisable. Systems such as cross wall construction found in some hotel

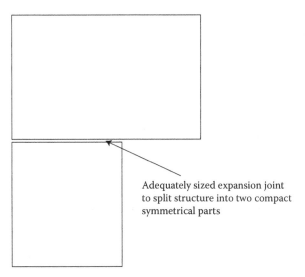

Adequately sized expansion joint to split structure into two compact symmetrical parts

Figure 4.1 Introduction of joints to achieve uniformity and symmetry in plan.

buildings, where there are many partition walls along the short direction but fewer in the long direction, work well for wind loading, which is greatest in the short direction, but tend to be unsatisfactory for seismic loads.

4.2.5 Torsional resistance and stiffness

Pure torsional excitation in an earthquake may arise in a site across which there are significantly varying soils, but significant torsional excitations on buildings are unusual. However, coupled lateral-torsional excitation, arising from an eccentricity between centres of mass and stiffness, is common and is found to increase damage in earthquakes. Such response may be inadequately represented by a linear dynamic analysis, because yielding caused by lateral-torsional response can reduce the stiffness on one side of a building structure and further increase the eccentricity between mass and stiffness centres.

Minimising the eccentricity of mass and stiffness is one important goal during scheme design, and achieving symmetry and uniformity should help to satisfy it. However, some eccentricity is likely to remain, and may be significant due to a number of effects, which may be difficult for the structural design to control; they may arise from uneven mass distributions, uneven stiffness contributions from non-structural elements or non-uniform stiffness degradation of structural members during a severe earthquake. Therefore, achieving good torsional strength and stiffness is an important goal. Stiff and resistant elements on the outside the building, for example in the form of a perimeter frame, will help to achieve this, while internal elements, such as a central core, contribute much less. Quantified rules are provided later in Section 4 of EC8 Part 1, as discussed in Section 4.5 of this chapter.

4.2.6 Adequacy of diaphragms at each storey level

Floor diaphragms perform several vital functions. They distribute seismic inertia loads at each floor level back to the main vertical seismic resisting elements, such as walls or frames. They act as a horizontal tie, preventing excessive relative deformations between the vertical elements, and so helping to distribute seismic loads between them. In masonry buildings, they act to restrain the walls laterally. At transfer levels, for example between a podium and a tower structure, they may also serve to transfer global seismic forces from one set of elements to another.

Floor diaphragms which have very elongated plan shapes, or large openings, are likely to be inefficient at the distribution of seismic loads to the vertical elements. Pre-cast concrete floors need to have adequate bearing to prevent the loss of bearing and subsequent floor collapse observed in a number of earthquakes. In masonry buildings, it is especially important to ensure a good connection between floors and the masonry walls they bear onto in order to provide lateral stability for the walls.

4.2.7 Adequate foundations

EC8 Part 1 Section 4.2.1.6 states that 'the design and construction of the foundations and of the connection to the superstructure shall ensure that the whole building is subjected to a uniform seismic excitation'. To achieve this, it recommends that a rigid cellular foundation should usually be provided where the superstructure consists of discrete walls of differing stiffnesses. Where individual piled or pad foundations are employed, they should be connected by a slab or by ground beams, unless they are founded on rock.

The interaction of foundations with the ground, in addition to interaction with the superstructure, is of course vital to seismic performance. Part 5 of EC8 gives related advice on conceptual seismic design of foundations, and this is further discussed in Chapters 8 and 9.

4.3 SITING CONSIDERATIONS

The regional seismic hazard is not the only determinant of how strongly a building may be shaken. (Regional seismic hazard is defined here as the ground shaking expected on a rock site as a function of return period.) Within an area of uniform regional hazard, the level of expected ground shaking is likely to vary strongly, and so is the threat from other hazards related to seismic hazard, such as land sliding or fault rupture, for reasons described in the next paragraph. Choice of the exact location of a building structure may not always be within a designer's control, but sometime even quite small changes in siting can make a dramatic difference to the seismic hazard.

The most obvious cause of local variation in hazard arises from the soils overlying bedrock, which affect the intensity and period of ground motions. It is not only the soils immediately below the site that affect the hazard; the horizontal profiles of soil and rock can also be important, due to 'basin effects'. Soil amplification effects are discussed in Chapters 8 and 9. Topographic amplification of motions may be significant near the crest of steep slopes. Fault rupture, slope instability, liquefaction, and shakedown settlement are other hazards associated with seismic activity which may also need to be considered. Figure 4.2 shows just a few examples where a failure to assess these phenomena has impinged on the performance of structures during a major earthquake.

Section 3 of EC8 Part 1 addresses soil amplification, Annex A of EC8 Part 5 addresses topographical amplification and Section 4 of EC8 Part 5 addresses the other siting considerations. By ensuring these potential hazards at a site are identified, the designer can take

Figure 4.2 Examples of poorly sited structures. (a) Fault Rupture – Luzon, Philippines 1990. (b) Liquefaction – Adapazari, Turkey 1999. (c) Slope Instability – Pacific Palisades, Los Angeles, 1994.

appropriate actions to minimise those hazards. In some cases, choice of a different site may be the best (or indeed only satisfactory) choice, for example to avoid building on an unstable slope or crossing a fault assessed as potentially active. If the hazard cannot be avoided, appropriate design measures must be taken to accommodate or mitigate it. For example, ground improvement measures may be one option for a site assessed as susceptible to liquefaction, and suitable articulation to accommodate fault movements may be possible for extended structures such as pipelines and bridges.

4.4 CHOICE OF STRUCTURAL FORM

The most appropriate structural material and form to use in a building is influenced by a host of different factors, including relative costs, locally available skills, environmental, durability, architectural considerations and so on. Some very brief notes on the seismic aspects are given below; further discussion is given in text books such as Booth (2014) and, for concrete structures, Fardis et al. (2015).

Steel has high strength to mass ratio, a clear advantage over concrete because seismic forces are generated through inertia. It is also easy to make steel members ductile in both flexure and shear. Steel moment frames can be highly ductile, although achieving adequate seismic resistance of connections can be difficult, and deflections may govern the design rather than strength. Braced steel frames are less ductile, because buckling modes of failure lack ductility, but braced frames possess good lateral strength and stiffness, which serves to protect non-structural as well as structural elements. Eccentrically braced frames (EBFs), where some of the bracing members are arranged so that their ends do not meet concentrically on a main member, but are separated to meet eccentrically at a ductile shear link, possess some of the advantages of both systems. More recently, buckling-restrained braces (also known as unbonded braces) have found more favour than EBFs in California; these consist of concentrically braced systems where the braces are restrained laterally but not longitudinally by concrete filled tubes, which results in a response in compression, which is as ductile as that in tension (Hamburger and Nazir 2003). Buckling-restrained braces combine ductility and stiffness in a similar way to EBFs.

Concrete has an unfavourably low strength to mass ratio, and it is easy to produce beams and columns which are brittle in shear, and columns which are brittle in compression. However, with proper design and detailing, ductility in flexure can be excellent, ductility in compression can be greatly improved by provision of adequate confinement steel and failure in shear can be avoided by 'capacity design' measures. Moreover, brittle buckling modes of failure are much less likely than in steel. Although poorly built concrete frames have an appalling record of collapse in earthquakes, well-built frames perform well. Concrete shear wall buildings do not typically suffer the 'pancake' collapses of framed buildings, and their inherent lateral stiffness provides good protection for non-structural elements such as cladding elements. Their performance in earthquakes is found to be generally good.

'Seismic isolation' involves the introduction of low lateral stiffness bearings to detune the building from the predominant frequencies of an earthquake; it has proved highly effective in the earthquakes of the past decade. 'Supplemental damping' involves the addition of damping elements to the structure, for example in the form of viscous dampers similar to the shock absorbers in cars. More recently, pre-stressed 'rocking systems', described by Pampanin (2012), have been built in New Zealand, the United States, Chile and elsewhere; a hospital with this system performed well in the Christchurch New Zealand earthquake of 2011. All these systems are discussed by Christopolous and Filiatrault (2006) but only seismic isolation is covered by the 2004 edition of EC8 Part 1; future editions are likely to be more comprehensive.

4.5 EVALUATING REGULARITY IN PLAN AND ELEVATION

4.5.1 General

EC8 Part 1 Section 4.2.3 sets out quantified criteria for assessing structural regularity, complementing the qualitative advice on symmetry and uniformity given in Section 4.2.1. Note that irregular configurations are allowed by EC8, but lead to more onerous design requirements.

A classification of 'non-regularity' in plan requires the use modal analysis, as opposed to equivalent lateral force analysis, and (generally) a 3-D as opposed to a 2-D structural model. For a linear analysis, a 3-D model would usually be chosen for convenience, even for regular structures. However, a non-linear static (push-over) analysis becomes much less straightforward with 3-D analysis models, and should be used with caution if there is plan irregularity, because of the difficulty in capturing coupled lateral-torsional modes of response. Other consequences of non-regularity in plan are the need to combine the effects of earthquakes in the two principal directions of a structure and for certain structures (primarily moment frame buildings) the q factor must be reduced by up to 13%. Moreover, in 'torsionally flexible' concrete buildings, the q value is reduced to 2 for medium ductility and 3 for high ductility, with a further reduction of 20% if there is irregularity in elevation. 'Torsionally flexible' buildings are defined in the next section.

A classification of 'non-regular' in elevation also requires the use of modal analysis, and leads to a reduced q factor, equal to the reference value for regular structures reduced by 20%.

The EC8 Manual (ISE/AFPS 2010) proposes some simplified methods of evaluating regularity, which are suitable for preliminary design purposes.

Section 4.9 of this book provides a worked example of assessing the regularity in plan and elevation of the demonstration building structure adopted for this book.

4.5.2 Regularity in plan

Classification as regular in plan requires the following.

1. 'Approximately' symmetrical distribution of mass and stiffness in plan.
2. A 'compact' shape, that is, one in which the perimeter line is always convex, or at least encloses not more than 5% re-entrant area (Figure 4.3).
3. The floor diaphragms shall be sufficiently stiff in-plane not to affect the distribution of lateral loads between vertical elements. EC8 warns that this should be carefully examined in the branches of branched systems, such as L, C, H, I and X plan shapes.
4. The ratio of longer side to shorter sides in plan does not exceed 4.
5. The torsional radius r_x in the x direction must exceed 3.33 times e_{ox} the eccentricity between centres of stiffness and mass in the x direction. Similarly, r_y must exceed 3.33 times e_{oy}. The terms r_x, r_y, e_{ox} and e_{oy} are defined below.
6. r_x and r_y must exceed the radius of gyration l_s, otherwise the building is classified as 'torsionally flexible', and the q values in concrete buildings are greatly reduced. The terms r_x, r_y and l_s are defined below.

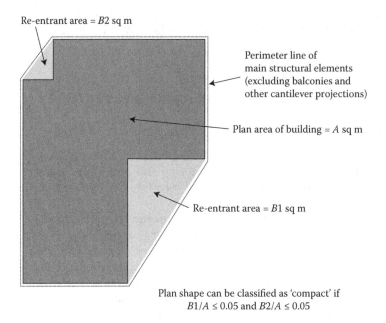

Re-entrant area = $B2$ sq m

Perimeter line of
main structural elements
(excluding balconies and
other cantilever projections)

Plan area of building = A sq m

Re-entrant area = $B1$ sq m

Plan shape can be classified as 'compact' if
$B1/A \le 0.05$ and $B2/A \le 0.05$

Figure 4.3 Definition of compact shapes.

The torsional radius r_x is the square root of the ratio of torsional stiffness (rotation per unit moment) to lateral stiffness in the x direction (deflection per unit force). A similar definition applies to r_y.

These are not exact definitions for a multi-storey building, since only approximate definitions of centre of stiffness and torsional radius are possible; they depend on the vertical distribution of lateral force and moment assumed. Approximate values may be obtained, based on the moments of inertia (and hence lateral stiffness) of the individual vertical elements comprising the lateral force resisting system; see Figure 4.4 and Equations 4.1 and 4.2. These equations are not reliable where the lateral load-resisting system consists of elements that assume different deflected shapes under lateral loading, for example unbraced frames combined with shear walls. Alternatively, using a computer analysis, values can be obtained from the deflections and rotations at each floor level found from the application of unit forces and torsional moments applied to a 3-D model of the structure; various vertical distributions of forces and moments may need to be considered. A worked example is provided in Section 4.9 below. Further advice is provided in the EC8 Manual (ISE/AFPS 2010).

$$x_{cs} \approx \sum \frac{(xEI_y)}{EI_y} \quad y_{cs} \approx \sum \frac{(yEI_x)}{EI_x} \tag{4.1}$$

$$r_x \approx \sqrt{\sum \frac{(x^2EI_y + y^2EI_x)}{\sum (EI_y)}} \quad r_y \approx \sqrt{\sum \frac{(x^2EI_y + y^2EI_x)}{\sum (EI_x)}} \tag{4.2}$$

The radius of gyration l_s is the square root of the ratio of the polar moment of inertia to the mass, the the polar moment of inertia being calculated about the centre of mass. For a

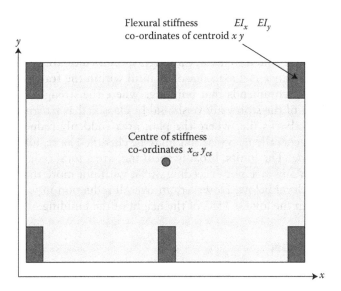

Figure 4.4 Approximate calculation of torsional radii (see also Equations 4.1 and 4.2).

rectangular building of side lengths l and b, and a uniform mass distribution, Equation 4.3 applies.

$$l_s = \sqrt{(l^2 + b^2)/12} \tag{4.3}$$

The requirement for torsional radius r_x to exceed 3.33 times the mass-stiffness eccentricity e_{ox} (item 5 on the list at the beginning of this section) relates the torsional resistance to the driving lateral-torsional excitation, correctly favouring configurations with stiff perimeter elements and penalising those relying on central elements for lateral resistance. It is very similar to a requirement that has appeared for many years in the Japanese code.

The requirement for r_x to exceed radius of gyration l_s (item 6 on the list at the beginning of this section) ensures that the first torsional mode of vibration does not occur at a higher period than the first translational mode in either direction, and demonstrating that this applies is an alternative way of showing that 'torsional flexibility' is avoided (EC8 Manual, ISE/AFPS 2010).

4.5.3 Regularity in elevation

Building must satisfy all the following requirements to be classified as regular in elevation.

1. All the vertical load-resisting elements must continue uninterrupted from foundation level to the top of the building or (where setbacks are present – see (4) below) to the top of the setback.
2. Mass and stiffness must either remain constant with height or reduce only gradually, without abrupt changes. Quantification is not provided in EC8; the EC8 Manual (ISE/ AFPS 2010) recommends that buildings where the mass or stiffness of any storey is less than 70% of that of the storey above or less than 80% of the average of the three storeys above should be classified as irregular in elevation.

3. In buildings with moment-resisting frames, the lateral resistance of each storey (i.e. the seismic shear initiating failure within that storey, for the code-specified distribution of seismic loads) should not vary 'disproportionately' between storeys. Generally, no quantified limits are stated by EC8, although special rules are given where the variation in lateral resistance is due to masonry infill within the frames. The EC8 Manual (ISE/AFPS 2010) recommends that buildings where the strength of any storey is less than 80% of that of the storey above should be classified as irregular in elevation.

4. Buildings with setbacks (i.e. where the plan area suddenly reduces between successive storeys) are generally irregular, but may be classified as regular if less than limits defined in the code. The limits, broadly speaking, are a total reduction in width from top to bottom on any face not exceeding 30%, with not more than 10% at any level compared to the level below. However, an overall reduction in width of up to half is permissible within the lowest 15% of the height of the building.

4.6 CAPACITY DESIGN

EC8 Part 1 Section 2.2.4 contains some specific design measures for ensuring that structures meet the performance requirements of the code. These apply to all structures, not just buildings, and a crucial requirement concerns capacity design, which determines much of the content of the material-specific rules for concrete, steel and composite buildings in Sections 5, 6 and 7 of EC8 Part 1.

Clause 2(P) of Section 2.2.4.1 states

> In order to ensure an overall dissipative and ductile behaviour, brittle failure or the premature formation of unstable mechanisms shall be avoided. To this end, where required in the relevant Parts of EN 1998, resort shall be made to the capacity design procedure, which is used to obtain the hierarchy of resistance of the various structural components and failure modes necessary for ensuring a suitable plastic mechanism and for avoiding brittle failure modes.

Professor Paulay's 'ductile chain' illustrates the principle of capacity design (see Figure 4.5). The idea is that the ductile link yields at load that is well below the failure load of the brittle links. Although most building structures are somewhat less straightforward than the chain used in Tom Paulay's example, one of the great strengths of the capacity design principle is that it relies on simple static analysis to ensure good performance, and is not dependent on the vagaries of a complex dynamic calculation.

Ensuring that columns are stronger than beams in moment frames, concrete beams are stronger in shear than in flexure and steel braces buckle before columns are three important examples of capacity design. A general rule for all types of frame building given in EC8 Part

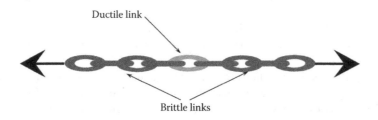

Figure 4.5 Capacity design – ensuring that ductile links are weaker than brittle ones.

1 Section 4.4.2.3 is that the moment strength of columns connected to a particular node should be 30% greater than the moment strength of the beams:

$$\sum M_{Rc} \geq 1.3 \sum M_{Rb} \tag{4.4}$$

The rule must be satisfied for concrete buildings, but the alternative capacity design rules given in EC8 Section 6.6.3 may apply to steel columns (see Chapter 6 of this book).

One feature of capacity design is that it ensures that designers identify clearly which parts of the structure will yield in a severe earthquake (the 'critical' regions) and which will remain elastic. An important related clause is given by clause 3(P) of Section 2.2.4.1.

> Since the seismic performance of a structure is largely dependent on the behaviour of its critical regions or elements, the detailing of the structure in general and of these regions or elements in particular, shall be such as to maintain the capacity to transmit the necessary forces and to dissipate energy under cyclic conditions. To this end, the detailing of connections between structural elements and of regions where non-linear behaviour is foreseeable should receive special care in design.

4.7 OTHER BASIC ISSUES FOR BUILDING DESIGN

4.7.1 Load combinations

Basic load combinations are given in EN1990: Basis for design, and for seismic load combinations are as follows:

$$\underset{\text{Design action effect}}{E_d} = \underset{\text{Permanent}}{\sum G_{kj}} + \underset{\text{Earthquake}}{A_{Ed}} + \underset{\text{Reduced variable load}}{\sum \psi_{2i} Q_{ki}} \tag{4.5}$$

ψ_{2i} is the factor defined in EN1990, which reduces the variable (or live) load from its characteristic (upper bound) value to its 'quasi-permanent' value, expected to be present for most of the time. It is typically in the range 0.0–0.8, depending on the variability of the loading type.

4.7.2 'Seismic' mass

The mass taken when calculating the earthquake loads should comprise the full permanent (or dead) load plus the variable (or live) load multiplied by a factor ψ_{Ei}. EC8 Part 1 Section 4.2.4 quantifies this as the factor ψ_{2i} defined in Section 4.7.1 above multiplied by a further reduction factor φ, which allows for the incomplete coupling between the structure and its live load:

$$\psi_{Ei} = \psi_{2i} \varphi \tag{4.6}$$

Typical values of φ are in the range 0.5–1, depending on the loading type.

4.7.3 Importance classes and factors

Four importance classes are recognised, as shown in Table 4.1, which also shows the recommended γ_I factor; this is, however, a 'nationally determined parameter' (NDP) which may be varied in the National Annex.

Table 4.1 Importance classes

Importance class	Buildings	γ_I
I	Buildings of minor importance for public safety, for example, agricultural buildings, etc.	0.8
II	Ordinary buildings, not belonging in the other categories.	1.0 (NB: not an NDP)
III	Buildings whose seismic resistance is of importance in view of the consequences associated with a collapse, for example, schools, assembly halls, cultural institutions etc.	1.2
IV	Buildings whose integrity during earthquakes is of vital importance for civil protection, for example, hospitals, fire stations, power plants, etc.	1.4

Note that whereas in US practice, the importance factors are applied to the seismic loads, in EC8 they are applied to the input motions. This makes an important difference when non-linear analysis is employed, since increasing the ground motions by X% may cause an increase of less than X% in forces, due to yielding of elements, but (possibly) more than X% in deflections, due to plastic strains and P-delta effects.

4.7.4 Primary and secondary members

EC8 Part 1 Section 4.2.2 distinguishes between primary and secondary elements. Primary elements are those which contribute to the seismic resistance of the structure. Some structural elements can, however, be designated as 'secondary' elements, which are taken as resisting gravity loads only. Their contribution to seismic resistance must be neglected. These elements must be shown to be capable of maintaining their ability to support the gravity loads under the maximum deflections occurring during the design earthquake. This may be done by showing that the actions (moments, shears, axial forces) that develop in them under the calculated seismic deformations do not exceed their design strength, as calculated in EC2. Otherwise no further seismic design or detailing requirements are required.

An example of the use of secondary elements occurs in a frame building is the following arrangement (Figure 4.6). The perimeter frame is considered as the primary seismic resisting element, and is designed for high ductility while the internal members are considered secondary. This gives considerable architectural freedom for the layout of the internal spaces; the column spacing can be much greater than would be efficient in a moment resisting frame, while close spaced columns on the perimeter represents much less obstruction.

4.7.5 Other design measures in EC8 Part 1 Section 2.2.4

The need for an adequate structural model for analysis is identified, and where necessary, soil deformability, the influence of non-structural elements and adjacent structures should be included in the analysis (clause 2.2.4 1(4)P). More detailed advice on analysis is given in the EC8 Manual (ISE/AFPS 2010).

The need for quality control is discussed and in particular, a formal quality system plan is specified for areas of high seismicity and structures of special importance (Section 2.2.4.3).

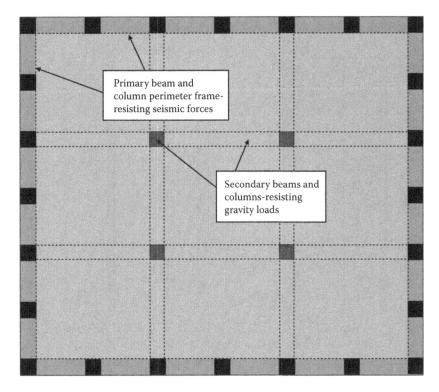

Primary beam and
column perimeter frame-
resisting seismic forces

Secondary beams and
columns-resisting
gravity loads

Figure 4.6 Building with external primary perimeter frame and internal secondary members.

Where a formal quality plan is applied to concrete buildings, a reduction in q values (and hence lateral strength requirements) is permitted (see Section 5.2.2.2(10)).

4.8 WORKED EXAMPLE FOR SITING OF STRUCTURES

4.8.1 Introduction

For this example, four sites (A, B, C and D) are postulated to be available for construction of the demonstration hotel structure. Preliminary site investigation was carried out at all the sites. Borehole data and SPT and field vane shear tests were carried out at each site. This information is shown in Figure 4.7.

4.8.2 Notes on key aspects of each site

SITE A: Loose sands below water table imply a high liquefaction risk. Piled foundations are likely to be necessary; piling through liquefiable material poses serious design problems associated with ensuring pile integrity and/or pile settlements.

SITE B: Strong stiffness contrast between top 5 m of soft clay and stiff clay below implies a high amplification of ground motions, especially around the 0.5 second ($= 4H/V_s$) period. Founding would need to be on to stiff clay, via piles or a deep basement.

SITE C: Stiff materials throughout form good foundation material with lowered potential for ground motion amplification. Shallow foundation is feasible.

Figure 4.7 Example borehole logs for possible sites for the building.

SITE D: A 6 m strata of soft clay may give rise to significant amplification of ground motions. Piling would be likely to be necessary into sandstone layer; relatively high shear strain differential between soft clay strata and stiffer strata above and below would probably result in plastic hinge formation in the piles.

4.8.3 Site selected for the hotel

Choose 'Site C' for shallow foundation design.
  Reasons:

1. Good, dense sand layer with 16 m thickness with high SPT numbers, overlying stiff clay
2. Angle of internal friction is 36°
3. Above the water table

4.9 WORKED EXAMPLE FOR ASSESSING STRUCTURAL REGULARITY

4.9.1 Introduction

The structural layout shown in Figure 4.8 is now checked for regularity in plan and elevation. A concrete frame and shear wall scheme has been adopted.

4.9.2 Regularity in plan

All the following conditions must be met.

1. 'Approximately' symmetrical distribution of mass and stiffness in plan.
   By inspection, it can be seen that a symmetrical distribution of stiffness has been achieved in plan, and there is no indication from the brief that significantly asymmetrical distributions of mass are to be expected.
2. A 'compact' shape, that is, one in which the perimeter line is always convex, or at least encloses not more than 5% re-entrant area (Figure 4.1).
   There are no re-entrant corners.
3. The floor diaphragms shall be sufficiently stiff in-plane not to affect the distribution of lateral loads between vertical elements. EC8 warns that this should be carefully examined in the branches of branched systems, such as L, C, H, I and X plan shapes.
   The floor slabs in the tower are rectangular, without branches, and have an aspect ratio in the tower (see 4 below) of 56 m/20 m = 2.8, which is relatively compact. Given the uniform distribution of mass and lateral load-resisting elements (i.e. the frames and shear walls) in the long direction, a continuous concrete solid slab or topping slab over pre-cast elements of at least 70 mm would not be expected to give rise to uneven load distributions, unless there were substantial openings in the slabs.
4. The ratio of longer side to shorter sides in plan does not exceed 4.
   The ratio in the tower is 2.8 (see above).
5. The torsional radius r_x in the X (short) direction must exceed 3.33 times e_{ox} the eccentricity between centres of stiffness and mass in the X direction. Similarly, r_y must exceed 3.33 times e_{oy}.

The EC8 Manual (ISE/AFPS 2010) gives conservative but simplified rules for satisfying this condition for some standard cases, but does not cover that of a uniform space frame with isolated shear walls, as here. The well distributed layout of shear walls and frames suggests that the structure should possess adequate torsional stiffness. A 3-D computer analysis was carried out to perform a detailed check, as follows.

Top deflection at top of building in X (short) direction under 1,000 kN load applied at stiffness centre in X direction:	7.35 mm
Top deflection at top of building in Y (long) direction under 1,000 kN load applied at stiffness centre in Y direction:	7.14 mm
Top rotation at top of building about Z (vertical) axis under 1,000 kNm moment about Z-axis	8.18 E–6 radians

(a)

3.0 m

13.0 m   5.4 m   11.6 m   6.9 m

8.0 m

All walls 3,500 mm by 350 mm

Transverse primary beams in tower 450 mm by 600 mm deep   Transverse primary beams in terrace 450 mm by 750 mm deep

All columns 750 mm square reducing to 600 mm square above level 4

All longitudinal primary beams: 450 mm by 600 mm deep

Single storey terrace   8 storey tower   Single storey terrace

Secondary beams, spanning in the transverse direction onto longitudinal beams at 8 m centres between primary beams, are omitted for clarity.

All floor and roof slabs are 150 mm thick solid concrete.

Figure 4.8 Structural layout taken for regularity checks. (a) Plan and (b) elevation.   *(Continued)*

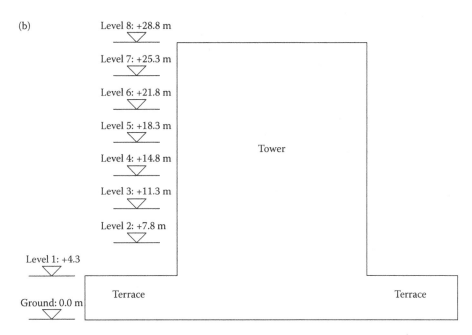

(b)

Level 8: +28.8 m

Level 7: +25.3 m

Level 6: +21.8 m

Level 5: +18.3 m

Level 4: +14.8 m

Tower

Level 3: +11.3 m

Level 2: +7.8 m

Level 1: +4.3

Ground: 0.0 m

Terrace

Terrace

Figure 4.8 (Continued) Structural layout taken for regularity checks. (a) Plan and (b) Elevation.

NB: The building is taken as perfectly symmetrical, and so the geometric centre, the centre of stiffness and the centre of mass all coincide. For cases where the stiffness and mass centre do not coincide with the geometric centre, see the example calculation in Appendix A of the EC8 Manual (ISE/AFPS 2010).

X stiffness = $1,000/(7.35E{-}3) = 136E3$ kN/m
Y stiffness = $1,000/(7.14E{-}3) = 140E3$ kN/m
Torsional stiffness = $1,000/(8.18E{-}6) = 122E6$ kN m/radian

$r_x = (122E6/140E3)^{1/2} = 29.5$ m
$0.3r_x = 0.3*29.5 = 8.9$ m

$r_y = (61.7E6/137E3)^{1/2} = 30.0$ m
$0.3r_y = 0.3*30 = 9.0$ m

Therefore, the separation between centres of mass and stiffness needs to be less than about 9 m.

6. r_x and r_y must exceed the radius of gyration l_s, otherwise the building is classified as 'torsionally flexible', and the q values in concrete buildings are greatly reduced.
The radius of gyration, assuming a uniform mass distribution, is calculated as follows. It can be seen that the requirement for regularity is satisfied.

$l_s = [(56^2 + 20^2)/12]^{1/2} = 17.2$ m $< r_x(= 29.5$ m$)$ and $< r_y(= 30$ m$)$ – ok.

The EC8 Manual (ISE/AFPS 2010) notes that an alternative demonstration that this condition is satisfied is to show that the first predominantly torsional mode has a lower period than either of the first predominantly translational modes in the two principal directions. A 3-D computer analysis, which assumed that the mass and stiffness centres coincided, gave the following values, confirming that this applies to the present structure. The period of the first torsional mode is well below that of the first two translational modes, reflecting the large excess of r_x and r_y over l_s calculated previously.

Period of first Y translational mode	0.90 s
Period of first X translational mode	0.88 s
Period of first torsional mode	0.62 s

Hence all the conditions for regularity in plan are satisfied.

4.9.3 Regularity in elevation

The following conditions must be met:

1. All the vertical load-resisting elements must continue uninterrupted from foundation level to the top of the building or (where setbacks are present – see (4) below) to the top of the setback.
   Satisfied by inspection.
2. Mass and stiffness must either remain constant with height or reduce only gradually, without abrupt changes. Quantification is not provided in EC8; the EC8 Manual (ISE/ AFPS 2010) recommends that buildings where the mass or stiffness of any storey is less than 70% of that of the storey above or less than 80% of the average of the three storeys above should be classified as irregular in elevation.
   The ground floor has a storey height of 4.3 m, compared with 3.5 m for the upper storeys, which tends to reduce stiffness by a factor of approximately $(3.5/4.3)^2 = 66\%$, which is a bit less than the 70% or 80% proposed above. However, there are more columns in the ground floor – an additional 50% – which offsets this, as does the base fixity of the ground floor columns and shear walls. Overall, this suggests that the stiffness ratio is within limits.
   A 3-D computer analysis shows that under earthquake loading, the ground floor storey drift is significantly less than that of the first floor, confirming that the stiffness check is satisfied. There is a stiffness change where the columns reduce in section at the fifth floor, but this is a reduction in stiffness so the regularity condition is met.
   The assumption that there is similar use of the floors in the tower at all levels above ground level leads to the conclusion that the mass at one level is always less that of the level below.
   Hence the 'soft storey' check is satisfied.
3. In buildings with moment-resisting frames, the lateral resistance of each storey (i.e. the seismic shear initiating failure within that storey, for the code-specified distribution of seismic loads) should not vary 'disproportionately' between storeys. Generally, no quantified limits are stated by EC8, although special rules are given where the variation in lateral resistance is due to masonry infill within the frames. The ISE Manual on EC8 (ISE 2008) recommends that buildings where the strength of any storey is less than 80% of that of the storey above should be classified as irregular in elevation.
   It is unlikely that any viable design would violate this condition. It cannot of course be checked without knowledge of the reinforcement in the beams, columns and walls.

4. Buildings with setbacks (i.e. where the plan area suddenly reduces between successive storeys) are generally irregular, but may be classified as regular if less than limits are defined in the code. The limits broadly speaking are a total reduction in width from top to bottom on any face not exceeding 30%, with not more than 10% at any level compared to the level below. However, an overall reduction in width of up to half is permissible within the lowest 15% of the height of the building.

The reduction in building width between the ground and first floors, as the tower rises above the podium, constitutes a setback. Since the ground floor height, at 4.3 m, is less than 15% of the total height of 28.8 m (28.8 times 0.15 = 4.32 m), and the reduction in width is from 40 m to 20 m (= 50% reduction) the setback remains (just) within 'regular' limits.

Hence all the conditions for regularity in elevation are satisfied.

REFERENCES

Booth E. 2014. *Earthquake Design Practice for Buildings*. ICE Publishing, London.

Christopoulos C. and Filiatrault A. 2006. *Principles of Passive Supplemental Damping and Seismic Isolation*. IUSS Press, Pavia Italy.

Fardis M., Carvalho E., Fajfar P. and Pecker A. 2015. *Seismic Design of Concrete Buildings to Eurocode 8*. CRC Press, Boca Raton, FL.

Hamburger R. and Nazir N. 2003. Seismic design of steel structures. In: Chen and Scawthorn (editors), *Earthquake Engineering Handbook*. CRC Press, Boca Raton, FL.

ISE/AFPS. 2010. *Manual for the Seismic Design of Steel and Concrete Buildings to Eurocode 8*. Institution of Structural Engineers, London.

Pampanin S. 2012. Reality-check and renewed challenges in earthquake engineering: Implementing low-damage structural systems – From theory to practice. In: *Proc. 15th World Conf. on Earthquake Engineering*, Lisbon.

Design of concrete structures

Andy Campbell and Mário Lopes

CONTENTS

5.1 INTRODUCTION

As noted in earlier chapters, EC8 aims to ensure life safety in a large earthquake together with damage limitation following a more frequent event. Whilst the code allows these events to be resisted by either dissipative (ductile) or non-dissipative (essentially elastic) behaviour, there is a clear preference for resisting larger events through dissipative behaviour. Hence, much of the code is framed with the aim of ensuring stable, reliable dissipative performance in pre-defined 'critical regions', which limit the inertial loads experienced by other parts of the structure. The design and detailing rules are formulated to reflect the extent of the intended plasticity in these critical regions, with the benefits of reduced inertial loads being obtained through the penalty of more stringent layout, design and detailing requirements.

This is particularly the case for reinforced concrete structures where such performance can only be achieved if strength degradation during hysteretic cycling is suppressed by appropriate detailing of these critical zones to ensure that stable plastic behaviour is not undermined by the occurrence of brittle failure modes such as shear or compression in the concrete or buckling of reinforcing steel.

With this in mind, three dissipation classes are introduced:

- Low (DCL) in which virtually no hysteretic ductility is intended and the resistance to earthquake loading is achieved through the strength of the structure rather than its ductility.
- Medium (DCM) in which quite high levels of plasticity are permitted and corresponding design and detailing requirements are imposed.
- High (DCH) where very large inelastic excursions are permitted accompanied by even more onerous and complex design and detailing requirements.

In this chapter, the primary focus is on DCM structures, which are likely to form the most commonly used group in practice. However, the limited provisions for DCL structures and the additional requirements for DCH structures are briefly introduced. Only the design of in-situ reinforced concrete buildings to EC8 Part 1 is addressed here. Rules for the design of pre-cast concrete structures are included in Section 5.11 of the code and guidance on their use in standard building structures is given in the Institution of Structural Engineers' manual on the application of EC8 (ISE/AFPS, 2010). Pre-stressed concrete structures, although not explicitly excluded from the scope of EC8 Part 1, are implicitly excluded as dissipative structures since the rules for detailing of critical regions are limited to reinforced concrete elements. Pre-stressed components could still be used within dissipative structures but should then be designed as protected elements, as discussed later.

5.2 DESIGN CONCEPTS

5.2.1 Energy dissipation and ductility class

EC8 is not a stand-alone code but relies heavily on the material Eurocodes to calculate resistance to seismic actions. Eurocode 2 (BS EN1992-1-1:2004 in the UK) fulfils this function for concrete structures. For DCL structures, EC8 imposes very limited material requirements in addition to the Eurocode 2 provisions, whereas for DCM and DCH structures, increasingly more onerous material requirements are imposed, together with geometrical constraints, capacity design provisions and detailing rules tied to local ductility demand.

These rules are aimed at the suppression of brittle failure modes, provision of capacity to withstand non-linear load cycles without significant strength degradation and improving the ability of defined critical regions to undergo very high local rotational ductility demands in order to achieve the lower global demands. Typically, this includes

- Ensuring flexural yielding prior to shear failure
- Providing stronger columns than beams to promote a more efficient beam side-way mode of response and avoid soft storey failure
- Retention of an intact concrete core within confining links
- Prevention of buckling of longitudinal reinforcement
- Limiting flexural tension reinforcement to suppress concrete crushing in the compression zone

These detailed requirements build upon the guidelines in Section 4 of EC8 Part 1 on

- Regularity of structural arrangement, aiming to promote an even distribution of ductility demand throughout the structure.
- Providing adequate stiffness, both to limit damage in events smaller than the design earthquake and to reduce the potential for significant secondary $P\text{-}\delta$ effects.

5.2.2 Structural types

EC8 Part 1 classifies concrete buildings into the following structural types:

- Frame system
- Dual system, which may be either frame or wall equivalent
- Ductile wall system
- System of large lightly reinforced walls

- Inverted pendulum system
- Torsionally flexible system

Apart from torsionally flexible systems, buildings may be classified as different systems in the two orthogonal directions.

Frame systems are defined as those systems where moment frames carry both vertical and lateral loads and provide resistance to 65% or more of the total base shear.

Conversely, buildings are designated as wall systems if walls resist 65% or more of the base shear. Walls may be classed as either ductile walls, which are designed to respond as vertical cantilevers yielding just above a rigid foundation, or as large lightly reinforced walls. Ductile walls are further sub-divided into coupled or uncoupled walls. Coupled walls comprise individual walls linked by coupling beams, shown in Figure 5.1, resisting lateral loads through moment and shear reactions in the individual walls together with an axial tensile reaction in one wall balanced by an axial compressive reaction in the other to create a global moment reaction. The magnitude of these axial loads is limited by the shear forces that can be transferred across the coupling beams. In order to qualify as a coupled wall system, the inclusion of coupling beams must cause at least a 25% reduction in the base moments of the individual walls from that which would have occurred in the uncoupled case. As coupled walls dissipate energy, not only in yielding at the base but also in yielding of the coupling beams, buildings with coupled walls may be designed for lower inertial loads than buildings with uncoupled walls to reflect their greater ductility and redundancy.

Large lightly reinforced walls are a category of structure introduced in EC8 and not found in other national or international seismic codes. These walls are assumed to dissipate energy, not through hysteresis in plastic hinges, but by rocking and uplift of the foundation, converting kinetic energy into potential energy of the structural mass and dissipating this through radiation damping. The dimensions of these walls or their fixity conditions or the presence of stiff orthogonal walls effectively prevent plastic hinging at the base. These provisions are likely to find wide application in heavy concrete industrial structures. However,

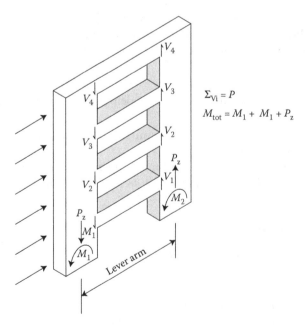

$$\Sigma_{Vi} = P$$
$$M_{tot} = M_1 + M_1 + P_z$$

Figure 5.1 Coupled wall system.

since this book is concerned primarily with conventional building structures, this type of structure is not considered further here.

Dual systems are structural systems in which vertical loads are carried primarily by structural frames but lateral loads are resisted by both frame and wall systems. From the earlier definitions, it is clear that, to act as a dual system, the frame and wall components must each carry more than 35% but less than 65% of the total base shear. When more than 50% of the base shear is carried by the frames, it is designated a frame-equivalent dual system. Conversely, it is termed a wall-equivalent dual system when walls carry more than 50% of the base shear.

Torsionally flexible systems are defined as those systems where the radius of gyration of the floor mass exceeds the torsional radius in one or both directions. An example of this type of system is a dual system of structural frames and walls with the stiffer walls all concentrated near the centre of the building on plan.

Inverted pendulum systems are defined as systems where 50% of the total mass is concentrated in the upper third of the height of the structure or where energy dissipation is concentrated at the base of a single element. A common example would normally be one storey frame structures. However, single-storey frames are specifically excluded from this category provided the normalised axial load, υ_d, does not exceed 0.3.

$$\upsilon_d = N_{Ed}/(A_c * f_{cd}) \tag{5.1}$$

where N_{Ed} is the applied axial load in the seismic design situation, A_c is the area of the column and f_{cd} is the design compressive strength of the concrete (i.e. the characteristic strength divided by the partial material factor, which can usually be taken as 1.5).

The treatment of both torsionally flexible and inverted pendulum systems within EC8 is discussed further in Section 5.4.

5.2.3 q factors for concrete buildings

Table 5.1 shows the basic values of q factors for reinforced concrete buildings. These are the factors by which the inertial loads derived from an elastic response analysis may be reduced to account for the anticipated non-linear response of the structure, together with associated aspects such as frequency shift, increased damping, over-strength and redundancy. The factor, α_u/α_1, represents the ratio between the lateral load at which structural instability occurs and that at which first yield occurs in any member. Default values of α_u/α_1 between 1.0 and 1.3 are given in the code with an upper limit of 1.5. Higher values than the default figures may be utilised but need to be justified by push-over analysis.

For walls or wall-equivalent dual systems, the basic value of the behaviour factor then needs to be modified by a factor, k_w, which accounts for the prevailing failure mode of the

Table 5.1 Basic value of the behaviour factor, q_0, for systems regular in elevation

Structural type	DCM	DCH
Frame system, dual system, coupled wall system	$3.0\alpha_u/\alpha_1$	$4.5\alpha_u/\alpha_1$
Uncoupled wall system	3.0	$4.0\alpha_u/\alpha_1$
Torsionally flexible system	2.0	3.0
Inverted pendulum system	1.5	2.0

wall, the q factors being reduced on squat walls where more brittle shear failure modes tend to govern the design.

$$k_w = (1 + \alpha_0)/3 \qquad (5.2)$$

where α_0 is the prevailing aspect ratio, h_w/l_w, of the walls.

A lower limit of 0.5 is placed on k_w for walls with an aspect ratio of 0.5 or less, with the basic q factor being applied unmodified to walls with an aspect ratio of 2 or more.

The basic q_0 factors tabulated are for structures which satisfy the EC8 regularity criteria, the basic factors needing to be reduced by 20% for structures which are irregular in elevation according to the criteria given earlier in Chapter 4.

5.2.4 Partial factors

In checking the resistance of concrete elements, the partial factors for material properties, γ_c and γ_s, for concrete and reinforcement respectively, are generally taken as those for the persistent and transient design situation rather than for the accidental design situation, which may initially appear to be more in keeping with an infrequent event, such as the design earthquake. Hence, a value of 1.5 is adopted for γ_c and 1.15 for γ_s in the UK, the values being defined in the National Annex to Eurocode 2 for each country. This practice is based upon an implicit assumption that the difference between the partial factors for the persistent and transient situation and those for the accidental situation are adequate to cater for possible strength degradation due to cyclic deformations. The use of these material factors has the added benefit to the design process that standard Eurocode 2 design charts can be used.

5.3 DESIGN CRITERIA

5.3.1 Capacity design

Capacity design is the basic concept underpinning the EC8 design philosophy for ductile structures (DCM and DCH). Therefore, it is important to fully understand this basic principle in order to place in context the design rules aimed at implementing it.

This concept can be exemplified considering the chain, introduced by Paulay (1993) and represented in Figure 5.2, in which link 1 is ductile and all other links are brittle.

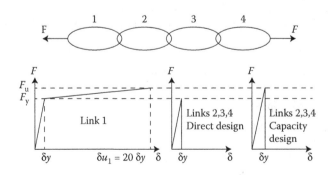

Figure 5.2 Ductility of chain with brittle and ductile links.

According to standard design procedures for quasi-static loading (termed direct design), as the applied force is the same for all links, the design force is also equal on all links. Therefore, assuming there are no reserve strengths, the yield capacity of all links is the same. In this situation, the system cannot resist any force above F_y, at which rupture of the brittle links takes place. Therefore with direct design the overall increase in length of the chain at rupture is

$$\delta u = 4\delta y \qquad (5.3)$$

According to capacity design principles, to maximise the ductility of the chain, some links have to be chosen to have ductile behaviour and be designed with that purpose. The rest of the structure must be designed with excess strength in order to remain elastic during the plastic deformations of the ductile links. For this purpose, the design force of the brittle links must be equal to the maximum resistance of the ductile links after yielding, this is, a force equal or above F_u. The ductile link behaves like a fuse, which does not allow the applied force acting on the brittle links to increase above their maximum resistance. Therefore, the force applied on the chain can increase above F_y up to the value F_u, but cannot exceed this value. At this stage, the chain collapses at a displacement much higher than the chain designed with the direct design methodology, as follows:

$$\delta u = 3\delta y + \delta u_1 = 3\delta y + 20\delta y = 23\delta y \qquad (5.4)$$

Hence, the brittle links must be designed for a force different from the ductile link, which is a function not of the notional applied load but of the capacity of the ductile link, in order to prevent the premature failure of the brittle links before the deformation capacity of the ductile links is exhausted. The fact that the design action-effects in pre-defined 'protected' elements are a function of the resistance of other key elements is a basic characteristic of capacity design, and is an important difference to standard design procedures for quasi-static loading.

This highlights the fact that the indiscriminate provision of excess strength, which is usually considered positive according to standard design procedures, may adversely affect the non-linear behaviour of a structural system, as it may prevent an intended ductile link from acting like a fuse. Hence, if after designing a ductile frame, the flexural reinforcement of beams or of the base section of walls is increased this is not necessarily a 'safe' change since it may increase the forces transmitted to other parts of the structure.

While capacity design is an important concept for seismic design in all materials, it is included here because it is particularly relevant to reinforced concrete structures, which can potentially exhibit brittle failure modes unless attention is paid to suppressing these modes in the design and detailing.

In the case of reinforced concrete elements, the best way to dissipate energy is by flexural yielding, as shear and axial forces tend to induce brittle behaviour. Therefore, the ductility of a structure can generally be optimised by enforcing flexural yielding at specific locations (ductile links), called plastic hinge zones, avoiding any type of shear or axial compressive failure (brittle links) and designing the rest of the structure to remain elastic throughout the development of the plastic hinges.

The approach adopted by EC8 to promote capacity design of reinforced concrete structures is to choose critical regions of the structure (the plastic hinge zones referred to above) that are designed to yield in flexure when subject to the design earthquake loading, modified by the q factor appropriate to the structural system. These critical regions are then detailed to undergo large, inelastic cyclic deformations and fulfil the role of structural 'fuses', limiting the inertial loads that can be transferred to the remaining 'protected' parts of the structure, which can then be designed to normal Eurocode 2 provisions.

The capacity design rules in EC8 are discussed in more detail later but primarily cover:

- Derivation of shear forces in members from the flexural capacity of their critical regions.
- Promotion of the strong column/weak beam hierarchy in frame structures, evaluating column moments as a function of the capacity of the beams framing into them.

In both cases, in the design of notionally elastic parts of the structure, an allowance for over-strength of the critical regions is made, a greater allowance being made for DCH than for DCM structures.

5.3.2 Local ductility provisions

The EC8 design rules take account of the fact that, to achieve the global response reductions consistent with the q factor chosen, much greater local ductility has to be available within the critical regions of the structure. Design and detailing rules for these critical regions are therefore formulated with the objective of ensuring that

- Sufficient curvature ductility is provided in critical regions of primary elements.
- Local buckling of compressed steel within plastic hinge regions is prevented.

This is fulfilled by special rules for confinement of critical regions, particularly at the base of columns, within beam/column joints and in boundary elements of ductile walls, which depend, in part, on the local curvature ductility factor μ_Φ. This is related to the global q factor as follows:

$$\mu_\Phi = 2q_0 - 1 \quad \text{if } T_1 \geq T_C \tag{5.5}$$

$$\mu_\Phi = 1 + 2(q_0 - 1)T_C/T_1 \quad \text{if } T_1 < T_C \tag{5.6}$$

where q_0 is the basic behaviour factor given in Table 5.1 before any reductions are made for lack of structural regularity or low aspect ratio of walls. T_1 is the fundamental period of the building and T_C is the period at the upper end of the constant acceleration zone of the input spectrum as described in Section 3.2 of EC8 Part 1.

Additionally, if Class B reinforcement is chosen rather than Class C in DCM structures, the value of μ_Φ should be at least 1.5 times the value given by Equations 5.5 and 5.6, whichever is applicable.

5.3.3 Primary and secondary members

Primary elements are specified as being those elements that contribute to the seismic resistance of the structure and are designed and detailed to the relevant provisions of EC8 for the designated ductility class. Elements that are not part of the main system for resisting seismic loading can be classed as secondary elements. They are assumed to make no contribution to seismic resistance and secondary concrete elements are designed to Eurocode 2 to resist gravity loads together with imposed seismic displacements derived from the response of the primary system. In this case, no special detailing requirements are imposed upon these elements.

A common problem in seismic design is that of unintentional stiffening of the designated seismic load-resisting system by secondary or non-structural elements (e.g. masonry partition walls), leading to a higher frequency of response and generally increased inertial loads. To guard against this, EC8 specifies that the contribution of secondary elements to the lateral stiffness should be no more than 15% of that of the primary elements. If secondary elements do not meet this criterion, one option is to provide flexible joints to prevent stiffening of the primary system by these elements.

Whilst this stiffness limit protects against the global effects of unintentional stiffening, the designer also needs to be aware of potentially adverse local effects such as

- Local changes to the intended load paths, potentially leading to increased loads on members not designed to cater for them or introducing a lack of regularity into discrete areas of the structure, modifying their dynamic response.
- Stiffening of parts of individual members (e.g. columns restrained by masonry panels over part of their height) preventing the intended ductile flexural response from occurring and resulting in a brittle shear failure.

Guidance on local stiffening issues associated with the most common case of masonry infill panels is given in Sections 4.3.6 and Section 5.9 of the code.

5.3.4 Stiffness considerations

Apart from its major influence in determining the magnitude of inertial loads, dealt with in earlier chapters, structural stiffness is important in meeting the damage limitation provisions of EC8 Part 1 (Clause 4.4.3) and in assessing the significance of P-δ effects as per Clause 4.4.2.2 (2) to (4).

Both effectively place limits on storey drift, the former explicitly albeit for a lower return period earthquake, and the latter implicitly through the inter-storey drift sensitivity coefficient, θ. In both cases, the relative displacements between storeys, d_r, if obtained from a linear analysis, should be multiplied by a displacement behaviour factor, q_d, to obtain the plastic relative displacements, d_r. When the period of response of the structure is greater than T_C (i.e. on the constant displacement or constant velocity portion of the response spectrum), q_d is equal to the behaviour factor q, so that the plastic displacement is equal to the elastic displacement obtained from the unreduced input spectrum. However, q_d exceeds q at lower periods as defined in Appendix B of the code.

In calculating displacements, EC8 requires that the flexural and shear stiffness of concrete structures reflect the effective stiffness consistent with the level of cracking expected at the initiation of yield of the reinforcement. If the designer does not take the option of calculating the stiffness reduction directly through push-over analysis, for example, the code allows the effective stiffness to be based upon half of the gross section stiffness [Clause 4.3.1 (7)] to account for softening of the structure at the strain levels consistent with reinforcement yield. It is acknowledged that the true stiffness reduction would probably be greater than this but the value chosen is a compromise; lower stiffness being more onerous for P-δ effects but less onerous for calculation of inertial loading on the structure. The EC8 approach, whilst similar to performance-based methodologies elsewhere, differs in applying a uniform stiffness reduction independent of the type of element considered. Paulay and Priestley (1992) and

Priestley (2003) propose greater stiffness reductions in beams than in columns, reflecting the weak beam/strong column philosophy and the beneficial effects of compressive axial loads.

Checks on damage limitation aim to maintain the maximum storey drifts below limiting values set between 0.5% and 1% of the storey height, dependent upon the ductility and fixity conditions of the non-structural elements. The amplified displacements for the design earthquake are modified by a reduction factor, ν, of either 0.4 or 0.5, varying with the importance class of the building, to derive the displacements applicable for the more frequent return period earthquake considered for the damage limitation state.

The inter-storey drift sensitivity coefficient, θ, used to take account of P-δ effects, is defined in the equation:

$$\theta = P_{tot} * d_r / (V_{tot} * h) \tag{5.7}$$

P_{tot} is the total gravity load at and above the storey, V_{tot} the cumulative seismic shear force acting at each storey and h the storey height. If the maximum value of θ at any level is less than 0.1, then P-δ effects may be ignored. If θ exceeds 0.3, then the frame is insufficiently stiff and an alternative solution is required.

For values of θ between 0.1 and 0.2, an approximate allowance for P-δ effects may be made by increasing the analysis forces by a factor of $1/(1 - \theta)$ whilst, for values of θ of between 0.2 and 0.3, a second order analysis is required.

5.3.5 Torsional effects

A simplified approach towards catering for the increase in seismic forces due to accidental eccentricity in regular structures is given in Clause 4.3.3.2.4 of EC8 Part 1. Loads on each frame are multiplied by a factor, δ, equal to $[1 + 0.6(x/L_e)]$ where x is the distance of the frame from the centre of mass and L_e is the distance between the two outermost load resisting elements.

Hence, for a building where the mass is uniformly distributed, the forces and moments on the outermost frames are increased by 30%. Fardis et al. (2005) note that this simplified method is conservative by a factor of 2 on average for structures with the stiffness uniformly distributed in plan. Where this is judged to be excessive, the general approach of Clause 4.3.2 may be applied within a 3-D analysis.

However, as the expression for δ was derived for structures with the stiffness uniformly distributed in plan, it may produce unsafe results for structures with a large proportion of the lateral stiffness concentrated at a single location. Therefore, it should not be applied to torsionally flexible systems.

5.4 CONCEPTUAL DESIGN

As already referred to in Chapter 4, EC8 provides guidance on the basic principles of good conception of building structures for earthquake resistance. These principles apply to all types of buildings and are qualitative and not mandatory. However, in Section 5 'Specific Rules for Concrete Buildings', besides providing guidance and rules for the design of several types of reinforced concrete building structures, EC8 clearly encourages designers to choose the most adequate structural types. Next the most important quantitative aspects and clauses of Section 5 of EC8 which condition the choice of structural types are high-lighted. These are

- Reduction of the q-factors assigned to the less adequate structural types and to irregular structures.
- The control of inter-storey drifts, which tends to penalise more flexible structures.

The reduction of the q-factors is apparent in Table 5.1. The torsionally flexible system and the inverted pendulum system are clearly penalised with q-factors that can be less than half of the ones prescribed for the frame, dual or coupled wall systems. Buildings with walls may fall under the classification of torsionally flexible if the walls are concentrated at a single location in plan. Buildings with several walls closer to the periphery of the floor plans tend to meet the criterion that avoids this type of classification.

Buildings with irregularities in plan or along the height are penalised as the irregularities tend to induce concentration of ductility demands at some locations of the structure opposed to the more uniform spread of ductility demands in regular buildings. In particular, the interruption of vertical elements which are important for the resistance to horizontal inertia forces (including both columns and walls, the latter being particularly important) before reaching the foundations, is a type of irregularity that the observation of past earthquakes shows is more likely to lead to catastrophic failure. EC8 only includes a moderate reduction of 20% in the behaviour factor for structures with this type of irregularity, but designers are cautioned to avoid it if at all possible.

The above means there is a price to pay for the use of these systems and buildings, both in terms of increasing amounts of reinforcement and dimensions of structural elements, which in regions of medium and high seismicity may create problems of compatibility with architectural requirements.

The control of inter-storey drifts may render frame or frame equivalent dual structures not stiff enough to meet these requirements, especially in the cases of tall buildings in regions of medium and high seismicity, prompting designers to conceive coupled wall or wall equivalent dual structures. These structural types generally present a better combination of stiffness and ductility characteristics, important for the seismic behaviour of reinforced concrete buildings.

Other requirements of design, related to the application of capacity design, may influence the overall structural conception of the buildings in order to make it possible to satisfy those requirements. The most important is the weak-beam/strong-column design of frames, referred to in Section 5.6.2, aimed at preventing the formation of soft-storey mechanisms. For this purpose, it is necessary that the sum of the flexural capacity of the columns converging at a joint is greater than the flexural capacity of the beams converging at the same joint. In practical terms, this implies that the dimension of the columns in the bending plane must not be much smaller than the dimension of the beams. This is not too difficult to enforce in a single plan direction, but its implementation in two orthogonal plan directions simultaneously may imply that both dimensions of the columns have to be large. This is likely to create difficulties in compatibility with architectural requirements as, in many cases, architects wish columns to protrude as little as possible from inner partition walls and exterior walls. However, the weak-beam/strong column requirement is not mandatory in all ductile structures with frames: the inclusion of walls with reasonable stiffness and strength in the horizontal resisting system of reinforced concrete buildings is enough for the prevention of the formation of soft-storey mechanisms, associated with hinging of both extremities of all columns at one floor. This is because there is a kinematic incompatibility between the wall deformation and the deformation of the frames at the formation of the soft-storey

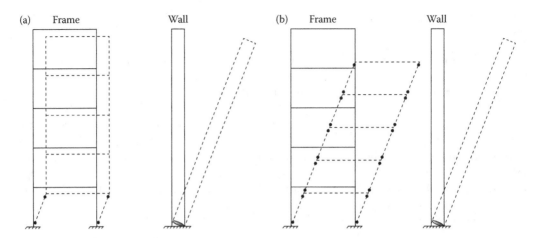

Figure 5.3 Kinematic incompatibility between wall deformation and soft-storey.

mechanism, as illustrated in Figure 5.3a. Figure 5.3b shows that in dual systems if the hinges develop at the columns a mechanism can only develop if plasticity spreads throughout the height of the building.

In order to quantify how 'reasonable' the wall stiffness and strength is, EC8 establishes that for the above purpose, the walls must absorb at least 50% of the total base shear in the seismic design situation. Therefore, in wall or wall equivalent dual structures, the walls are considered stiff enough to prevent the formation of the soft-storey mechanisms, regardless of frame design.

This allows designers to solve the above-mentioned problem of compatibility between the weak-beam/strong column design with architectural requirements by providing at least in one direction walls stiff enough to take at least half of the global seismic shear in that direction. And the fact that there is no need to enforce the weak-beam/strong column requirement simplifies the design process. This makes the choice of wall or wall equivalent dual structures even more advantageous.

The control of inter-storey drifts due to the presence of walls in dual or wall structures, in particular the ones not designed for the lower levels of lateral stiffness and strength (DCH), also helps to limit the possible consequences of effects associated with the presence of secondary structural elements or non-structural elements, as already referred to in Section 5.3.3. For this reason, the additional measures prescribed by EC8 to account for the presence of masonry infills apply only to frame or frame equivalent dual structures (clause 4.3.6.1) of DCH structures.

Another requirement, with possible implications for all vertical structural elements in all types of structures, is the limitation of the axial force, aiming at restricting the negative effects of large compressive forces on the available ductility. In order to meet this requirement, in some cases, designers may be forced to conceive columns with cross section areas larger than desirable for compatibility with architectural requirements.

5.5 DESIGN FOR DCL

As noted earlier, EC8 permits the design of structures for non-dissipative behaviour. If this option is taken, then standard concrete design to Eurocode 2 should be carried out, the only additional requirement being that reasonably ductile reinforcing steel, Class B or C as defined in Eurocode 2, must be used. A q factor of up to 1.5 is permitted, this being regarded as effectively an over-strength factor. However, other than for design of secondary elements, the DCL option is only recommended for areas of low seismicity as defined by Clause 3.2.1(4) of EC8 Part 1.

5.6 FRAMES: DESIGN FOR DCM

5.6.1 Material and geometrical restrictions

There are limited material restrictions for DCM structures. In addition to the requirement to use Class B or C reinforcement, as for DCL, only ribbed bars are permitted as longitudinal reinforcement of critical regions and concrete of class C16/20 or higher must be used.

Geometrical constraints are also imposed on primary elements.

Beams	In order to promote an efficient transfer of moments between columns and beams, and reduce secondary effects, the offset of the beam centre line from the column centre line is limited to less than a quarter of the column width.
	Also, to take advantage of the favourable effect of column compression on the bond of reinforcement passing through the beam/column joint:
	Width of beam ≤ (column width + depth of beam) ≤ twice column width if less
	This requirement makes the use of flat slabs in ductile frames inefficient as the slab width that contributes to the stiffness and strength of the frame is reduced. Their use as primary elements is further discouraged by Clause 5.1.1(2).
Columns	The cross-sectional dimension should be at least 1/10th of the distance between point of contraflexure and the end of the column, if the inter-storey drift sensitivity coefficient θ is larger than 0.1.

5.6.2 Calculation of action effects

Action effects are calculated initially from analytical output, for elements and effects associated with non-linear ductile behaviour, and then from capacity design principles for effects that are to be resisted in the linear range.

In frame structures, the starting point is the calculation of beam flexural reinforcement to resist the loads output from the analysis for the relevant gravity load and seismic combination with the seismic loads reduced by the applicable q factor and factored as appropriate to account for $P\text{-}\delta$ effects and accidental eccentricity.

The shear actions on the beam should then be established from the flexural capacity for the actual reinforcement arrangement provided.

The shear force is calculated from the shear that develops when plastic hinges develop in the critical regions at each end of the beam. This equates to the sum of the negative yield moment capacity at one end and the positive yield moment capacity at the other, divided by the clear span, to which the shear due to gravity loads should be added. The yield moment is calculated from the design flexural strength, multiplied by an over-strength factor γ_{Rd}, but this is taken as 1.0 in DCM beams. In calculating the hogging capacity of the beam, the slab reinforcement within an effective flange width, defined in Clause 5.4.3.1.1(3), needs to be included. If the reinforcement differs at opposite ends of the beam, the calculation must be repeated to cater for sway in both directions.

The beam shear forces may be reduced in cases where the sum of the column moment strengths at the joint being considered is less than the sum of the beam moment strengths. This will not generally apply because of the provisions encouraging a strong column/weak beam mechanism, and only usually occurs in the top storeys of multi-storey frames, or in single storey frames.

The principle is illustrated in Figure 5.4, following the rules of EC8.

The moments that should then be applied to the columns are also calculated from capacity design principles to meet the strong column/weak beam requirement.

$$\sum M_{Rc} \geq 1.3 \; \sum M_{Rb} \tag{5.8}$$

where

$\sum M_{Rc}$ is the sum of the column strengths provided at the face of the joint.

$\sum M_{Rb}$ is the sum of the beam strengths provided at the face of the joint.

This rule need not be observed in the top storeys of multi-storey frames, or in single-storey frames. It is also not necessary to apply this rule in frames belonging to wall or wall-equivalent dual structures.

Therefore, if a structure is classified as a frame system or frame-equivalent dual system in only one vertical plane of bending, there is no need for this rule to be satisfied in the orthogonal plane.

The proportion of the summed beam moments to be resisted by the column sections above and below the beam/column joint should be allocated in accordance with the relative stiffnesses. Fardis et al. (2005) suggest that for columns of equal proportions and spans, 45% of

$$\sum M_{Rb1} < \sum M_{Rc1}$$
$$\gamma \sum M_{Rc1} * (\sum M_{Rb}/\sum M_{Rc})$$

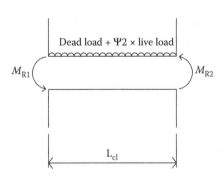

Usual case: $\sum M_{Rb1} < \sum M_{Rc1}$
$$M_{Ri} = \gamma M_{Rd} M_{Rbi}$$
Infrequent case: $\sum M_{Rb} > \sum M_{Rc}$
$$M_{Ri} = \gamma_{Rd} M_{Rbi} * (\sum M_{Rc}/\sum M_{Rb})$$

$$V = \frac{M_{R1} + M_{R2}}{L_{cl}} + V\text{static}$$

Figure 5.4 Capacity design values of shear forces in beams.

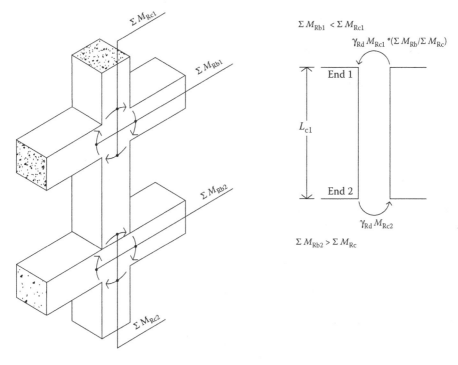

Figure 5.5 Capacity design values of shear forces in columns.

the total moment should be assigned to the column above the joint and 55% to that below. This aims at constant column reinforcement, allowing for the flexural capacity of the column generally increasing with axial compression.

Having obtained the flexural demand on the column, its capacity to carry combined flexure and axial load can be checked against standard Eurocode 2 interaction charts.

The shear load to be applied to the column is then derived from the flexural capacity in a similar way to that described above for beams, as shown in Figure 5.5. Generally, there will be significant axial loads in the column, which affects the moment strength. Also, there will not usually be significant lateral loading within the length of the column, so there is no additional term analogous to the gravity loading applied to the beams. However, in all cases, the moment strength at each end of the column is factored by γ_{Rd}, (equal to 1.1 for DCM columns) and may also be factored by $\Sigma M_{Rb}/\Sigma M_{Rc}$ provided this ratio is less than 1. In most cases, following the 'strong column/weak beam' rule, the ratio of column to beam strength will be at least 1.3 so the capacity design shear could be reduced accordingly. However, it is not generally recommended to take advantage of this relaxation as it leads to a reduction in reserve strengths in shear as compared to flexure. As is the case for beams, the calculation must be done for sway in both directions; this will mainly affect the influence of axial load on the bending strength, since the column might be in tension for the positive direction of seismic load, and in compression for the negative direction, and this has a large influence on bending strength.

5.6.3 Strength verification

Having derived the design shear and bending actions in the structural members, the resistances are then calculated according to Eurocode 2. If the partial material factors are chosen as discussed in Section 5.2.4 to cater for potential strength degradation, then the design process is simplified. Standard design aids for strength such as Narayanan and Beeby (2005) or guidance available on the Internet (e.g. http://www.concretecentre.com/) can then be used for seismic design. However, EC8 allows National Authorities to choose more complex options.

An additional restriction in columns is that the normalised axial compression force v_d must be less than 0.65:

$$v_d = N_{Ed}/A_c f_{cd} \leq 0.65 \tag{5.9}$$

This is intended to limit the adverse effects of cover spalling and avoid the situation, characteristic of members subject to high levels of axial stress, where only limited ductility is available.

For DCM frames, biaxial bending is allowed to be taken into account in a simplified way, by carrying out the checks separately in each direction but with the uniaxial moment of resistance reduced by 30%.

5.6.4 Design and detailing for ductility

Special detailing is required in the 'critical' regions, where plastic hinges are expected to form. These requirements are a mixture of standard prescriptive measures outlining a set of rules to be followed for all structures in a given ductility class and numerically based measures, where the detailing rules are dependent upon the calculated local ductility demand. The latter are typically required at the key locations for assurance of ductile performance such as hinge regions at the base of columns, Beam–column joints and boundary elements of ductile walls, the detailing provisions becoming progressively more onerous as the ductility demand is increased.

In frame structures, specific requirements are outlined for beams, columns and beam/column joints, addressed in turn below.

5.6.4.1 Beams

5.6.4.1.1 Critical regions

The critical regions are defined as extending a length h_w away from the face of the support, and a distance of h_w to either side of an anticipated hinge position (e.g. where a beam supports a discontinued column), where h_w is the depth of the beam.

5.6.4.1.2 Main (longitudinal) steel

Although flexural response of reinforced concrete beams to seismic excitation is generally deformation-controlled, abrupt brittle failure can occur if the area of reinforcement provided is so low that the yield moment is lower than the concrete cracking moment. In this situation, when the concrete cracks and tensile forces are transferred suddenly to the reinforcement, the beam may be unable to withstand the applied bending moment. To guard against this, EC8 requires a minimum amount of tension steel equal to

$$\rho_{min} = 0.5\left(\frac{f_{ctm}}{f_{yk}}\right) \tag{5.10}$$

along the entire length of the beam (and not just in the critical regions). In this expression, f_{ctm} is the mean value of concrete tensile strength as defined in Table 3.1 of Eurocode 2 and f_{yk} is the characteristic yield strength of the reinforcement.

To ensure that yielding of the flexural reinforcement occurs prior to crushing of the compression block, the maximum amount of tension steel provided, ρ_{max}, is limited to

$$\rho_{max} = \rho' + \frac{0.0018}{\mu_\phi \varepsilon_{sy,d}} \cdot \frac{f_{cd}}{f_{yd}} \tag{5.11}$$

Here, $\varepsilon_{sy,d}$ is the design value of reinforcement strain at yield, ρ' is the compression steel ratio in the beam and f_{cd} and f_{yd} are the design compressive strength of the concrete and design yield strength of the reinforcement respectively.

The development of the required local curvature ductility, μ_ϕ, is also promoted by specifying that the area of steel in the compression zone should be no less than half of the steel provided in the tension zone in addition to any design compression steel.

Since bond between concrete and reinforcement becomes less reliable under conditions of repeated inelastic load cycles, no splicing of bars should take place in critical regions according to Clause 8.7.2(2) of EC2. All splices must be confined by specially designed transverse steel as defined in Equations 5.51 and 5.52 of EC8.

Another area where particular attention needs to be paid to bond stresses is in beam/column joints of primary seismic frames, due to the high rate of change of reinforcement stress, generally varying from negative to positive yield on either side of the joint. To cater for this, the diameter of bars passing through the beam/column joint region is limited according to Equations 5.50a and 5.50b of EC8 Part 1. For DCM structures, these become

$$\frac{d_{bL}}{h_c} \leq 7.5 \cdot f_{ctm}/f_{yd})(1 + 0.8\upsilon_d)/(1 + 0.5\rho'/\rho_{max}) \quad \text{for interior columns} \tag{5.12}$$

$$\frac{d_{bL}}{h_c} \leq 7.5 \cdot f_{ctm}/f_{yd})(1 + 0.8\upsilon_d) \quad \text{for exterior columns} \tag{5.13}$$

where d_{bL} is the longitudinal bar diameter and h_c is the depth of the column in the direction of interest.

5.6.4.1.3 Hoop (transverse) steel

Many of the detailing provisions in EC8 revolve around the inclusion of transverse reinforcement to provide a degree of triaxial confinement to the concrete core of compression zones and restraint against buckling of longitudinal reinforcement. As confinement increases the available compressive capacity, in terms of both strength and more pertinently strain, it has enormous benefits in assuring the availability of local curvature ductility in plastic hinge regions. Eurocode 2 gives relationships for increased compressive strength and available strain associated with triaxial confinement, illustrated in Figure 5.6. These indicate that for the minimum areas of confinement reinforcement required at column bases and in beam column joints, the ultimate strain available would be between about two and four times that of the unconfined situation, dependent on the effectiveness of the confinement arrangement, as defined later.

Figure 5.6 Stress–strain relationships for confined concrete.

Figure 5.7 Transverse reinforcement in beams, from Eurocode 8 Part 1.

The requirements set out in EC8 to achieve this through detailing of critical regions are briefly summarised below.

- Hoops of at least 6 mm diameter d_{bw} must be provided.
- The spacing, s, of hoops should be less than the minimum of

$$h_w/4; \quad 24d_{bw}; \quad 225 \text{ mm} \quad \text{or} \quad 8d_{BL}$$

- The first hoop should be placed not more than 50 mm from the beam end section as shown in Figure 5.7.

Hoops must have 10 bar diameters anchorage length into the core of the beam.

5.6.4.2 Columns

5.6.4.2.1 Critical regions

These are the regions closer to both end sections of all primary seismic columns. The length of the critical region (where special detailing is required) is the largest of the following:

h_c; $l_{cl}/6$;   0.45 m

where
   h_c is the largest cross section dimension of the column
   l_{cl} is the clear length of the column

The whole length of the column between floors is considered a critical region:

- If (h_c/l_{cl}) is less than 3
- For structures with masonry infills
  - If it is a ground floor column
  - If the height of adjacent infills is less than the clear height of the column
  - If there is a masonry panel on only one side of the column in a given plane

5.6.4.2.2 Main (longitudinal) steel

The longitudinal reinforcement ratio must be between 0.01 and 0.04.
   Symmetric sections must be symmetrically reinforced.
   At least one intermediate bar is required along each side of the column.
   Full tension anchorage lengths must always be provided, and 50% additional length supplied if the column is in tension under any seismic load combination.
   As for beams, no splicing of bars is allowed in the critical regions and where splices are made, they must be confined by specially designed transverse steel.

5.6.4.2.3 Transverse steel (hoops and ties)

The amount of transverse steel supplied in the critical regions at the base of columns must satisfy the following equation:

$$\alpha\omega_{wd} \geq 30\mu_\Phi\upsilon_d \cdot \varepsilon_{sy,d} \cdot b_c/b_0 - 0.035 \tag{5.14}$$

where ω_{wd} is (volume of confining hoops $* f_{yd}$)/(volume of concrete core $* f_{cd}$), b_0 is the minimum dimension of concrete core and α is a confinement effectiveness factor, depending on concrete core dimensions, confinement spacing and the arrangement of hoops and ties. It is defined in Equations 5.16a to 5.16c and 5.17a to 5.17c of the code for various cross-sections.
   In the critical region at the base of columns, a minimum value of ω_{wd} of 0.08 is specified. However, for structures utilising low levels of ductility (q of 2 or less) and subject to relatively low compressive stresses ($\upsilon_d < 0.2$), this requirement is waived and the normal Eurocode 2 provisions apply. In all other critical regions of columns, the following applies:

   Minimum hoop and tie diameter is 6 mm.
   Maximum spacing of hoops and ties in the critical region is the least of

   $b_0/2$;   175 mm; $8dB$

The maximum distance between restrained longitudinal bars should be 200 mm.

Hoops must have 10 bar diameter anchorage length into the core of the column. Typical detailing requirements in critical regions of columns are illustrated in Figures 5.8 and 5.9.

Beam/column
joint 'Special'
confinement to
clauses 5.4.3.3
(1) to (3)

Critical region for
full length of short
column (lc/hc < 3)

Critical region
prescriptive
confinement to
clauses 5.4.3.2.2
(1) to (4), (10)
and (11)

Reduced
confinement
away from
plastic hinge
zones. All
column splices
within this
region

Secondary
wall

Critical region
prescriptive
confinement to
clauses 5.4.3.2.2
(1) to (4), (10)
and (11)

Beam/column
joint 'Special'
confinement to
clauses 5.4.3.3
(1) to (3)

Figure 5.8 Typical column details – elevation.

5.6.4.3 Beam–column joints

The beam–column joints of frames represent a highly stressed region with quite complex reinforcement detailing. The design requirements in EC8 are much more straightforward for DCM than for DCH.

In DCM joints, no explicit calculation of shear resistance is required, provided the following rules are satisfied:

1. To ensure that there is adequate bond between reinforcement and concrete, the diameter of the main beam bars passing through the joint must be limited as given earlier in Equations 5.12 and 5.13.
2. At least one intermediate column bar is provided between each of the corners of the columns.

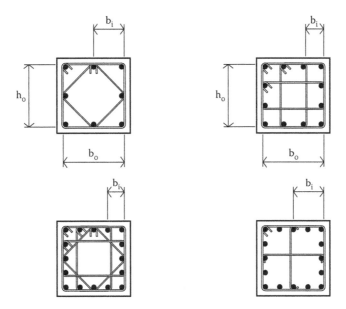

Figure 5.9 Typical column details – cross-section.

3. Hoops must continue unreduced through the joint from the critical region of the column, or must meet the confinement requirements of Equation 5.14 if greater, unless the joint is confined on all four sides by beams. In this case, the hoop spacing may be doubled (but must not exceed 150 mm).

5.7 DUCTILE WALLS: DESIGN FOR DCM

Reinforced concrete walls are defined as vertical elements in which one of the dimensions of its cross-section is at least four times the other dimension. In these elements, the flexural resistance is provided by two boundary elements at section extremities, where flexural reinforcement is concentrated and concrete is confined, with the web in between providing most of the shear resistance. The association of intersecting rectangular wall segments that develop in different directions may give rise to a three dimensional element, which must be analysed as an integral unit.

5.7.1 Geometrical restrictions

The thickness of the web of reinforced concrete walls, b_{wo}, must be larger than $h_s/20$ (in which h_s is the clear story height), with a minimum of 0.15 m. The width of the boundary elements b_w must not be less than 0.20 m. If the length of the boundary element, l_c, is restricted to no more than twice its thickness, b_w, and one-fifth of the wall cross-section length, l_w, then b_w must be greater than or equal to $h_s/15$. If the wall length l_w exceeds the above value, then the thickness of the boundary element must be higher than $h_s/10$. Figure 5.10 illustrates the above restrictions.

5.7.2 Calculation of action effects

Figure 5.3 highlights that for the best seismic performance the walls must act as vertical cantilevers and only be allowed to develop a single plastic hinge at the base. The formation

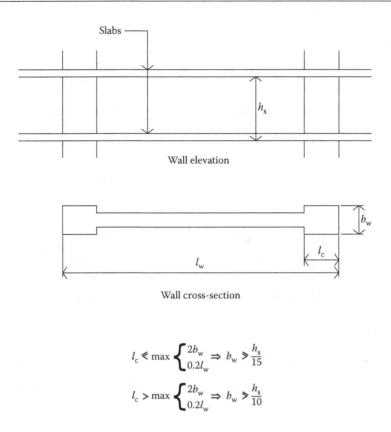

$$l_c \ll \max \begin{cases} 2b_w \\ 0.2l_w \end{cases} \Rightarrow b_w > \frac{h_s}{15}$$

$$l_c > \max \begin{cases} 2b_w \\ 0.2l_w \end{cases} \Rightarrow b_w > \frac{h_s}{10}$$

Figure 5.10 Minimum thickness of wall boundary elements, following the rules of EC8.

of this hinge is practically unavoidable, which can be explained as follows: both walls and frames have to withstand similar displacements at floor levels, and therefore both types of elements have to withstand similar curvatures. As shown in Figure 5.11, this induces higher axial strains in wall sections due to their larger cross-section dimensions. The fact that this derives only from kinematic compatibility, and is therefore independent of flexural design, implies that wall hinging is practically unavoidable for structures designed to resist earthquakes in the inelastic range and tend to occur before hinges develop in the frames.

However, besides avoiding the soft-storey mechanism there are other advantages in maintaining elastic behaviour in the rest of the wall, by preventing the formation of plastic hinges in the wall at the upper storeys: the elastic part of the wall tends to behave almost as a rigid body above the flexible zone of the hinge at the base, maintaining relatively uniform interstorey drifts throughout the height of the building. This tends to minimise the local ductility demand in the frames and hence the extent of non-structural damage, for the same global ductility of the structure.

In order that walls act as vertical cantilevers the length of their cross sections must be significantly greater than the height of the beams to which they are connected in the plane that contains the larger wall dimension. For this purpose, Fardis et al. recommend a minimum value of wall length $l_w = 1.5$ m in low rise buildings and $l_w = 2.0$ m in medium and high rise buildings.

The non-linear behaviour of an uncoupled reinforced concrete wall is governed by a single plastic hinge at its base. This section must be designed in flexure for the bending moment that results from the analysis for the seismic design situation. However, unlike the frames, the bending moment diagram does not change sign between successive floor levels, which

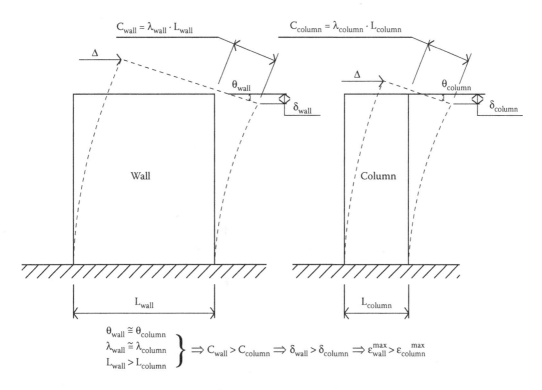

Figure 5.11 Unavoidability of wall hinging.

creates uncertainties in moment distribution along the wall. In order to avoid yielding above the base hinge, EC8 prescribes that the design bending moment diagram is based on an envelope derived from the bending moment diagram obtained from analysis, as shown in Figure 5.12.

$$\text{tension shift} \; a_1 = z \cdot \cot\theta$$

where
   z – effective depth at base of the wall
   θ – strut angle in shear strength calculation according to EC2

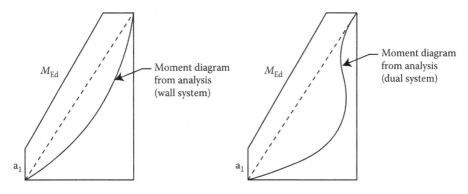

Figure 5.12 Design bending moment for RC walls (EC8).

Figure 5.13 Design envelope of shear forces of dual systems.

Turning to shear design, since the bending moment diagrams in the wall at the various stages of hinge development are not known it is not possible to derive shear forces based on equilibrium equations. In order to avoid shear failure considering (i) possible bending moments at the base section being greater than the design value used to derive the flexural reinforcement, (ii) possible variations of the distribution of inertia forces in the non-linear range and (iii) effects of higher vibration modes, EC8 prescribes that the evaluation of design shear forces should comprise the magnification of the shear forces from analysis by a factor of 1.5. Besides the over-strength of the plastic hinge, the flexural capacity can easily increase above the design value used to derive the flexural reinforcement if seismic action effects in the wall include a reasonably significant axial force. In this situation, the amount of flexural reinforcement at the base is probably conditioned by the situation in which the seismic axial force is tensile, a higher flexural capacity corresponding to the case in which the seismic axial force is compressive. The EC8 approach for DCM structures is clearly a simplified and generous one, as the magnifying factor is constant. Therefore, designers should be aware that it may not cover all the factors that may increase shear forces above the value obtained from analysis.

In the case of dual structures and due to the larger influence of the frames in the overall behaviour of the structure at the upper floors, EC8 prescribes an envelope of shear forces as the one shown in Figure 5.13, which accounts for higher shear forces than predicted at upper floor levels.

5.7.3 Strength verification

Just as for the frames, both flexural and shear verifications for walls of DCM structures follow standard Eurocode 2 procedures. Flexural reinforcement should be concentrated at the extremities of the section, in the zones furthest away from the neutral axis. This is also the most efficient distribution of the flexural reinforcement in terms of curvature ductility. The minimum longitudinal reinforcement ratio in the boundary elements is 0.005.

In what regards three dimensional elements, resulting from the intersection of rectangular wall segments, for the purpose of evaluating the flexural capacity, the effective flange width on each side of a web should be the minimum of

1. The actual flange width
2. Half the distance to an adjacent web
3. 25% of the height of the wall above that level

Just as for columns, EC8 imposes an upper limit to the normalised axial compression force on the walls, as follows:

$$\upsilon_d = N_{Ed}/A_c f_{cd} \leq 0.4 \tag{5.15}$$

where A_c is the cross-section area of the wall, and other symbols are as defined earlier for column design.

5.7.4 Design and detailing for ductility

It is not necessary to design the boundary elements for ductility if the normalised axial force is below 0.15 ($\upsilon_d \leq 0.15$). In this situation, the transverse reinforcement can be evaluated as prescribed in EC2. If $0.15 \leq \upsilon_d \leq 0.20$ the design of the transverse reinforcement can also follow EC2 if the q factor is reduced by 15%. If $\upsilon_d > 0.20$, the ductility of the rectangular wall plastic hinges is achieved by confinement of the wall boundary elements, according to EC8 prescriptions, as follows:

1. Height of confined boundary elements (h_{cr})

$$h_{cr} = \max[l_w, h_w/6] \tag{5.16}$$

l_w – length of wall section (largest dimension)
h_w – total height of the wall above the foundation or top basement floor
but h_{cr} need not be greater than

$$h_{cr} \leq \begin{cases} 2l_w \\ \begin{cases} h_s & \text{for } n \leq 6 \text{ storeys} \\ 2h_s & \text{for } n \geq 7 \text{ storeys} \end{cases} \end{cases} \tag{5.17}$$

h_s – clear storey height
2. Length of confined boundary element: The confined boundary element must extend throughout the zone of the section where the axial strain exceeds the code limit for unconfined concrete $\varepsilon_{cu2} = 0.0035$. Therefore, for rectangular sections, it must extend at least to a distance from the hoop centreline on the compressive side of

$$x_u(1 - \varepsilon_{cu2}/\varepsilon_{cu2,c}) \tag{5.18}$$

x_u – depth of compressive zone
$\varepsilon_{cu2,c}$ – maximum strain of confined concrete

The values of x_u and $\varepsilon_{cu2,c}$ can be evaluated as follows:

$$x_u = (\upsilon_d + \omega_v)\frac{l_w b_c}{b_0} \tag{5.19}$$

$$\varepsilon_{cu2,c} = 0.0035 + 0.1\alpha\omega_{wd} \tag{5.20}$$

$$\omega_v = (A_{sv}/h_c b_c)f_{yd}/f_{cd} \tag{5.21}$$

where b_c is the width of web, b_0 is the width of confined boundary element (measured to centrelines of hoops), h_c is the largest dimension of the web, and A_{sv} is the amount of vertical web reinforcement.

The value of $\alpha\omega_{wd}$ can be evaluated as follows:

$$\alpha\omega_{wd} \geq 30\mu_\phi(\upsilon_d + \omega_v)\,\varepsilon_{sy,d}\frac{b_c}{b_0} - 0.035 \tag{5.22}$$

in which μ_ϕ is the local curvature ductility factor, evaluated by Equation 5.5 or 5.6, with the basic value of the q factor q_0 replaced by the product of q_0 times the maximum value of the ratio M_{Ed}/M_{Rd} at the base of the wall.

Regardless of the above, EC8 specifies that the length of boundary elements should not be smaller than $0.15l_w$ or $1.5b_w$, with b_w being the width of the wall.

3. Amount of confinement reinforcement.

This is calculated from the mechanical volumetric ratio of confinement reinforcement ω_{wd}, evaluated according to Equation 5.22. The minimum value of $\omega_{wd} = 0.08$.

Sections with barbells, flanges or sections, consisting of several intersecting rectangular segments, can be treated as rectangular sections with the width of the barbell or flange provided that all the compressive zone is within the barbell or flange. If the depth of the compressive zone exceeds the depth of the barbell or flange the designer may: (1) increase the depth of the barbell or flange in order that all the zone under compression is within the barbell or flange, (2) if the width of the barbell or flange is not much higher than the width of the web, design the section as rectangular with the width of the web and confine the entire barbell or flange similarly to the web or (3) verify if the available curvature ductility exceeds the curvature ductility demand by non-linear analysis of the section, including the effect of confinement, after full detailing of the section.

In cases of three-dimensional elements consisting of several intersecting rectangular wall segments, parts of the section that act as the web for bending moments about one axis may act as flanges for the bending moment acting on an orthogonal axis. Therefore, it is possible that in some of these cases the entire cross-section may need to be designed for ductility as boundary elements.

General detailing rules regarding the diameter, spacing and anchorage of hoops and ties for wall boundary elements designed for ductility according to EC8 are the same as for columns. The maximum distance between longitudinal bars is also 200 mm.

5.8 DESIGN FOR DCH

The rules for DCH structures build upon those for DCM and, in certain instances, introduce additional or more onerous design checks. These are briefly introduced below. Additionally, the option of large lightly reinforced walls is removed, this type of system not being considered suitable for DCH performance.

5.8.1 Material and geometrical restrictions

The major differences from DCM are

- Concrete must be Class C20/25 or above.
- Only Class C reinforcement must be used.
- The potential over-strength of reinforcement is limited by requiring the upper characteristic (95% fractile) value of the yield strength to be no more than 25% higher than the nominal value.
- Additional limitations on the arrangement of ductile walls and minimum dimensions of beams and columns.

5.8.2 Derivation of actions

The capacity design approach used in DCM structures is reinforced as follows:

- Over-strength factors are increased to 1.2 on beams, 1.3 on columns and 1.2 on beam/column joints.
- An additional requirement for calculating the shear demand on beam/column joints is introduced in Clause 5.5.2.3.
- The shear demand on ductile walls is generally greater, the enhancement of the shear forces output from the analysis increasing from a constant factor of 1.5 to a factor of between 1.5 and q, determined from Equation 5.25 of EC8 Part 1. An over-strength factor of 1.2 is introduced for this purpose.
- Additional requirements are introduced for calculating the shear demand on squat walls.

5.8.3 Resistances and detailing

The main changes and additions are as follows:

- The assumed strut inclination in checking the shear capacity of beams to Eurocode 2 is limited to 45° and additional shear checks are introduced when almost full reversal of shear loading can occur.
- The maximum permissible normalised axial force is reduced from 0.65 to 0.55 in columns and from 0.4 to 0.35 in walls.
- Wall boundary elements need to be designed for ductility according to EC8, regardless of the level of the normalised axial force.
- The length (l_{cr}) of the critical regions of beams, columns and walls is increased and the spacing of confinement reinforcement reduced.
- Confinement requirements are extended to a length of 1.5 l_{cr} for columns in the bottom two storeys of buildings.
- The minimum value of ω_{wd} in the critical region at the base of columns and in the boundary elements of ductile walls is increased from 0.08 to 0.12.
- The maximum distance between column and wall longitudinal bars restrained by transverse hoops or ties is reduced from 200 to 150 mm.

- More comprehensive and complex checks for the shear resistance and confinement requirements at beam/column joints are introduced.
- Much more stringent checks on the resistance to shear by diagonal tension and diagonal compression are introduced, namely the limitation of the strut inclination to 45° and the reduction of the resistance to diagonal compression of the web in the critical region to 40% of the resistance outside the critical region. A different verification is also introduced on the resistance against shear failure by diagonal tension in walls with shear ratio $\alpha_s = M_{Ed}/(V_{Ed}\, l_w)$ below 2. The need for verification against sliding shear is also introduced.
- Special provisions for short coupling beams ($l/h < 3$) are included effectively comprising confined diagonal reinforcement cages as proposed by Park and Paulay (1974).

5.9 CONCRETE DESIGN EXAMPLE: WALL EQUIVALENT DUAL STRUCTURE

5.9.1 Introduction

The concrete design example is based on a dual frame solution for the eight-storey hotel introduced in earlier chapters. For clarity, the example only considers the critical transverse direction with primary frames at 8 m spacing. The frames on GLs 1, 7, 9 and 15 incorporate structural walls whereas those on GLs 3, 5, 11 and 13 are moment frames.

In a typical frame, transverse beams support masonry cross-walls between GLs B & C and D & E, primary beams fulfilling this function on odd gridlines and secondary beams on even gridlines. The masonry walls are effectively isolated from the frames so as not to stiffen the primary structural system.

At all levels in the moment frames, primary columns are situated on gridlines B, C, D and E, whilst at Level 1 additional primary columns are located on GLs A and F.

Longitudinal beams are also continuous along these gridlines, the beams on GLs B and E supporting the external cavity wall, those on GLs C and D supporting internal corridor walls and those on GLs A and B supporting the external glazing.

The structure has been analysed for both gravity loads and seismic Equivalent lateral force loading derived as in Chapter 3 but with an additional allowance for the mass of the concrete frame.

The primary members are as follows:

Slabs	150 mm thick

Frames including structural walls: Walls 2 No – 3,500 mm × 350 mm, with the outer edge on GL B and E respectively.

Primary Beams	450 mm × 600 mm
Columns	600 mm square (upper four storeys)
	750 mm square (lower four storeys)

Moment Frames:

Primary Beams	450 mm × 750 mm
Columns	600 mm square (upper four storeys)
	750 mm square (lower four storeys)

Secondary transverse beams on intermediate frames	350 mm × 500 mm
Longitudinal beams	350 mm × 600 mm

All reinforcement is Class C with characteristic yield strength of 500 N/mm² and concrete is Class C30/37.

Cover to main bars is 45 mm in beams, columns and walls and 30 mm in slabs.

The slabs span longitudinally between the main transverse frames and secondary beams on the intermediate grid-lines. They have been designed for the dominant gravity load condition and the longitudinal reinforcement comprises Φ12-300 T&B (754 mm²/m).

For the purposes of the example, the structure can be assumed to be braced in the longitudinal direction and is to be designed and detailed for ductility class DCM. Sample checks are carried out on a structural wall and on typical frame members and are intended to illustrate the requirements for design and detailing of critical regions. It should be noted that references to Equations and Clauses all refer to EC8 Part 1 unless noted otherwise.

A dual system has been chosen in order to illustrate key aspects of the code for both frame and wall elements. Hence, compromises have been made to the member sizes to remain within the limitations for the proportion of total base shear carried by the frame and wall members. In particular, to reduce structural displacements for the damage limitation requirement, a wall equivalent dual system is provided but this has the potential disadvantage of increasing the level of acceleration applied to the stiffer structure. Thus, in the initial design, the spectral acceleration is derived based on a C_t coefficient of 0.05 (from EC8 Clause 4.3.3.2.2(3) for structures other than moment resistant space frames or eccentrically braced steel frames) as in Chapter 3. This results in a conservative estimate of the inertial loads for the relatively flexible dual structure and a comparison with an approach based on modal response spectrum analysis, which results in significantly lower inertial loads, is given later to justify the performance in the damage limitation case.

5.9.2 Layout

See Figure 5.14.

5.9.3 Evaluation of the q factor

$q = q_0 k_w$

According to Table 5.1 of EC8,

DCM + dual system $\Rightarrow q_0 = 3.0\alpha_u/\alpha_1$

The ratio α_u/α_1 depends on the classification of the structure. For multi-storey, multi-bay wall equivalent dual structures, and unless a more accurate value is obtained by pushover analysis, EC8 allows the assumption that $\alpha_u/\alpha_1 = 1.2$

$q_0 = 3 \times 1.2 = 3.6$

$k_w = (1 + h_w/l_w)/3 \leq 1$

$k_w = (1 + 28.8/3.5)/3 = 3.06$

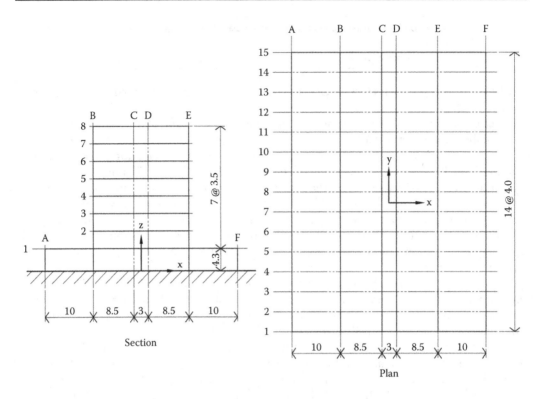

Figure 5.14 Structural layout.

Therefore: $k_w = 1$

$q = 3.6 \times 1.0 = 3.6$

The lateral loads imposed on the structure are based on those derived in Chapter 3 for structures other than moment frames. Thus, the spectral acceleration derived from the empirical period calculation is 2.32 m/s² as per Chapter 3 but the masses to which it is applied are increased to reflect the frame weight. Hence, the applied lateral loads at each level are as listed below. It will be shown in Section 5.9.6.8 that use of modal response spectrum analysis to calculate the structural frequencies results in significantly reduced inertial forces but the more conservative equivalent lateral force approach is retained initially for the preliminary design.

Level 8	3,448 kN
Level 7	4,592 kN
Level 6	3,952 kN
Level 5	3,328 kN
Level 4	2,720 kN
Level 3	2,112 kN
Level 2	1,456 kN
Level 1	1,384 kN

5.9.3.1 Part of the total base shear taken by the walls

$$V_{total} = 22{,}992 \text{ kN}$$

$$V_{frames} = 8{,}218 \text{ kN}$$

$$V_{walls} = 14{,}774 \text{ kN}$$

$$\frac{V_{wall}}{V_{total}} = \frac{14{,}774}{22{,}992} = 0.64 \geq 0.5 \leq 0.65$$

Based on the above the structure is classified as a wall equivalent dual structure.

5.9.3.2 Verification of P-δ effects and inter-storey drifts

The stability index, θ, needs to be checked to see if P-δ effects can be ignored or covered by an approximate method. [Clause 4.4.2.2 (2)]

$$\theta = \frac{P_{tot} \times d_r}{V_{tot} \times h} \qquad\qquad \text{(4.4.2.2(2), Equation 4.28)}$$

This should be based on the displacements of the structure that have been output from an analysis based on a stiffness of $0.5 * E * I_g$ [Clause 4.3.1 (7)]

$$E \text{ for grade } C30/37 \text{ concrete} = 33 \times 10^6 \text{ kN/m}^2 \qquad\qquad \text{[EC2 Table 3.1]}$$

Table 5.2 shows the values necessary to verify EC8 rules for P-δ effects and inter-storey drifts. The adequacy to meet the Damage Limitation case is addressed later.

Table 5.2 Horizontal displacements and inter-storey drift sensitivity coefficient

Level	d_e(mm)	d_{er}(mm)	d_r(mm)	$d_r * \upsilon$	P_{tot}	V_{tot}	h(mm)	θ
8	149	–	–	–	–	–		
7	130	19	68.4	34.2	12,660	3,448	3,500	0.072
6	110	20	72	36	25,482	8,040	3,500	0.065
5	88	22	79.2	39.6	38,304	11,992	3,500	0.072
4	65	23	82.8	41.4	51,126	15,320	3,500	0.079
3	44	21	75.6	37.8	64,317	18,040	3,500	0.077
2	24	20	72	36	77,508	20,152	3,500	0.079
1	9	15	54	27	90,699	21,608	3,500	0.065
	0	9	32.4	16.2	1,13,690	22,992	4,300	0.037

d_e – average horizontal displacement from analysis (based on the design response spectrum) in the transverse direction at each floor level.
d_{er} – relative displacement between storeys based on the design response spectrum.
$d_r = q_d * d_{er}$ – relative displacement between storeys accounting for the ductility-modified spectrum.
υ – reduction factor that accounts for the lower return period of the seismic action associated with the damage limitation requirement.
P_{tot} – total gravity load at and above the storey considered at the seismic design situation.
V_{tot} – total seismic storey shear.
As $\theta_{max} = 0.079 < 0.1 \Rightarrow$ no need to increase action-effects to cater for P-δ effects.

5.9.4 Design of wall elements

5.9.4.1 Allowance for torsion

In the preliminary design, it was decided to increase the action effects on the stiff wall elements using the simplified conservative allowance for torsion given in Section 4.3.3.2.4 of the code. Because the structure contains stiff perimeter elements that provide good resistance to torsional effects and would thus reduce the likelihood of significant torsional response, no increase is applied to the frame elements at this stage. It would be expected that the final actions taking full account of torsion introduced through accidental eccentricity would be derived in the final design using the more rigorous approach of Section 4.3.2 and that this would confirm the assumptions of the preliminary design.

$$\delta = 1 + 0.6\frac{x}{L} \qquad\qquad \text{(Equation 4.12)}$$

For GLs 7 and 9	For GLs 1 and 15
$x = 1 \times 4 = 4$	$x = 7 \times 4 = 28$
$L = 14 \times 4 = 56$	
$\delta = 1 + 0.6 \times \dfrac{4}{56} = 1.04$	$\delta = 1 + 0.6 \times \dfrac{28}{56} = 1.3$

Increase forces output from the analysis by a factor of 1.3 to account for torsional effects on GL 1 and 15, and by 1.04 on GL 7 and 9.

5.9.4.2 Design of the wall base section

5.9.4.2.1 Action-effects

Axial force at the base of the wall due to vertical actions $(G + 0.3Q)$. These were evaluated using the area of influence of the wall (Figure 5.15).

$$d = 3.5 + (8.5 - 3.5)/2 = 6.0 \text{ m}$$
$$A_1 = 6.0 \times 8 = 48 \text{ m}^2$$
$$A_2 = 5 \times 8 = 40 \text{ m}^2$$
$$A_{tower} = A_{terrace} = 56 \times 20 = 1,120 \text{ m}^2$$

General loading (from Chapter 3)

Tower Section $G_k = 5,152 + 7 * 8,862 = 67,186$ kN
Terrace Section $G_k = 5,482$ kN
Tower Section $0.3Q_k = 672 + 7 * 834 = 6,510$ kN
Terrace Section $0.3Q_k = 2,178 - 834 = 1,344$ kN

GL 1 and 15 (half-bay tributary area)

$$N(G_k) = (67,186) \times \frac{24}{1,120} + (5,482) \times \frac{20}{1,120} = 1,440 + 98 = 1,538 \text{ kN}$$

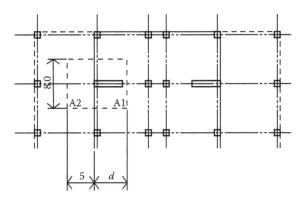

Figure 5.15 Area of influence of the walls.

$$N(G_k + 0.3Q_k) = (67,186 + 6,510) \times \frac{24}{1,120} + (5,482 + 1,344) \times \frac{20}{1,120} = 1,579 + 122 = 1,701 \text{ kN}$$

Allowance for weight of frame

Wall + transverse beams + longitudinal beams = 1,068 kN

GL 7 and 9 (full-bay tributary area)

$$N(G_k) = (67,186) \times \frac{48}{1,120} + (5,482) \times \frac{40}{1,120} = 2,879 + 196 = 3,075 \text{ kN}$$

$$N(G_k + 0.3Q_k) = (67,186 + 6,510) \times \frac{48}{1,120} + (5,482 + 1,344) \times \frac{40}{1,120} = 3,158 + 244 = 3,402 \text{ kN}$$

Allowance for weight of frame

Wall + transverse beams + longitudinal beams = 1,288 kN

GL 7 and 9	GL 1 and 15
$N(G_k)_{total} = 3,075 + 1,288 = 4,363 \text{ kN}$	$N(G_k)_{total} = 1,538 + 1,068 = 2,606 \text{ kN}$
$N(G_k + 0.3Q_k)_{total} = 3,402 + 1,288$ $= 4,690 \text{ kN}$	$N(G_k + 0.3Q_k)_{total} = 1,701 + 1,068$ $= 2,769 \text{ kN}$

Action-effects due to the seismic action (from analysis)

$$M = \pm 19,793 \text{ kN} \cdot \text{m}$$
$$V = \pm 1,847 \text{ kN}$$
$$N = \pm 1,217 \text{ kN}$$

To obtain the design action-effects due to the seismic action, the above values must be multiplied by $\xi = 1.3$ on GL 1 and 15 and 1.04 on GL 7 and 9.

GL 7 and 9	GL 1 and 15
$M_{Ed1} = 19,793 \times 1.04 = 20,585 \, kN \cdot m$	$M_{Ed1} = 19,793 \times 1.3 = 25,731 \, kN \cdot m$
$V_{Ed} = 1,847 \times 1.04 = 1,921 \, kN$	$V_{Ed} = 1,847 \times 1.3 = 2,401 \, kN$
$N_{Ed} = 1,217 \times 1.04 = 1,266 \, kN$	$N_{Ed} = 1,217 \times 1.3 = 1,582 \, kN$

Wall design – base section:

$$M_{Ed} = M_{Ed1} + \frac{h}{30} \times N_{Ed}$$

$\dfrac{h}{30} = \dfrac{3.5}{30} = 0.117$ m is the maximum eccentricity according to EC2 clause 6.1 (4)P

Maximum axial force:

GL 7 and 9	GL 1 and 15
$N_{Ed} = 4,690 + 1,266 = 5,956 \, kN$	$N_{Ed} = 2,769 + 1,582 = 4,351 \, kN$
$M_{Ed} = 20,585 + 0.117 \times 5,956$	$M_{Ed} = 25,731 + 0.117 \times 4,351$
$\quad = 21,282 \, kN \cdot m$	$\quad = 26,240 \, kN \cdot m$

Minimum axial force:

GL 7 and 9	GL 1 and 15
$N_{Ed} = 4,363 - 1,266 = 3,097 \, kN$	$N_{Ed} = 2,606 - 1,582 = 1,024 \, kN$
$M_{Ed} = 20,585 + 0.117 \times 3,097$	$M_{Ed} = 25,731 + 0.117 \times 1,024$
$\quad = 20,947 \, kN \cdot m$	$\quad = 25,851 \, kN \cdot m$

The proportion of live load to be included in the gravity load component in the seismic combination is an area that is open to judgement by the designer based on the use of the building, the make-up of the live load and the potential consequences of failure. In this case, the minimum vertical load calculated above includes no live load in the gravity component although the lateral loads are based on 30% of the characteristic live load being present. The rationale for this is that whilst 30% of the live load may be present globally, this will not be distributed evenly around the floor slab and the tributary load local to individual elements may have more or less live load than this. The full characteristic live load is unlikely to be present during an earthquake whereas it is feasible that certain parts of the structure may be empty. Since it is the minimum axial load that tends to govern wall reinforcement design, this conservative approach has been adopted in deriving the minimum loads but the maximum loads have been based on only 30% of the characteristic live load consistent with the global horizontal loads.

5.9.4.3 Flexural design

- Use design charts or design programme
- Assume symmetric reinforcement $d_1/d = 0.1$ (d_1 being the distance from the centre of tensile reinforcement to the edge of wall section)

- Steel constitutive relationship with a horizontal top branch (EC2 3.2.7(2))
- Use partial factors for the persistent and transient design situations. [Clause 5.2.4(2)]

$$\gamma_s = 1.15 \quad \gamma_c = 1.5$$

Check normalised axial load for N_{max}:

$$\upsilon = \frac{N}{bhf_{cd}} = \frac{5{,}956 \times 10^3}{350 \times 3{,}500 \times 20} = 0.243 \tag{5.4.3.4.1(2)}$$

$\upsilon_{máx} < 0.4 \Rightarrow$ the design axial force does not exceed the maximum limit for DCM structures.

$\upsilon_{máx} < 0.2 \Rightarrow$ it is necessary to design the boundary elements explicitly for ductility according to EC8 (5.4.3.4.2(12))

Situation with N_{max}: Design using concrete centre charts (from http://www.concretecentre.com/):

Note: These are based on characteristic concrete strength f_{ck} rather than design strength f_{cd}.

GL 7 and 9	GL 1 and 15
$\dfrac{N}{bhf_{ck}} = \dfrac{5{,}956 \times 10^3}{350 \times 3{,}500 \times 30} = 0.162$	$\dfrac{N}{bhf_{ck}} = \dfrac{4{,}351 \times 10^3}{350 \times 3{,}500 \times 30} = 0.118$
$\dfrac{M}{bh^2 f_{ck}} = \dfrac{21{,}282 \times 10^6}{350 \times 3{,}500^2 \times 30} = 0.165$	$\dfrac{M}{bh^2 f_{ck}} = \dfrac{26{,}240 \times 10^6}{350 \times 3{,}500^2 \times 30} = 0.204$

Situation with N_{min}:

GL 7 and 9	GL 1 and 15
$\dfrac{N}{bhf_{ck}} = \dfrac{3{,}097 \times 10^3}{350 \times 3{,}500 \times 30} = 0.084$	$\dfrac{N}{bhf_{ck}} = \dfrac{1{,}024 \times 10^3}{350 \times 3{,}500 \times 30} = 0.028$
$\dfrac{M}{bh^2 f_{ck}} = \dfrac{20{,}947 \times 10^6}{350 \times 3{,}500^2 \times 30} = 0.163$	$\dfrac{M}{bh^2 f_{ck}} = \dfrac{25{,}851 \times 10^6}{350 \times 3{,}500^2 \times 30} = 0.201$
$\dfrac{A_{S,tot} f_{yk}}{bhf_{ck}} = 0.58$	

A_S = total area of flexural reinforcement in the boundary elements of the wall section

$$A_{S,tot} = bh \frac{f_{ck}}{f_{yk}} 0.58 = 350 \times 3{,}500 \times \frac{30}{500} \times 0.58 = 42{,}630 \text{ mm}^2 \Rightarrow 2 \times 21{,}315 \text{ mm}^2$$

Before detailing the number and diameter of the flexural reinforcement bars, the length of the boundary elements will be evaluated. This is because the flexural reinforcement on the boundary elements cannot be distributed arbitrarily. Even though it is convenient to concentrate it near the extremities, in practice, it is necessary to spread part of it along the faces of the boundary element because the minimum diameter of longitudinal bars is related to the spacing of the confinement reinforcement (hoops and ties) according to clauses 5.4.3.4.2(9) and 5.4.3.2.2(11), and because the spacing of flexural steel bars and of confinement reinforcement is relevant for the evaluation of the effectiveness of confinement, according to Equations 5.16a and 5.17a.

Minimum length of the boundary elements:

$$0.15l_w = 0.15 \times 3.5 = 0.525 \text{ m} \tag{5.4.3.4.2(6)}$$

where l_w – length of wall cross-section.

$$1.5b_w = 1.5 \times 0.35 = 0.525 \text{ m} \tag{5.4.3.4.2(6)}$$

where b_w – width of wall cross-section.

The length of the boundary elements (h_0) may be evaluated as follows:

$$h_0 = x_u \times \left(\frac{1 - \varepsilon_{cu2}}{\varepsilon_{cu2,c}} \right) \tag{5.4.3.4.2(6)}$$

$$\varepsilon_{cu2} = 0.0035 \tag{5.4.3.4.2(6)}$$

$$\varepsilon_{cu2,c} = 0.0035 + 0.1\alpha\omega_{wd} \tag{5.4.3.4.2(6)}$$

$$x_u = (\upsilon_d + \omega_v) \frac{l_w b_c}{b_0} \tag{5.4.3.4.2(5)a, Equation 5.21}$$

$$\alpha\omega_{wd} \geq 30\mu_\phi(\upsilon_d + \omega_v)\varepsilon_{sy,d} \frac{b_c}{b_0} - 0.035 \tag{Equation 5.20}$$

b_0 – minimum dimension of concrete core, measured to centreline of the hoops
x_u – depth of the compressive zone
ε_{cu2} – maximum strain of unconfined concrete
$\varepsilon_{cu2,c}$ – maximum strain of confined concrete
α – confinement effectiveness factor

ω_{wd} – mechanical ratio of confinement reinforcement $\omega_{wd} = \dfrac{V_{conf\,reinf}}{V_{concrete\,core}} \dfrac{f_{yd}}{f_{cd}}$

Assuming a concrete cover of 45 mm to the main flexural reinforcement and $\phi = 10$ mm hoops

$$b_0 = 350 - 2 \times 45 + 10 = 270 \text{ mm}$$
$$b_c = 350 \text{ mm}$$

ρ_v – $A_{s,v}/A_c$ ratio of vertical web reinforcement
Minimum amount of vertical web reinforcement (EC2, 9.6.2(1))

$$A_{s,v\,min} = 0.002A_c = 0.002 \times 350 \times 1,000 = 700 \text{ mm}^2/\text{m} \Rightarrow 2 \text{ legs } T10 \text{ at } 200 \text{ mm}$$

$$\text{spacing} = 785 \text{ mm}^2/\text{m}$$

$$\mu_\phi = 2q_0 - 1 \quad (T > T_C)$$

$$\mu_\phi = 2 \times 3.6 - 1 = 6.2 \text{ (assuming that } M_{Rd} \approx M_{Ed}) \qquad \text{(Equation 5.4) and}$$

$$\text{[Clause 5.4.3.4.2(2)]}$$

$$\varepsilon_{sy,d} = \frac{435}{2,00,000} = 0.002175$$

ω_v – mechanical ratio of vertical web reinforcement:

$$\omega_v = \rho_v \frac{f_{yd,v}}{f_{cd}} = \frac{785}{350 \times 1,000} \frac{435}{20} = 0.049$$

$$\alpha\omega_{wd} = 30 \times 6.2 \times (0.243 + 0.049) \times 0.002175 \frac{350}{270} - 0.035 = 0.118$$

$$\varepsilon_{cu2,c} = 0.0035 + 0.1 \times 0.118 = 0.0153$$

$$x_u = (0.243 + 0.049) \frac{3,500 \times 350}{270} = 1,325 \text{ mm}$$

Length of boundary elements:

$$h_0 = 1,325 \times (1 - 0.0035/0.0153) = 1,022 \text{ mm}$$

Knowing the amount of flexural reinforcement and the length of the boundary elements, it is possible to make a first detail of the boundary elements. Figure 5.16 shows a possible solution:

$$h_0 = 7 \times 80 + 5 \times 85 + \frac{10}{2} + \frac{32}{2} + \frac{10}{2} + \frac{25}{2} = 1,024 \text{ mm}$$

In the evaluation of the dimension of the boundary elements, it was assumed the diameter of the stirrups and hoops is $\phi = 10$ mm.

The proposed detail of the boundary elements meets EC8 and EC2 requirements. According to clause 5.3.4.3.2(9) of EC8, it is only necessary that 'every other longitudinal bar is engaged by a hoop or cross-tie' and according to clause 5.4.3.2.2(11) 'the distance between consecutive longitudinal bars engaged by hoops and cross-ties does not exceed 200 mm'. EC2 states that 'No bar within a compressive zone should be further than 150 mm from a restrained bar' (clause 9.5.3(6)). However, it should be noted that there are several options to meet EC2 and EC8 requirements for the design of the boundary elements, as will be discussed at a later stage.

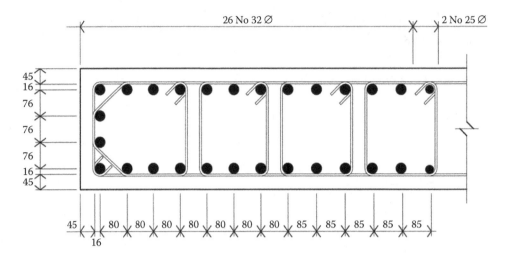

Figure 5.16 Possible detailing of wall boundary elements.

Minimum concrete cover to main vertical reinforcement (EC2, 8.1(2))

$$c_{nom} = c_{min} + \Delta c_{dev}$$ (EC2, Equation 4.1)

$$c_{min} = \phi = 32 \text{ mm}$$

$$\Delta c_{dev} = 10 \text{ mm}$$

$$c_{nom} = 42 \text{ mm} < 45 \text{ mm}$$

Minimum distance between flexural bars (EC2, 8.2(2)):

$$\text{Max. of} \begin{cases} k_1\phi = 1 \times 32 = 32 \text{ mm} \\ d_g + k_2 = 25 + 5 = 30 \text{ mm} \\ 20 \text{ mm} \end{cases}$$

d_g = maximum aggregate size

5.9.4.4 Shear design

Shear failure associated with compressive failure of diagonal struts:

$$V_{Rd,máx} = \frac{\alpha_{cw} b_w z \upsilon_1 f_{cd}}{\cot g\theta + tg\theta}$$ (EC2, Equation 6.9)

$\alpha_{cw} = 1$ for non pre-stressed structures

$$b_w = 0.35 \text{ m}$$

$$z = 0.9d = 0.9 \times (0.9 \times 3.5) = 2.835 \text{ m} \quad d = 0.9 \times 3.5$$

$$\nu_1 = 0.6 \times \left[1 - \frac{f_{ck}}{250}\right] = 0.6 \times \left[1 - \frac{30}{250}\right] = 0.528 \qquad \text{(EC2, Equation 6.6N)}$$

According to clause 6.2.3(2) of EC2, the limiting values for use in each country can be found in the respective National Annex of EC2. The limiting values recommended in EC2 are $1 \leq \cot g\theta \leq 2.5$.

$$\cot g\theta = 2.5 \quad tg\theta = 0.4$$

$$V_{Rd,máx} = \frac{1 \times 350 \times 2,835 \times 0.528 \times 20}{2.5 + 0.4} = 36,13,158 \text{ N} \approx 3,613 \text{ kN}$$

$$\cot g\theta = 1 \quad tg\theta = 1$$

$$V_{Rd,máx} = \frac{1 \times 350 \times 2,835 \times 0.528 \times 20}{1 + 1} = 52,39,080 \text{ N} \approx 5,239 \text{ kN}$$

The design value of the shear force must be obtained multiplying the shear force obtained from the global structural analysis by the magnification factor referred to in 5.7.2, as follows:

$$V_{Ed} = 1.5 \times 2,401 = 3,602 \text{ kN} \qquad \text{(5.4.2.4(7))}$$

If $V_{Ed} > V_{Rd,max}$ (associated with the $\cot g\theta = 2,5$) it would be necessary to adopt a lower value of $\cot g\theta$ until $V_{Ed} \leq V_{Rd,max}$. This would obviously lead to a larger amount of stirrups, according to Equation 6.8.

Shear resistance associated with failure in shear by diagonal tension

$$V_{Rd,s} = \frac{A_s}{s} z \cdot f_{ywd} \cdot \cot g\theta \qquad \text{(EC2, Equation 6.8)}$$

(Equation 6.8 allows the evaluation of the amount of stirrups)

According to Equation 6.8 of EC2, the higher the value of $\cot g\theta (\theta -$ inclination of diagonal compressive struts) the lower is the necessary amount of stirrups.

Evaluate the amount of stirrups (assume $\cot g\theta = 2.5$ and apply Equation. 6.8 of EC2)

$$\frac{A_s}{s} = \frac{V_{Ed}}{z \cdot f_{ywd} \cdot \cot g\theta} = \frac{3,602 \times 10^3}{2,835 \times 435 \times 2.5} = 1.17 \text{ mm}^2/\text{mm} = 1,170 \text{ mm}^2/\text{m}$$

$\Rightarrow$ 2 legs Φ10 at 125 mm spacing (1,256 mm²/m)

Verification of minimum wall horizontal reinforcement

$$A_{sh,min} \begin{cases} 0.25 \times A_{sv,min} = 0.25 \times 785 = 197 \text{ mm}^2/\text{m} \\ 0.001A_c = 0.001 \times 350 \times 1,000 = 350 \text{ mm}^2/\text{m} \end{cases}$$

The above design represents the most economic design that respects the limits for cot $g\theta$ recommended by EC8. However the shear capacity of RC members depends on factors that are not explicitly accounted for in Equation 6.8, namely the level of axial force and the formation and development of plastic hinges. It may be considered that in some situations, Equation 6.8 does not provide enough protection against shear failure. Considering the potentially catastrophic consequences of brittle shear failure of RC walls, if designers want to adopt a more conservative approach in shear wall design the following suggestion is offered: in the zones outside the plastic hinge adopt $\theta \geq 30°$, and in the plastic hinge adopt $\theta \geq 38°$ if the design axial force is compressive and $\theta = 45°$ if the design axial force is tensile. This is less stringent that what is required for DCH structures but it reduces the gap between DCM and DCH requirements for shear design. This gap may be considered excessive in particular for RC walls, as these elements are more prone to shear failure than beams and columns.

If this suggestion had been adopted, and since the design axial force is always compressive, the necessary amount of shear reinforcement would be

$$\frac{A_s}{s} = \frac{3,602 \times 10^3}{2,835 \times 435 \times \cot g38} = 2.282 \text{ mm}^2/\text{mm} = 2,282 \text{ mm}^2/\text{m}$$

(e.g. 2 legs of $\Phi16@175$ or $\Phi12@100$)

5.9.4.5 Detailing for local ductility

The height of the plastic hinge above the base of the wall for the purpose of providing confinement reinforcement

$$h_{cr} = \max[l_w, h_w/6] \qquad \text{(Equation 5.19a)}$$

$$h_{cr} = \max[3.5, 28.8/6] = 4.8 \text{ m}$$

$$\begin{cases} h_{cr} \leq 2 \cdot l_w = 2 \times 3.5 = 7 \text{ m} \\ h_{cr} \leq 2 \cdot h_s = 2 \times 4.3 = 8.6 \text{ m} \end{cases} \quad h_{cr} \leq 7 \text{ m}$$

$$h_{cr} = 4.8 \text{ m}$$

Evaluation of confinement reinforcement in the boundary elements.
According to Equation 5.20

$$\alpha \cdot w_{wd} \geq 0.118$$

$$\alpha = \alpha_n \alpha_s$$

$$\alpha_n = 1 - \sum_i b_i^2 / (6 \cdot b_0 h_0)$$

(Equation 5.16a)

$$\alpha_s = \left(1 - \frac{s}{2 \cdot b_0}\right) \times \left(1 - \frac{s}{2 \cdot h_0}\right) \qquad \text{(Equation 5.17a)}$$

All distances (b_i, b_0, h_0, s) are measured to centrelines of hoops or flexural reinforcement. The values b_i are based on the detail of the edge members and represent the distance between consecutive engaged bars. The reason for this is that confining stresses are transferred from the steel cage to the concrete essentially at the intersection of flexural engaged bars with the hoops and cross-ties that engage them. These are the points at which the outward movement of the steel cage is strongly restricted. The points where the flexural reinforcement is only connected to sides of rectangular hoops, as shown in Figure 5.17, are restricted against outward movement in a much less efficient manner. This is because rectangular hoops work efficiently under tension and not under flexure and therefore restrict the outward deformation of flexural bars and transfer confining stresses to concrete essentially at the corners and not along straight sides, as illustrated in Figure 5.18.

Since with straight hoops confinement stresses are transferred to the concrete at discrete locations (with circular or spiral hoops the distribution of confinement stresses takes place continuously along the length of the hoops) in between those locations, the effect of confinement is felt essentially by arch action. This effect takes place both on the vertical and horizontal planes, leading to a reduction of the zone effectively confined between hoops layers, as shown in Figure 5.17.

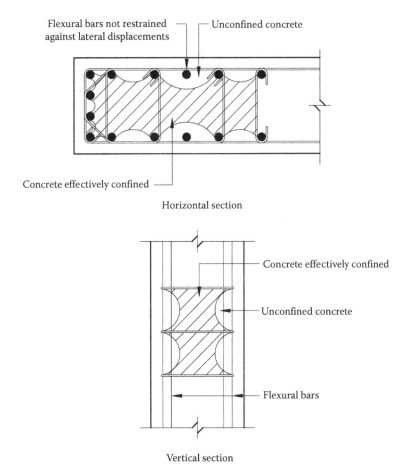

Figure 5.17 Effect of confinement between layers by arch action.

Deformation of hoops   Confining stresses

Figure 5.18 Efficiency of rectangular hoops.

The reduction of the zone effectively confined, away from the points in which most of the confining stresses are transferred to the concrete, is considered by means of the confinement effectiveness factor, α, which corresponds to the ratio of the smallest area effectively confined by the area of the concrete core, of rectangular shape with dimensions $b_0 \times h_0$ in this case. Therefore the factor α is evaluated as $\alpha = \alpha_n \cdot \alpha_s$ in which the term α_s accounts for the loss of confined area due to arch action in the vertical plane and α_n for the loss of confined area due to arch action in the horizontal plane. Therefore, both the spacing between flexural engaged bars, as well as the vertical spacing between hoop layers, are critical parameters for the effectiveness of the confinement. For this reason, both these spacings must be kept below the smallest dimension of the confined concrete core. In the case of circular hoops or spirals, arch action only takes place in the vertical plane, therefore $\alpha_n = 1$ and the spacing between flexural bars is irrelevant.

The longitudinal bars pointed out with arrows at Figure 5.17 are not considered for the evaluation of the effectiveness of confinement, as they are not engaged by hoops or cross-ties (5.4.3.2.2(8)). This led to values of $b_i = 160$ mm being adopted instead of pairs of values of $b_i = 80$ mm. According to Figure 5.16 the value of α_n must be evaluated as follows:

$$\alpha_n = 1 - \frac{[2 \times (80^2 + 160^2 + 80^2 + 160^2 + 80^2 + 170^2 + 85^2 + 170^2) + 3 \times 76^2]}{(6 \times 270 \times 1{,}024)} = 0.826$$

According to Equation 5.18, hoop spacing should not exceed any of the following values:

$$s = \min\left(\frac{b_0}{2}; 175; 8d_{BL}\right) \qquad \text{(Equation 5.18)}$$

$b_0 = 270$ mm	(width of confined boundary element)

d_{bl} – diameter of flexural reinforcement

$$s = \min\left(\frac{270}{2}; 175; 8 \times 20\right) = 135 \text{ mm}$$

In order to match the stirrup spacing, $s = 0.125$ m can be adopted. Note that with the adopted hoop spacing of 125 mm, the minimum diameter of the longitudinal flexural bars within the boundary elements would be 16 mm. If a spacing of 150 mm had been adopted, the minimum diameter of the longitudinal bars would be 20 mm. The need to minimise the spacing of the longitudinal bars as far as practicable, coupled with the need to avoid longitudinal bars with small diameters in the boundary elements, forces the spread of a reasonable amount of flexural reinforcement along the faces of the boundary elements, as previously mentioned.

Assuming $s = 0.125$ m initially

$$\alpha_s = \left(1 - \frac{125}{2 \times 270}\right) \times \left(1 - \frac{125}{2 \times 1{,}024}\right) = 0.72$$

$$\alpha = 0.826 \times 0.72 = 0.59$$

$$\alpha \cdot w_{wd} = 0.118 \Rightarrow 0.59 \times w_{wd} = 0.118$$

$$w_{wd} = 0.20$$

$$w_{wd} \geq w_{wd,min} = 0.08 \tag{5.4.3.2.2(9)}$$

Evaluation of ω_{wd} for the adopted detail of the boundary elements.
Length of confining hoops:

Exterior hoops $= 270 + 2 \times 1{,}024 = 2{,}318$ mm

$$\text{Interior hoops} = 2 \times (2 \times 80 + 32 + 10) + (2 \times 85 + 32 + 10) + (2 \times 85 + 28.5 + 10) + 7 \\ \times 270 + 76 + 2 \times \sqrt{(76 + 32 + 10/2)^2 + (80 + 32/2/ + 10/2)^2} = 3{,}094 \text{ mm}$$

Exterior hoops ($=$ stirrups, which also contribute to confine the concrete) $\phi = 10$ mm
Assuming inner hoops $\phi = 10$ mm
Volume of hoops/m

$$V = \frac{1}{0.125} \times (2{,}318 + 3{,}094) \times 78.54 = 33{,}98{,}736 \text{ mm}^2/\text{m}$$

$$w_{wd} = \frac{33{,}98{,}736}{1{,}024 \times 270 \times 1{,}000} \frac{435}{20} = 0.267$$

If the diameter of the inner hoops is reduced to $\phi_i = 8$ mm

$$V = \frac{1}{0.125}(2{,}318 \times 78.5 + 3{,}094 \times 50) = 26{,}93{,}304 \text{ mm}^2/\text{m}$$

$$w_{wd} = 0.21$$

Adopt exterior hoops (stirrups from one edge of the wall section to the other) 2 legs $\Phi 10$ at 125 mm spacing.

Adopt inner hoops (according to the detail of the boundary elements) $\Phi 8$ at 125 mm spacing.

5.9.4.6 Improvements to the detail of the boundary elements

Designers will generally have several options for the design of walls' boundary elements. In this section, some possible improvements of the detail of the boundary elements are analysed.

5.9.4.6.1 Hoops and cross-ties

It was noted previously that flexural bars such as the ones pointed out in Figure 5.17 are not engaged and are inefficient from the point of view of confinement. Besides improving the effectiveness of confinement, engaging these bars with corner hoops or cross ties would provide additional restraint against buckling of those flexural bars.

Therefore, the detailing of the boundary elements can be improved by additional hoops or cross ties that engage the eight flexural bars not engaged in the detail of Figure 5.16 to increase the efficiency of the confinement and give additional restraint against buckling of these bars. One simple way of doing this would be by adding cross ties, as shown in Figure 5.19.

With this detail the value of factor α_n (Equation 5.16a) would be as follows:

$$\alpha_n = 1 - (2 \times 7 \times 80^2 + 3 \times 76^2 + 2 \times 5 \times 85^2)/(6 \times 270 \times 1{,}024) = 0.89$$

This represents an increase of around 8% in the efficiency of confinement.

The layout of the inner stirrups is also less efficient than it could be. Figure 5.20 shows that the concrete between the inner hoops is less confined than the concrete within these hoops: the expansion of the concrete within the inner hoops is restricted in the direction of the largest dimension of the wall cross-section by the stirrups (2T10) and the inner hoops (2T8), while the expansion of the concrete between the inner hoops in the same direction is restricted only by the stirrups. This is not consistent with the underlying EC8 design philosophy for the provision of confinement. Note that the boundary elements are analysed as integral units since the amount of confinement reinforcement is evaluated for the whole boundary element (by means of a single value of ω_{wd}) and not parts of it.

Even though EC8 does not account for situations with different levels of confinement within the edge members of the walls, the relative importance of the above situation decreases if the zone with less confinement is closer to the neutral axis, as the strain demand on the concrete is less than near the section extremity. Anyway the inconvenience of having

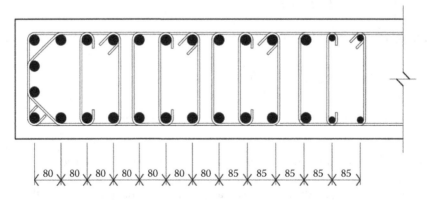

Figure 5.19 Detail of boundary element with all flexural bars engaged by hoops or cross-ties.

Figure 5.20 Zones with different confinement within the wall edge member.

Figure 5.21 Detail of boundary element with overlapping hoops.

zones with different levels of confinement near the extremity of the wall can be avoided by overlapping the inner hoops, as shown in Figure 5.21.

This last detail is equivalent to having four hoops instead of two in the largest dimension of the wall cross-section throughout the boundary element, increasing the confining stress in the weaker zones, shown in Figure 5.20, and thus providing a uniform distribution of the available strain ductility in the edge member. This is an improvement as compared with the detail of Figure 5.20, but its relative importance and efficiency will vary from case to case.

In rectangular or elongated sections, the confining stress in two orthogonal directions may be different, but it is good design practice to make them similar, as the concrete is only properly confined if it is confined in all directions. This is illustrated in Annex E of EC8-Part 2, according to which in situations with different confining stresses (σ_x, σ_y) in orthogonal directions, an effective confining stress can be evaluated as $\sigma_e = \sqrt{\sigma_x \cdot \sigma_y}$. In order that the orthogonal confining stresses are similar, the ratio of confinement reinforcement should be similar in both directions. In the case of the details of the boundary elements previously referred to, these values are as follows:

$$\rho_{swx} = \frac{A_{swx}}{s \cdot b_y}$$

$$\rho_{swy} = \frac{A_{swy}}{s \cdot b_x}$$

where s – longitudinal spacing of hoops and cross ties.

For detail shown in Figure 5.16:

$$A_{swx}(2T10) = 157 \text{ mm}^2 \quad \rho_{swx} = \frac{157}{125 \times 270} = 0.00465$$

$$A_{swy}(1T10 + 8T8) = 481 \text{ mm}^2 \quad \rho_{swy} = \frac{481}{125 \times 1,024} = 0.00376$$

For detail shown in Figure 5.19:

$$A_{swx}(2T10) = 157 \text{ mm}^2 \quad \rho_{swx} = 0.00465$$

$$A_{swy}(1T10 + 12T8) = 681 \text{ mm}^2 \quad \rho_{swy} = \frac{681}{125 \times 1,024} = 0.00532$$

For detail shown in Figure 5.21:

$$A_{swx}(2T10 + 2T8) = 258 \text{ mm}^2 \quad \rho_{swx} = \frac{258}{125 \times 270} = 0.00764$$

$$A_{swy}(1T10 + 12T8) = 681 \text{ mm}^2 \quad \rho_{swy} = 0.00532$$

The overlapping hoops, as shown in Figure 5.22, are clearly a better detail than non-overlapping hoops, and the efficiency is higher in the cases in which it increases the smallest confining stress, according to the equation for the effective stress. Therefore the recommended

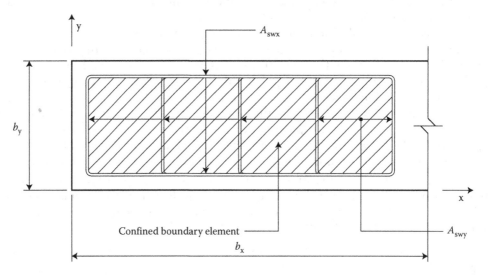

Figure 5.22 Confined boundary elements.

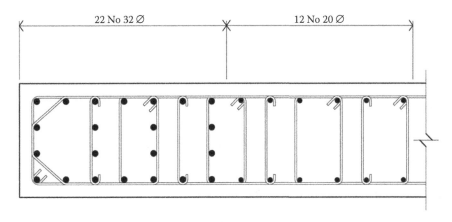

Figure 5.23 Detail with flexural reinforcement closer to the extremity of the section.

detail for the hoops and cross-ties of the boundary elements of the example wall would be the detail of Figure 5.21.

Flexural reinforcement: The distribution of the flexural reinforcement within the edge member shown in Figure 5.16 was done with steel bars with diameters 32 and 25 mm distributed along the periphery of the boundary element. However, this can be optimised by concentrating the reinforcement closer to the extremity of the wall section, leading to higher flexural capacity (for the same amount of reinforcement) and higher curvature ductility. This is due to the fact that the concentration of the flexural reinforcement at section extremities leads to the reduction of the dimension of the compressive zone, a feature of behaviour not accounted for in Equation 5.21.

The concentration of the flexural reinforcement closer to the wall extremities can be achieved for instance by placing some flexural reinforcement in the middle of the boundary elements, as shown in Figure 5.23.

The inner vertical bars can be maintained in their position during casting by tying them to the hoops and cross-ties. In order to maintain the spacing between flexural bars, to retain the effectiveness of confinement, the position of the ϕ32 bars that were moved closer to the extremity of the section, is taken by smaller flexural bars. In order not to increase the total amount of steel, four of the ϕ32 bars and the two ϕ25 bars were replaced by ϕ20 bars. This meets the requirement that the maximum spacing of confinement reinforcement should not be higher than eight times the diameter of flexural bars (according to Equation 5.18). For the chosen spacing of hoops and cross-ties of 125 mm, the minimum diameter of the flexural reinforcement is 16 mm.

5.9.5 Design of the wall above the plastic hinge

The design of the wall above the plastic hinge at the base is different from the design of the plastic hinge in two main features:

1. It is based on the provisions of EC2, since all these zones are supposed to remain in the elastic range throughout the seismic action. There is no need to provide confinement reinforcement.
2. In order to ensure that the wall remains elastic above the base hinge considering the uncertainties in the structure dynamic behaviour, the design bending moments and shear forces obtained from analysis are magnified.

From the bending moment diagram (Figure 5.24) obtained from analysis, the following linear envelope can be established:

The above diagram must be shifted upwards by a distance a_1, designated tension shift in EC8 (5.4.2.4(5)), consistent with the strut inclination adopted in the ULS verification for shear.

$$a_1 = d \cdot \cot g\theta = 3,150 \times 2.5 = 7,875 \text{ mm} = 7.875 \text{ m}$$

The design bending moment diagram (M_{Sd}) in Figure 5.24b is obtained for the design of the wall above the plastic hinge.

$$M_{top} = 19,793 - 669.1 \times (28.8 - 7.875) = 5,792 \text{ kN} \cdot \text{m}$$

The values above are the basic values prior to applying the factor accounting for torsional effects since these are dependent on location. Both the base and design moments need to be increased by the appropriate factor before being used in the design (e.g. for GL 7 and 9, $M_{base} = 19,793 * 1.04 = 20,585 \text{ kN} \cdot \text{m}$ and $M_{top} = 5,792 * 1.04 = 6,024 \text{ kN} \cdot \text{m}$.

Shear force design diagrams

$$V_{wall,top} = \frac{V_{wall,base}}{2} = \frac{3,602}{2} = 1,801 \text{ kN}$$

(a)

First bending moment diagram

(b)

Second bending moment diagram

Figure 5.24 Bending moment diagrams.

Shear diagram (GL 1 and GL 15)

Figure 5.25 Shear force diagrams.

The approach to design of the elastic sections at the higher levels is

- Choose a level at which first curtailment of flexural reinforcement would be appropriate (say at the third floor level in this instance)
- Carry out the design for moment and shear as described previously using the values from Figures 5.24 and 5.25 and the axial load appropriate for the level chosen
- There is no requirement to detail boundary elements above the height of the critical region other than EC2 prescriptions.

5.9.6 Design of frame elements

5.9.6.1 Torsional effects

The forces applied to the shear walls have been increased by a factor δ to account for accidental eccentricity [Clause 4.3.3.2.4]

$$\delta = 1 + 0.6\,x/L$$

However, as noted earlier, no increase has been applied to the frame elements in this preliminary analysis since torsional effects due to accidental eccentricity will tend to be controlled by the stiff perimeter walls and the simplified allowance for accidental eccentricity is considered to be quite conservative.

As previously, since the inter-storey drift sensitivity coefficient θ is less than 0.1 at all levels, no increase is required for P-δ effects.

5.9.6.2 Design forces

The beam flexural design is based on the maximum moments in the lower four storeys. The remainder of the design then follows from capacity design principles.

From the analysis output summarised in Figure 5.26 and Table 5.3.

$M_{hogging}$ (max) = 1,241 kN · m (Level 3)

$M_{sagging}$ (max) = 1,187 kN · m (Level 3)

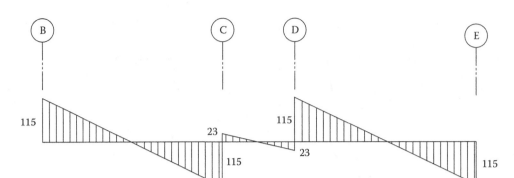

Figure 5.26 Shear force diagram for gravity sub-frame analysis ($1.0G_k + 0.3Q_k$).

Table 5.3 Gravity and seismic combinations – selected analysis output

(a) Sample applied moments

Element	Level/location[a]	Applied moment (kN · m)		Combined static and seismic moment (kN · m)	
		Static	Seismic	Hogging	Sagging
114	1 Outer	191	±482	673	−291
115	1 Inner	30	±751	781	−721
118	2 Outer	158	±799	957	−641
119	2 Inner	29	±1,069	1,098	−1,040
121	3 Outer	160	±931	1,091	−771
122	3 Inner	26	±1,215	1,241	−1,189
124	4 Outer	158	±925	1,083	−767
125	4 Inner	34	±1,015	1,049	−981
127	5 Outer	149	±825	974	−676
128	5 Inner	45	±659	704	−614

(b) Sample axial loads

Element	Level/location[b]	Axial load (kN)		Combined static and seismic axial load (kN)	
		Static	Seismic	Maximum	Minimum
80	1 Outer	3,362	±1,303	4,665	2,059
88	1 Inner	3,420	±2,558	5,978	862
81	2 Outer	2,677	±1,265	3,942	1,412
89	2 Inner	2,977	±2,164	5,141	813
82	3 Outer	2,266	±1,083	3,349	1,183
90	3 Inner	2,527	±1,638	4,165	889
83	4 Outer	1,855	±870	2,725	985
91	4 Inner	2,078	±1,041	3,119	1,037
84	5 Outer	1,443	±660	2,103	783
92	5 Inner	1,629	±575	2,204	1,054

[a] Location 'Outer' refers to the beams between GL B and C or D and E; 'Central' refers to the span between GL C and D.

[b] Location: 'Outer' refers to the columns on GL B and E; 'Inner' refers to the columns on GL C and D.

5.9.6.3 Beam design

Initially, treat as rectangular – add flange reinforcement later for capacity design.

Because of the shape of the bending moment diagram in the short span, it is assumed that no redistribution will take place.

M_1 (Hogging) = 1,241 kN · m
M_2 (Sagging) = 1,189 kN · m

Design for DCM – bending and shear resistances from EC2 [Clause 5.4.3.1.1(1)]

As the example is aimed at demonstrating the application of the seismic engineering principles of EC8, reference is made to design aids where standard design to EC2 is carried out as part of the verification. The EC2 design aids referenced here are the 'How to' sheets produced by the Concrete Centre and downloadable from http://www.concretecentre.com/.

$$K = M/bd^2f_{ck}$$

$$z = 0.5 * d\{1 + (1 - 3.53\,K)^{1/2}\}$$

5.9.6.3.1 Hogging

Assume two layers of 32 mm diameter bars, 45 mm cover to main reinforcement

$$d = 750 - 45 - 32 - (32/2) = 657 \text{ mm}$$

$$K = 1,241 \times 10^6 /(450 * 657^2 * 30) = 0.212$$

No redistribution – $K' = 0.205$
$K > K'$. Therefore, compression reinforcement is required.

$$z = 0.5 * 657\{1 + (1 - 3.53 * 0.205)^{1/2}\}$$

$$z = 501 \text{ mm}$$

Use partial factors for the persistent and transient design situations. [clause 5.2.4(2)]

$$\gamma_s = 1.15 \quad \gamma_c = 1.5$$

Compression reinforcement

$$x = 2(d - z) = 2(657 - 501) = 312 \text{ mm}$$

$$A_{s2} = \frac{(K - K')f_{ck}bd^2}{f_{sc}(d - d_2)}$$

$$f_{sc} = \frac{700(x - d_2)}{x} = \frac{700(312 - 93)}{312} = 491 \text{ N/mm}^2 > f_{yd}$$

$$f_{sc} = f_{yd} = 500/1.15 = 434.8 \text{ N/mm}^2$$

$$A_{s2} = \frac{(0.212 - 0.205)30 * 450 * 657^2}{434.8(657 - 93)} = 166 \text{ mm}^2$$

Nominal – will be enveloped by reinforcement provided for sagging moment in reverse cycle.

Tension Reinforcement

$$A_{s1} = \frac{K'f_{ck}bd^2}{f_{yd}z} + \frac{A_{s2}f_{sc}}{f_{yd}} = \frac{0.205 * 30 * 450 * 657^2}{434.8 * 501} + \frac{166 * 434.8}{434.8} = 5,650 \text{ mm}^2$$

Use $7 - \Phi 32$ (5,628 mm² – 1.9%) plus longitudinal slab reinforcement within effective width (see later).

Note: It is often recommended in the UK that K' is limited to 0.168 to ensure a ductile failure. If the calculation above were to be repeated with this limit applied, the resulting areas of tension and compression reinforcement would be

$$A_{s1} = 5,540 \text{ mm}^2 \quad \text{and} \quad A_{s2} = 1,046 \text{ mm}^2$$

(i.e. a similar area of tension reinforcement and significantly increased compression reinforcement is required. In this case, because of the much larger area of reinforcement provided in the bottom face to cater for the reverse loading cycle, it has no practical effect on the solution).

5.9.6.3.2 Sagging

$K = 1,189 \times 10^6/(450 * 657^2 * 30) = 0.204 < 0.205$, therefore, singly reinforced.
$z = 0.5 * d\{1 + (1 - 3.53 \text{ K})^{1/2}\} = 0.5 * 657\{1 + (1 - 3.53 * 0.204)^{1/2}\} = 502 \text{ mm}$
$A_{s2} = 1,189 \times 10^6/502 * 434.8 = 5,447 \text{ mm}^2$
$7 - \Phi 32$ (5,628 mm² – 1.9%)
Spacing $= (450 - (2 * 45) - 32)/4 = 82 \text{ mm}$
Clear space between bars $= 82 - 32 = 50 \text{ mm}$
Minimum clear space between bars = bar diameter OR aggregate size + 5 mm OR 20 mm
    OK.

$\rho \text{ min} = 0.5 * (f_{ctm}/f_{yk}) = 0.5 * 2.9/500 = 0.0029$ (0.29% cf 1.9% provided)

f_{ctm} from EC2 Table 3.1.

5.9.6.4 Derive shear demand from flexural capacity

Internal column connection framed by orthogonal beams
    Calculate hogging capacity:

$$b_{eff} = b_w + 8 * h_f \qquad \text{[Clause 5.4.3.1.1(3)]}$$

Slab width to be considered = $8 \times 0.15 = 1.2$ m
Slab reinforcement = $\Phi12 - 300$ T and B (754 mm²/m in total)

$$A_{s1} = 5{,}628 + 754 * 1.2 = 6{,}533 \text{ mm}^2$$

A_{s1} (required) = 5,650 mm² for an applied moment of 1,241 kN.m
Hogging Capacity = $1{,}241 * 6533/5650 = 1{,}435$ kN.m
Calculate sagging capacity:

$$A_{s2} = 5{,}628 \text{ mm}^2$$

A_{s2} (required) = 5,447 mm² for an applied moment of 1,189 kN.m
Sagging Capacity = $1{,}189 * 5{,}628/5{,}447 = 1{,}229$ kN.m

$$V_{E,d} = \gamma_{Rd} * [M_{Rd}(\text{top}) + M_{Rd}(\text{bottom})]/l_{cl} + V_g$$

1. Long Outer Spans
   For DCM structures, $\gamma_{Rd} = 1.0$ in beams
   From gravity load analysis $V_g = 115$ kN

   $$l_{cl} = 8.5 - 0.75 = 7.75$$

   $$V_{E,d} = 115 + (1{,}435 + 1{,}229)/7.75 = 459 \text{ kN}$$

2. Short central span
   From gravity load analysis $V_g = 23$ kN

   $$l_{cl} = 3.0 - 0.75 = 2.25$$

   $$V_{E,d} = 23 + (1{,}435 + 1{,}229)/2.25 = 1{,}207 \text{ kN}$$

5.9.6.4.1 Check shear resistance to EC2 for demand based on flexural capacity

As previously, where standard design to EC2 forms part of the verification, reference is made to the design aids downloadable from http://www.concretecentre.com/.

1. Outer spans

   $$v_{Ed} = 459 * 10^3/450 * 657 = 1.55 \text{ N/mm}^2$$

   Assume Cot $\theta = 2.5$
   $v_{Rd,max} = 3.64$ N/mm² (from 'How to' Sheet 4: Beams – Table 7)

   $$A_{sw}/s = V_{E,d}/(z * f_{ywd} * \text{Cot}\,\theta) = 459 \times 10^3/501 * 434.8 * 2.5 = 0.84$$

   Assume 8 mm links.

Figure 5.27 Reinforcement arrangement in critical region of beam.

In a critical region, $s = \min\{h_w/4 = 188: 24d_{bw} = 192; 225; 8d_{BL} = 256\}$ {Equation 5.13]
Use 175 mm spacing of links

$$A_{sw} = 0.84 * 175 = 147 \text{ mm}^2$$

Use 4 legs of $\Phi8$ (201 mm²) as shown on Figure 5.27
ρ_w (min) is OK from EC2 Equations 9.4 and 9.5.
2. Short Central Span

$$\upsilon_{Ed} = 1207 * 10^3/450 * 657 = 4.08 \text{ N/mm}^2 > 3.64$$

Cot θ is less than 2.5

$$\theta = 0.5 * \text{Sin } e^{-1}[\upsilon_{Ed}/0.2f_{ck} * (1 - f_{ck}/250)] \quad \text{[Concrete Centre How to Guide 4 – Beams]}$$

$$\theta = 0.5 * \text{Sin } e^{-1}[4.08/0.2 * 30 * (1 - 30/250)] = 25.3°$$

$$\text{Cot}\,\theta = 2.1$$

$$A_{sw}/s = V_{E,d}/(z * f_{ywd} * \text{Cot}\,\theta) = 1,207 \times 10^3/501 * 434.8 * 2.1 = 2.64$$

Assume links at 150 mm spacing.
$A_{sw} = 2.64 * 150 = 396 \text{ mm}^2$ – Use 4 legs of T12 (452 mm²)

5.9.6.5 Check local ductility demand

$$\rho_{max} = \rho' + 0.0018 * f_{cd}/(\mu_\phi * \varepsilon_{syd} * f_{yd}) \qquad \text{[EC8 Part 1 Equation 5.11]}$$

$\mu_\phi = 2 * q - 1$
$\mu_\phi = 2 * 3.6 - 1 = 6.2$
$f_{cd} = 30/1.5 = 20 \text{ N/mm}^2$
$\varepsilon_{syd} = 434.8/200E3 = 0.0022$

Area of reinforcement in the compression zone $= 5,628 \text{ mm}^2$

$\rho' = 5,628/450 * 657 = 0.019$

$\rho_{max} = \rho' + 0.0018 * 20/(6.2 * 0.0022 * 434.8) = \rho' + 0.006$

By inspection, because ρ only exceeds ρ' by the nominal slab reinforcement, the expression is satisfied.

Check maximum diameter of flexural bars according to Equation 5.50a:

$\rho_{max} = \rho' + 0.006 = 0.025$

At level 4, just above the critical node, $N_{static} = 2,078 \text{ kN}$ and $N_{seismic} = \pm 1,041 \text{ kN}$:

$N_{min} = 1,037 \text{ kN}$

$$\upsilon_d = \frac{1,037 \times 10^3}{750^2 \times 20} = 0.092$$

Equation 5.50a $\quad \dfrac{d_{bL}}{h_c} \leq \dfrac{7.5 f_{ctm}}{\gamma_{Rd} f_{yd}} \dfrac{1 + 0.8\upsilon_d}{1 + 0.75 k_D (\rho'/\rho_{max})}$

According to clause 5.6.2.2.b, for DCM structures:

$k_D = 2/3; \quad \gamma_{Rd} = 1.0$

$$\frac{d_{bL}}{750} \leq \frac{7.5 \times 2.9}{1.0 \times 434.8} \frac{1 + 0.8 \times 0.092}{1 + 0.75 \times \dfrac{2}{3} \times \dfrac{0.0190}{0.025}} \Rightarrow d_{bL} \leq 29.2 \text{ mm}$$

Hence, the bond requirements across the column joint are not satisfied by the reinforcement arrangement proposed in the preliminary design, illustrating the difficulty in meeting the EC8 bond provisions. In the final design, this could be addressed through

- Modification of the reinforcement arrangement (providing 12–25 mm diameter bars in two layers would satisfy spacing requirements). This would be reduced further (to only 8 bars) if the reduced inertial loads from the response spectrum analysis are considered rather than the equivalent lateral force approach (see the later calculations on the damage limitation requirement).
- Increasing the concrete grade to C35/45 (f_{ctm} would become 3.2 N/mm² which would result in a permitted bar diameter of 32.2 mm) or
- Increasing the column size.

5.9.6.6 Column design

If the frame was to be designed as a moment frame in both directions,

Clause 5.4.3.2.1 allows the frame to be designed for uniaxial bending about each direction in turn rather than considering biaxial bending, provided the uniaxial capacity is reduced by 30% [Clause 5.4.3.2.1].

In this case, the frame is assumed braced in the longitudinal direction. Therefore, no reduction in capacity is taken.

$$0.01 < \rho_l < 0.04 \qquad\qquad \text{[Clause 5.4.3.2.2 (1)]}$$

Definition of critical regions:

$$l_{cr} = \max\{h_c : l_{cl}/6; 0.45\} \qquad\qquad \text{[EC8 Part 1 – Equation 5.14]}$$

$$l_{cr} = \max\{0.75; \ 2.75/6 = 0.46; \ 0.45\}$$

$$l_{cr} = 0.75 \text{ m}$$

Consider the position of maximum moment at level 3.

In frame structures or frame-equivalent dual structures, it is necessary to design for a strong column/weak beam mechanism and satisfy EC8 Part 1 Equation 4.29:

$$\sum M_{Rc} > 1.3 \sum M_{Rb} \qquad\qquad \text{[EC8 Part 1 Equation 4.29]}$$

However, since the walls carry more than 50% of the base shear and the structure is therefore classified as a wall-equivalent dual system, this requirement is waived. Thus, the designer may design the columns for the moments output from the analysis. Even though it is implicit within the code that soft storey mechanisms are prevented by the presence of sufficient stiff walls in wall-equivalent dual systems, their inelastic behaviour is more uncertain than pure wall or frame systems, as noted by Fardis et al. (2005). To cater for this, the designer may decide to reduce the probability of extensive plasticity in the columns by continuing to relate the column moments to the capacities of the beams framing into them. In this case, the beam capacities need not be increased by the 1.3 factor of Equation 4.29. The output from the analysis shows a maximum value of 1,389 kN · m and a value of 1,465 kN · m is derived from the beam capacities. These values are similar and the calculations proceed using the higher value derived from the beam capacities.

Assume 45%/55% split between the column sections above and below the joint.

Design lower section.

$$\sum M_{Rb} = (1,435 + 1,229) = 2,664 \text{ kN} \cdot \text{m}$$

$$M_{Rc1} = 0.55 * 2,664 = 1,465 \text{ kN} \cdot \text{m}$$

Axial load from analysis:

$$N_{static} = 2,527 \text{ kN} \quad N_{seismic} = \pm 1,638 \text{ kN}$$

| Maximum compression | $N = 2{,}527 + 1{,}638 = 4{,}165$ kN |
| Minimum compression | $N = 2{,}527 - 1{,}638 = 889$ kN |

5.9.6.6.1 Check normalised axial compression

$\upsilon_d < 0.65$ for DCM   [Clause 5.4.3.2.1(3)]

$\upsilon_d = 4{,}16{,}510^3/750 * 750 * (30/1.5) = 0.37$ OK

5.9.6.6.2 Check column resistances

Design resistances to EC2 [Clause 5.4.3.2.1(1)]:
  From the concrete centre, 'how to' sheet 5–columns:

Using design chart for C30/37 concrete and $d_2/h = 0.1$
Assume 32 mm diameter main steel; $d_2 = 45 + 32/2 = 61$ mm
$d_2/h = 61/750 = 0.08$ Chart for $d_2/h = 0.1$ is most appropriate.
Maximum compression: $N/(b * h * f_{ck}) = 4{,}165 \times 10^3/750 * 750 * 30 = 0.25$
Minimum compression: $N/(b * h * f_{ck}) = 889 \times 10^3/750 * 750 * 30 = 0.05$
Flexure: $M/(b * h^2 * f_{ck}) = 1{,}465 \times 10^6/(750^3 * 30) = 0.12$
Maximum compression: $A_s * f_{yk}/b * h * f_{ck} = 0.2$
Minimum compression: $A_s * f_{yk}/b * h * f_{ck} = 0.3$

$A_s/b * h = 0.3 * 30/500 = 0.018(1.8\% -$ within prescribed limits$)$

$A_s = 0.018 * 750 * 750 = 10{,}125$ mm^2

Use 16 Φ32 – (5T32 in EF + 3 in each side) – [12,864 mm^2]

Check capacity for maximum compression

$$\frac{A_s f_{yk}}{bh f_{ck}} = \frac{12{,}864 * 434.8}{750 * 750 * 30} = 0.33$$

For $N_{max} = 4{,}165$ kN, $M/bh^2 f_{ck} = 0.18$
$M_{cap} = 0.18 * 750^3 * 30 * 10^{-6} = 2{,}278$ kN · m

5.9.6.6.3 Check shear: approach as for beams but without lateral load between supports

For a conservative design, the column shear could be based upon the flexural capacity at maximum compression calculated above. However, EC8 Equation 5.9 allows the column flexural capacities to be multiplied by the ratio $\Sigma M_{R,b}/\Sigma M_{R,c}$ on the basis that yielding may develop initially in the beams and hence does not allow the development of the column over-strength moments.

$$V_{E,d} = \gamma_{Rd} * (\Sigma M_{R,b} / \Sigma M_{R,c}) * (M_{c,top} + M_{c,bottom})/l_{cl}$$

For DCM columns, $\gamma_{Rd} = 1.1$

$l_{cl} = 3.5 - 0.75 = 2.75$ m

$$V_{E,d} = 1.1 * \frac{2,664}{2*2,278} * 2 * 2,278 * \frac{1}{2.75} = 1,066 \text{ kN}$$

$d = 75\ 0 - 45 - 32/2 = 689$
$v_{E,d} = 1,066 - 10^3/689 * 750 = 2.06 \text{ N/mm}^2 < 3.64 \text{ N/mm}^2$

As previously, Cot $\theta = 2.5$ and $f_{ywd} = f_{yk}/1.15$

$$A_{sw}/s = V_{E,d}/(z * f_{ywd} * \cot\theta)$$

z can be taken as $0.9d$ for a steel couple:

$$A_{sw}/s = 1,066 \times 10^3/(0.9*689*434.8*2.5) = 1.58 \text{ mm}^2/\text{mm}$$

Although the structural analysis shows that the flexural demand is lower at the lower levels, check normalised axial compression at the position of maximum axial load (on GL C and D at the base).

$N_{static} = 3,420 \text{ kN}$
$N_{seismic} = \pm 2,558 \text{ kN}$
$N_{max} = 5,978 \text{ kN}$
$v_d = 5,978 * 10^3/750 * 750 * 20 = 0.53 < 0.65$

Therefore, the normalised axial compression is satisfactory.

5.9.6.6.4 Detailing

For the critical regions of DCM columns:

$$s_{max} = \min\{b_0/2;\ 175;\ 8d_{BL}\} \qquad \text{[EC8 Part 1 Equation 5.18]}$$

For columns, take 45 mm cover to the main reinforcement

$b_0 = h_0 = 750 - 2 * 45 + 10 = 670$ mm (centre to centre of link)
$b_0/2 = 335$ mm $8 * d_{BL} = 8 * 32 = 256$ mm

$s_{max} = 175$ mm
$A_{sw} = 1.58 * 175 = 277$ mm^2 (5 legs of T10 – 392 mm^2)

Provide 5 legs of T10 hoops/ties at 175 mm spacing within the critical region, 750 mm from the underside of the beam as shown in Figure 5.28.
Distance between restrained main bars $= (750 - 2 * 45 - 32)/4 = 157$ mm < 200 mm OK [EC8 Part 1 Clause 5.4.3.2.2(11)]

5.9.6.7 Confinement reinforcement

For DCM, confinement reinforcement within a beam/column joint and in the critical regions at the base of a column must meet the provisions of Clauses 5.4.3.2.2(8) to (11).

Figure 5.28 Arrangement of column reinforcement.

Clauses 5.4.3.2.2(10) and (11) are satisfied by the detailing requirements outlined above for all critical regions of the column.

Therefore, the additional requirements of Clauses 5.4.3.2.2(8) and (9) need to be checked.

$$\alpha * \omega_{wd} \geq 30\mu_\Phi\upsilon_d\varepsilon_{sy,d} * (b_c/b_0) - 0.035 \qquad \text{[Clause 5.4.3.2.2(8)]}$$

and

$$\omega_{wd} \geq 0.08 \qquad \text{[Clause 5.4.3.2.2(9)]}$$

$$\omega_{wd} = [(A_{svx}/b_0 * s) + (A_{svy}/h_0 * s)] * (f_{yd}/f_{cd}) \qquad \text{[EC8 Part 1 – Clause 5.4.3.2.2(8)]}$$

$$\alpha = \alpha_n * \alpha_s$$

$$\alpha_n = 1 - \Sigma(b_i)^2/6b_0h_0 \qquad \text{[EC8 Part 1 Equation 5.16a]}$$

$$\alpha_s = (1 - s/2b_0) * (1 - s/2h_0) \qquad \text{[EC8 Part 1 Equation 5.17a]}$$

$$b_0 = h_0 = 670 \text{ mm}$$

Since the normalised axial compression is greatest at the base of the column, consider the detailing of this region to check the feasibility of the design.

All main column bars are equally spaced:

$$b_i = (660 - 32)/4 = 157 \text{ mm}$$

$$\alpha_n = 1 - (16 * 157^2)/6 * 670 * 670 = 0.85$$

$$\alpha_s = (1 - s/1,340) * 1 - s/1,340)$$

For $s = 100$ mm, $\alpha_s = 0.85$
For $s = 125$ mm, $\alpha_s = 0.82$
For $s = 150$ mm, $\alpha_s = 0.78$

As before

$$\mu_\phi = 2 * q - 1 = 2 * 3.6 - 1 = 6.2$$

$$\varepsilon_{syd} = 434.8/200E3 = 0.0022$$

$$\upsilon_d = 0.53$$

$$b_c/b_0 = 750/670 = 1.12$$

$$30\mu_\Phi\upsilon_d\varepsilon_{sy,d} * (b_c/b_0) - 0.035 = 30 * 6.2 * 0.53 * 0.0022 * 1.12 - 0.035 = 0.208$$

Try hoops/ties at 100 mm spacing: $\alpha = 0.85 * 0.85 = 0.72$

$\omega_{wd} = 0.208/0.72 = 0.29 >$ minimum of 0.08   [EC8 Part 1 – Clause 5.4.3.2.2(9)]
$\omega_{wd} = [(A_{svx}/b_0 * s) + (A_{svy}/b_0 * s)] * (f_{yd}/f_{cd})$
$\omega_{wd} = [2 * (392/670 * 100)] * 434.8/20 = 0.25 < 0.29$ Not sufficient

Consider 12 mm diameter hoops and ties $A_s = 565$ mm^2

$$\omega_{wd} = [2 * (565/670 * 100)] * 434.8/20 = 0.37 > 0.29 \quad \text{OK}$$

For beam–column joints, the density of confinement reinforcement may be reduced up the height of the building as the normalised axial compression reduces.

Also, the internal 600 mm square columns in the upper four storeys have beams of three-quarters of the column width that frame into them on all four sides. In these cases, the calculated confinement spacing may be doubled but may not exceed a limit of 150 mm [EC8 Clause 5.4.3.3(2)].

5.9.6.8 Damage limitation case

From Table 5.2, it can be seen that the maximum value of storey drift in the damage limitation event is $d_r \times \upsilon = 41.4$ mm.

This is above the maximum inter-storey drift for buildings having non-structural elements fixed in a way so as not to interfere with structural deformations, which is 0.01 h = 0.01 × 3,500 = 35 mm (h–storey height).

However, as noted earlier, the lateral loads on the structure were initially calculated based on a standard formula, which is applicable to a wide range of structures and, by necessity, this is quite conservative in its calculation of the period of response. Although it is wall-equivalent, the dual structure chosen is relatively flexible compared to typical shear wall structures and therefore might be expected to attract lower inertial loads. Therefore, a more realistic approach was adopted calculating the period using modal analysis with the stiffness of the structure based on $0.5 * E_c * I_g$ as per the deflection calculation.

The modal analysis gives a fundamental period of 1.2 seconds with 67% mass participating (compared to 0.62 seconds using the generic formula) together with significant secondary modes at periods of 0.32 seconds (18% mass participating) and 0.14 seconds (9% mass participating).

The spectral acceleration associated with the fundamental mode is only 1.35 ms^{-2} rather than 2.32 ms^{-2} previously obtained. Also, despite the higher spectral accelerations of the higher modes, their low mass participation means that the effective acceleration consistent with the SRSS combination of the individual modal inertial loads is lower than taking the fundamental mode acceleration with 100% mass participation in this case.

Hence, inertial loads would be less than 60% of those used in the initial analysis. This gives a maximum storey drift of $0.6 * 41.4 = 24.8$ mm, well within the EC8 limit. It can therefore be seen that the structure possesses adequate stiffness and the feasibility of the design is confirmed. The final design should proceed on the basis of these lower inertial loads, resulting in reduced quantities of reinforcement but the member sizes should remain unaltered to meet damage limitation requirements.

REFERENCES

Fardis, M.N., Carvalho, E., Elnashai, A., Faccioli, E., Pinto, P. and Plumier, A. 2005. *Designers' Guide to EN 1998-1 and EN 1998-5 Eurocode 8: Design of Structures for Earthquake Resistance. General rules, Seismic Actions, Design Rules for Buildings, Foundations and Retaining Structures*. Thomas Telford, London.

ISE/AFPS 2010. *Manual for the Seismic Design of Steel and Concrete Buildings to Eurocode 8*. Institute of Structural Engineers, London.

Narayanan, R.S. and Beeby, A. 2005. Designers' Guide to EN 1992-1-1 and EN 1992-1-2 Eurocode 2: Design of concrete structures. General rules and rules for buildings and structural fire design. Thomas Telford, London.

Park, R. and Paulay, T. 1974. *Reinforced Concrete Structures*. Wiley, New York.

Paulay, T. 1993. Simplicity and Confidence in Seismic Design. The Fourth Mallet-Milne Lecture. SECED/John Wiley and Sons.

Paulay, T. and Priestley, M.J.N. 1992. *Seismic Design of Reinforced Concrete and Masonry Buildings*. Wiley, New York.

Priestley, M.J.N. 2003. *Revisiting Myths and Fallacies in Earthquake Engineering*. The Mallet-Milne Lecture, 2003. IUSS Press, Pavia, Italy.

Chapter 6

Design of steel structures

Ahmed Y. Elghazouli and José Miguel Castro

CONTENTS

6.1 INTRODUCTION

In line with current seismic design practice, steel structures may be designed to EC8 according to either non-dissipative or dissipative behaviour. The former, through which the structure is dimensioned to respond largely in the elastic range, is normally limited to areas of low seismicity or to structures of special use and importance; it may also be feasible if vibration reduction devices are incorporated. Otherwise, codes aim to achieve economical design by employing dissipative behaviour in which considerable inelastic deformations can be accommodated under significant seismic events. In the case of irregular or complex structures, detailed non-linear dynamic analysis may be necessary. However, dissipative design of regular structures is usually performed by assigning a structural behaviour factor (i.e. force reduction or modification factor), which is used to reduce the code-specified forces, resulting from idealised elastic response spectra. This is carried out in conjunction with the capacity design concept, which requires an appropriate determination of the capacity of the structure based on a pre-defined plastic mechanism, often referred to as failure mode, coupled with the provision of sufficient ductility in plastic zones and adequate over-strength factors for other regions.

This chapter focuses on the dissipative seismic design of steel frame structures according to the provisions of EN1998-1 (2004), particularly Section 6 (Specific Rules for Steel Buildings). After giving an outline of common configurations and the associated behaviour factors, the seismic performance of the three main types of steel frame is discussed. Brief notes on material requirements and control of design and construction are also included. This chapter concludes with illustrative examples for the use of EC8 in the preliminary design of lateral resisting frames for the eight storey building dealt with in previous chapters of this book.

6.2 STRUCTURAL TYPES AND BEHAVIOUR FACTORS

There are essentially three main structural steel frame systems used to resist horizontal seismic actions, namely moment resisting, concentrically braced and eccentrically braced frames (EBFs). Other systems such as hybrid and dual configurations can be used and are referred to in EC8, but are not dealt with in detail herein. It should also be noted that other configurations such as those incorporating buckling restrained braces or special plate shear walls, which are covered in the most recent North American Provisions (AISC, 2010a,b), are not directly addressed in the current version of EC8.

As noted before, unless the complexity or importance of a structure dictates the use of non-linear dynamic analysis, regular structures are designed using the procedures of capacity design and specified behaviour factors. These factors (also referred to as force reduction factors) are recommended by codes of practice based on background research involving extensive analytical and experimental investigations. Before discussing the behaviour of each type of frame, it is useful to start by indicating the structural classification and reference behaviour factors (q) stipulated in EC8 as this provides a general idea about the ductility and energy dissipation capability of various configurations. Table 6.1 shows the main structural types together with the associated dissipative zones according to the provisions and classification of EC8 (described in Section 6.3 of EN1998-1). The upper values of q allowed for each system, provided that regularity criteria are met, are also shown in Table 6.1. The ability of the structure to dissipate energy is quantified by the behaviour factor; the higher the behaviour factor, the higher is the expected energy dissipation as well as the ductility demand on critical zones.

Table 6.1 Structural types and behaviour factors

Structural type	q-factor	
	DCM	DCH
Moment-resisting frames	4	$5\alpha_u/\alpha_1$

$\alpha_u/\alpha_1 = 1.1$

$\alpha_u/\alpha_1 = 1.2$ (1 bay)
$\alpha_u/\alpha_1 = 1.3$ (multi-bay)

Dissipative zones in beams and column bases

Concentrically-braced frames	4	4

Dissipative zones in tension diagonals

V-braced frames	2	2.5

Dissipative zones in tension and compression diagonals

Frames with K-bracings		Not allowed in dissipative design

Eccentrically-braced frames	4	$5\alpha_u/\alpha_1$

$\alpha_u/\alpha_1 = 1.2$
Dissipative zones in bending or shear links

(Continued)

Table 6.1 (*Continued*) Structural types and behaviour factors

| | q-factor | |
Structural type	DCM	DCH
Inverted pendulum structures	2	$2\alpha_u/\alpha_1$

$$\alpha_u/\alpha_1 = 1.0 \qquad\qquad \alpha_u/\alpha_1 = 1.1$$

Dissipative zones in column base, or column ends ($N_{Ed}/N_{pl,Rd} < 0.3$)

Moment-resisting frames with concentric bracing	4	$4\alpha_u/\alpha_1$

$$\alpha_u/\alpha_1 = 1.2$$

Dissipative zones in moment frame and tension diagonals

Moment frames with infills		DCM	DCH
	Unconnected concrete or masonry infills, in contact with the frame	2	2
	Connected reinforced concrete infills	See concrete rules	
	Infills isolated from moment frame	4	$5\alpha_u/\alpha_1$

Structures with concrete cores or walls	See concrete rules	

The multiplier α_u/α_1 depends on the full-to-first plasticity resistance ratio of the structure. A reasonable estimate of this value may be determined from conventional non-linear 'push-over' analysis, but should not exceed 1.6. In the absence of detailed calculations, the approximate values of this multiplier given in Table 6.1 may be used. If the building is irregular in elevation, the listed values should be reduced by 20%.

The values of the structural behaviour factor given in the code should be considered as an upper bound even if in some cases non-linear dynamic analysis indicates higher q factors. For regular structures in areas of low seismicity having standard structural systems with sections of standard sizes, a behaviour factor of 1.5–2.0 may be adopted (except for K-bracing) by satisfying only the resistance requirements of EN1993-1 (Eurocode 3, 2005).

Although a direct code comparison between codes can only be reliable if it involves the full design procedure, the reference q factors in EC8 appear generally lower than R values in the US provisions (ASCE/SEI, 2010) for similar frame configurations. It is also important to note that the same force-based behaviour factors (q) are proposed as displacement amplification factors (q_d), on the basis of the 'equal-displacement approximation' (Elghazouli et al., 2014). This is not the case in the US provisions where specific seismic drift amplification factors (C_d) are suggested; these values are generally lower than the corresponding R factors for all frame types.

6.3 DUCTILITY CLASSES AND RULES FOR CROSS-SECTIONS

To achieve some consistency with other parts of the code, the most recent version of EC8 explicitly addresses three ductility classes namely DCL, DCM and DCH referring to low, medium and high dissipative structural behaviour, respectively. For DCL, global elastic analysis and the resistance of the members and connections may be evaluated according to EC3 without any additional requirements. The recommended reference 'q' factor for DCL is 1.5–2.0. For buildings that are not seismically isolated or incorporating effective dissipation devices, design to DCL is only recommended for low-seismicity situations. In contrast, structures in DCM and DCH need to satisfy specific requirements primarily related to ensuring sufficient ductility in the main dissipative zones. Some of these requirements are general rules that apply to most structural types whilst others are more relevant to specific configurations.

The application of a behaviour factor larger than 1.5–2.0 must be coupled with sufficient local ductility within the critical dissipative zones. For elements in compression or bending (under any seismic loading scenario), this requirement is ensured in EC8 by restricting the width-to-thickness (b/t) ratios to avoid local buckling. An increase of b/t ratio results in lower element ductility due to the occurrence of local buckling (as illustrated in Figure 6.1), leading to a reduction in the energy dissipation capacity, which is expressed by a lower q factor. The classification used in EC3 is adopted but with restrictions related to the value of q factor as given in Table 6.2 (Section 6.5.3 and Table 6.3 of EN1998-1). It is worth noting that the seismic cross-section requirements in US practice imply more strict limits for certain section types.

The cross-section requirements apply to all types of frame considered in EC8. These provisions implicitly account for the relationship between local buckling and rotational ductility of steel members, which has been extensively investigated by several researchers (e.g. Lay and Galambos, 1967; Kato and Akiyama, 1982; Kato, 1989).

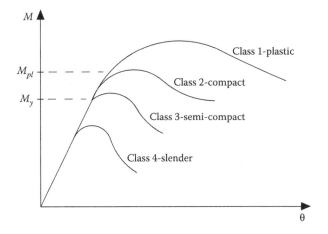

Figure 6.1 Moment–rotation characteristics for different cross-section classes.

Table 6.2 Cross-section requirements based on ductility class and reference q-factor

Ductility class	Reference q-factor	Cross-section class
DCM	$1.5 < q \leq 2$	Class 1, 2 or 3
	$2.0 < q \leq 4$	Class 1 or 2
DCH	$q > 4$	Class 1

In subsequent sections, the behaviour of the three main configurations of steel frame structure, namely moment resisting, concentrically and eccentrically braced frames is discussed. Whereas moment frames exhibit relatively ductile behaviour under earthquake loading, their low transverse stiffness may, in some situations, result in high-storey drifts, thus leading to unacceptable damage to non-structural components and possible stability problems. On the other hand, concentrically braced frames (CBF) may provide relatively higher stiffness, but can often suffer from reduced ductility once the compression braces buckle. EBFs have the potential of providing adequate ductility as well as stiffness, provided that the shear or bending links are carefully designed and detailed to withstand the substantial inelastic demands that are imposed on these dissipative zones.

6.4 MOMENT-RESISTING FRAMES

6.4.1 Frame characteristics

Moment-resisting frames are designed such that plastic hinges occur predominantly in beams rather than in columns (weak beam/strong column design) as shown in Figure 6.2. This provides favourable performance, compared to strong beam/weak column behaviour through which significant deformation and second order effects may arise in addition to the likelihood of premature storey collapse mechanisms. The only exception to this requirement is at the base of the ground floor columns, where plastic hinges may form.

Due to the spread of plasticity through flexural plastic hinges, moment-resisting frames usually possess high ductility as reflected in the high reference 'q' assigned in EC8. Nevertheless, due to their inherent low stiffness, lateral deformation effects need careful consideration.

6.4.2 Capacity design requirements

In EC8, the 'weak beam/strong column' concept is typically required, with plastic hinges allowed at the base of the frame, at the top floor of multi-storey frames and for single-storey frames. The most recent version of EC8 also allows dissipative zones to be located in the connections provided adequate behaviour can be demonstrated. Rules for moment-resisting frames are described mainly in Section 6.6 of EN1998-1.

To obtain ductile plastic hinges in the beams, checks are made that the full plastic moment resistance and rotation is not reduced by coexisting compression and shear forces. To satisfy this for each critical section, the applied moment (M_{Ed}) should not exceed the design plastic moment capacity ($M_{pl,Rd}$) (i.e. $M_{Ed}/M_{pl,Rd} \leq 1.0$), the applied axial force (N_{Ed}) should not exceed 15% of the plastic axial capacity ($N_{pl,Rd}$) (i.e. $N_{Ed}/N_{pl,Rd} \leq 0.15$). Also, the shear force (V_{Ed}) due to the application of the plastic moments with opposite signs at the extremities of the beam should not exceed 50% of the design plastic shear resistance ($V_{pl,Rd}$) of the section (i.e. $V_{Ed}/V_{pl,Rd} \leq 0.5$, in which $V_{Ed} = V_{Ed,G} + V_{Ed,M}$), where $V_{Ed,G}$ and $V_{Ed,M}$ are the shear forces due to the gravity and moment components on the beam, respectively.

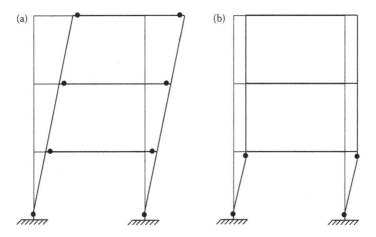

Figure 6.2 Weak-beam/strong-column and weak-column/strong-beam behaviour in moment-resisting frames. (a) Beam mechanism. (b) Storey column mechanism.

According to Section 6.6.3 of EN1998-1, columns should be verified for the most unfavourable combination of bending moments M_{Ed} and axial forces N_{Ed}, based on

$$M_{Ed} = M_{Ed,G} + 1.1\gamma_{ov}\Omega M_{Ed,E} \tag{6.1}$$

$$N_{Ed} = N_{Ed,G} + 1.1\gamma_{ov}\Omega N_{Ed,E} \tag{6.2}$$

where Ω is the minimum over-strength in the connected beams ($\Omega_i = M_{pl,Rd}/M_{Ed,i}$). The parameters $M_{Ed,G}$ and $M_{Ed,E}$ are the bending moments in the seismic design situation due to the gravity loads and lateral earthquake forces, respectively, as shown in Figure 6.3 (Elghazouli, 2007); the same subscripts also apply for axial and shear actions. Additionally, the most unfavourable shear force (V_{Ed}) of the column due to seismic combination actions must be less than 50% of the ultimate shear resistance of the section.

The beam over-strength parameter ($\Omega = M_{pl,Rd}/M_{Ed}$) as adopted in EC8 involves an approximation as it does not account accurately for the influence of gravity loads on the behaviour (Elghazouli, 2007). This issue becomes particularly pronounced in gravity-dominated frames (i.e. with large beam spans) or in low-rise configurations (since the initial column sizes are relatively small), in which the beam over-strength may be significantly underestimated. The extent of the problem depends on the interpretation of the code and whether Ω is used in isolation or in combination with an additional capacity design criterion based on a limiting ratio of 1.3 on the column-to-beam capacity (i.e. Equation 4.4 of Chapter 4). It is also important to note that whilst codes aim to achieve a 'weak-beam/strong-column' behaviour, some column hinging is often unavoidable. In the inelastic range, points of contra-flexure in members change and consequently the distribution of moments vary considerably from idealised conditions assumed in design. The benefit of meeting code requirements is to obtain relatively strong columns such that beam rather than column yielding dominates over several stories, hence achieving adequate overall performance.

6.4.3 Stability and drift considerations

Deformation-related criteria are stipulated for all building types in EC8 but, as expected, they are particularly important in steel moment frames due to their inherent flexibility,

Gravity loading ($M_{Ed,G}$)   Lateral loading ($M_{Ed,G}$) $M_{Ed,G} + M_{Ed,E}$

Figure 6.3 Moments due to gravity and lateral loading components in the seismic situation.

which often governs the design. Two deformation-related requirements, namely 'second-order effects' and 'inter-storey drifts', are stipulated in Sections 4.4.2.2 and 4.4.3.2 of EN1998-1. The former is associated with ultimate state whilst the latter is included as a damage-limitation (serviceability) condition.

Second-order (P-Δ) effects are specified through an inter-storey drift sensitivity coefficient (θ) given as

$$\theta = \frac{P_{tot}d_r}{V_{tot}h} \tag{6.3}$$

where P_{tot} and V_{tot} are the total cumulative gravity load and seismic shear, respectively, at the storey under consideration; h is the storey height and d_r is the design inter-storey drift (product of elastic inter-storey drift from analysis and q, i.e. $d_e \times q$). Instability is assumed beyond $\theta = 0.3$ and is hence considered as an upper limit. If $\theta \leq 0.1$, second-order effects could be ignored, whilst for $0.1 < \theta \leq 0.2$, P-Δ may be approximately accounted for in seismic action effects through the multiplier $1/(1-\theta)$.

For serviceability, 'd_r' is limited in proportion to 'h' such that

$$d_r\nu \leq \psi h \tag{6.4}$$

where ψ is suggested as 0.5%, 0.75% and 1.0% for brittle, ductile or non-interfering non-structural components, respectively; ν is a reduction factor, which accounts for the smaller more-frequent earthquakes associated with serviceability, recommended as 0.4–0.5 depending on the importance class.

Assessment of other codes, including US provisions, suggests that drift-related requirements in EC8 can be relatively more stringent, depending on the limit selected and the importance category under consideration. As a result, direct application of the specific rules for moment frames in EC8, followed by inter-storey drift and second-order stability checks, often results in an overall lateral capacity, which is notably different from that assumed in design (Elghazouli, 2007; Sanchez-Ricart and Plumier, 2008; Peres and Castro, 2010). Significant levels of lateral frame over-strength can be present particularly when large q factors are used and/or when the spectral design accelerations are not high. This over-strength is also a function of spectral acceleration and gravity design. Whereas the presence of over-strength reduces the ductility demand in dissipative zones, it also affects forces imposed on frame and foundation elements. A rational application of capacity design necessitates a realistic assessment of lateral capacity after the satisfaction of all provisions, followed by a re-evaluation of global over-strength and the required 'q'. Although high 'q' factors are allowed for moment frames, in recognition of their ductility and energy dissipation capabilities, such a choice may, in some cases, be unnecessary and undesirable (Villani et al., 2009; Elghazouli, 2010).

6.4.4 Beam-to-column connections

Steel moment frames have traditionally been designed with rigid full-strength connections, usually of fully welded or hybrid welded/bolted configuration. Typical design provisions ensured that connections are provided with sufficient over-strength such that dissipative zones occur mainly in the beams. However, the reliability of these commonly used forms of full-strength beam-to-column connection has come under question following poor performance in large events in the mid-1990s, particularly in the Northridge earthquake of 1994 (Bertero et al., 1994) and the Hyogo-ken Nanbu (Kobe) earthquake of 1995 (EERI, 1995), as illustrated in Figure 6.4. The extent and repetitive nature of damage observed in several types of welded and hybrid connections have prompted considerable research effort not only to repair methods for existing structures but also to alternative connection configurations to be incorporated in new designs.

The above-mentioned problems prompted the industry in the United States, in liaison with government agencies, professional institutions and academic establishments to create a joint venture (SAC) to respond to the questions raised by the extensive damage observed. Laboratory tests confirmed that connections designed and manufactured strictly to code requirements and conventional shop practice failed to provide the necessary levels of ductility (SAC, 1995–1996a,b). Observed damage was attributed to several factors, including defects associated with weld and steel materials, welding procedures, stress concentration, high rotational demands, scale effects, as well as the possible influence of strain levels and rates. In addition to the concerted effort dedicated to improving seismic design regulations for new construction, several proposals have been forwarded for the upgrading of existing connections (FEMA, 1995–2000; PEER, 2000). As shown schematically in Figure 6.5, this may be carried out by strengthening of the connection through haunches, cover or side plates or other means. Alternatively, it can be achieved by weakening of the beam by trimming the flanges (i.e. reduced beam section 'RBS' or 'dog-bone' connections), perforating the flanges or by reducing stress concentrations through slots in beam webs, enlarged access holes, etc. It should be noted, however, that most pre-qualification activities have focused on connections to open section columns, with comparatively less attention given to connections to tubular columns (Elghazouli and Packer, 2014). In general, the design can be based on either pre-qualified connections or on prototype tests. Pre-qualified connections have been proposed in the US (AISC, 2011), and a similar European activity is underway.

Another important aspect of connection behaviour is related to the influence of the column panel zone. This has direct implications on the ductility of dissipative zones as well as on the overall frame performance. Recent research studies (Castro et al., 2005; Castro et al.,

Figure 6.4 Examples of typical damage in connections of moment frames: (a) weld fracture at bottom flange; (b) fracture extending into structural section.

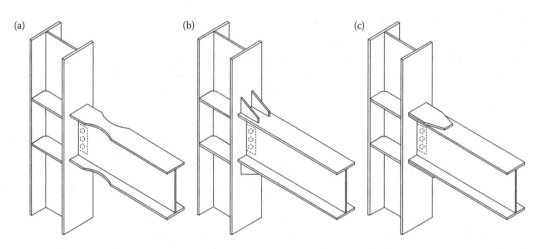

Figure 6.5 Schematic examples of modified connection configurations for moment frames. (a) Reduced beam section (RBS or dog-bone). (b) Connection with haunches. (c) Connection with cover plates.

2008) involved the development of realistic modelling approaches for panel zones within moment frames as well as assessment of current design procedures. One important issue is related to the treatment of the two yield points corresponding to the onset of plasticity in the column web and surrounding components, respectively. Another key design consideration is concerned with balancing the extent of plasticity between the panel zone and the connected beams, an issue which can be significantly affected by the level of gravity applied on the beams. On the one hand, allowing a degree of yielding in the panel reduces the plastic hinge rotations in the beams yet, on the other hand, relatively weak panel zone designs can result in excessive distortional demands which can cause unreliable behaviour of other connection components particularly in the welds.

Section 6.6.3 of EN1998-1 requires the web panel to be designed to ensure adequate shear and buckling resistance. The design shear assuming plasticity in the beams ($V_{wp,Ed}$) should not exceed the web panel plastic shear resistance ($V_{wp,Rd}$), (i.e. $V_{wp,Ed} \leq V_{wp,Rd}$). The design shear should also not exceed the buckling resistance of the web panel ($V_{wb,Rd}$), (i.e. $V_{wp,Ed} \leq V_{wb,Rd}$). If strengthening is required to the web panel, additional plates can be welded to the column panel zone.

Many of the drawbacks of fully rigid welded frames can be alleviated by bolted forms. To this end, the feasibility of using partial-strength bolted connections, which are usually semi-rigid as well, for seismic resistance has been the subject of a number of investigations (e.g. Nader and Astaneh, 1992; Elnashai and Elghazouli, 1994; Elghazouli, 1996; Mazzolani and Piluso, 1996; Faella et al., 2000; Elghazouli et al., 2009). Despite the economic advantages in fabrication and construction, this type of connection has not been traditionally employed for earthquake resistance due to two main reasons. The first is related to the semi-rigidity, which may lead to excessive deformation under static loads. It was shown in several investigations, however, that due to the relatively longer natural period of semi-rigid frames, the deflections may often not be higher under dynamic loads as compared to rigid frames. The second reason is that insufficient information has been available on the hysteretic behaviour and local ductility of partial strength connections. In general, semi-rigid partial-strength connections can be a viable alternative, particularly in moderate seismicity areas.

Revisions have also been introduced in the current version of EC8 to reflect recent research findings. If the structure is designed to dissipate energy in the beams, connections should

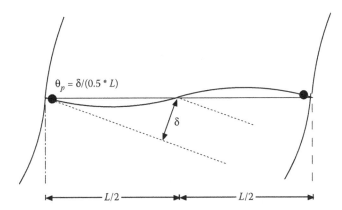

$$\theta_p = \delta/(0.5 * L)$$

Figure 6.6 Estimation of plastic hinge rotation θ_p.

be designed for the required degree of over-strength taking into account the plastic moment resistance of the attached beams. On the other hand, semi-rigid partial-strength connections are now permitted provided several conditions are satisfied (according to Section 6.6.4 of the code) including (i) all connections have rotation capacity consistent with global deformations, (ii) members framing into connections are stable at the ultimate limit state and (iii) connection deformation is accounted for through non-linear analysis.

For all connections, whether full- or partial-strength, design to EC8 should ensure sufficient plastic rotation (θ_p) of the plastic hinge region, such that $\theta_p \geq 35$ mrad for DCH and $\theta_p \geq 25$ mrad for DCM (with $q > 2$). The plastic rotation θ_p can be determined as $\delta/0.5L$, where δ is the beam deflection at mid-span and L is the beam span, as illustrated in Figure 6.6. In tests, it should be ensured that θ_p is achieved under cyclic loading with less than 20% degradation in stiffness and strength, and that the column web panel shear does not contribute more than 30% to θ_p, noting that the column elastic deformation is not included in θ_p. It is also important to note that if partial-strength connections are adopted, column capacity design checks need only to be verified based on the plastic capacity of the connections rather than that of the beams.

6.5 CONCENTRICALLY BRACED FRAMES

6.5.1 Frame characteristics

Because of their geometry, CBFs, such as those shown in Figure 6.7, provide truss action with members subjected largely to axial forces in the elastic range. However, during a moderate to severe earthquake, the bracing members and their connections undergo significant inelastic deformations into the post-buckling range, which has led to reported cases of damage in previous earthquakes (EERI, 1995).

The response of CBFs is typically dominated by the behaviour of its bracing members. This behaviour has been investigated previously by several researchers (e.g. Maison and Popov, 1980; Popov and Black, 1981; Ikeda and Mahin, 1986; Goel and El-Tayem, 1986), focusing mainly on the response under idealised cyclic loading conditions. A collaborative European project (Elghazouli, 2003; Broderick et al., 2005; Elghazouli et al., 2005) also examined the performance of bracing members through analytical studies, which were supported by monotonic and cyclic quasi-static axial tests as well as dynamic shake table tests.

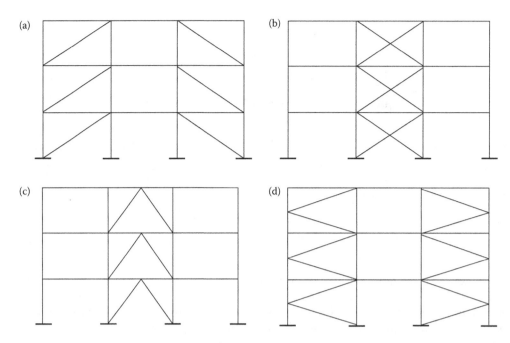

Figure 6.7 Typical idealised configurations of concentrically braced frames. (a) Non-intersecting braces, (b) cross-bracing, (c) V or inverted-V bracing and (d) K-bracing.

An example of the hysteretic axial load–deformation response of a bracing member is shown in Figure 6.8, in which the displacement and axial loads are normalised by the yield values. In compression, member buckling is followed by lateral deflection and the formation of a plastic hinge at mid-length, which leads to a gradual reduction in capacity. On reversing the load, elastic recovery occurs followed by loading in tension until yielding takes place. Subsequent loading in compression results in buckling at loads lower than the initial strength due to the residual deflections, the increase in length as well as the Bauschinger effect. Moreover, due to the accumulated permanent elongation, tensile yielding occurs at axial deformations that increase with each cycle of loading. Cyclic testing of diagonal bracing systems indicates that energy can be dissipated after the onset of global buckling if failures due to local buckling or at the connection are prevented.

Under the cyclic axial loading conditions applied on bracing members in seismic situations, failure can occur due to fracture of the cross-section following local buckling, provided that bracing connections are adequately designed and detailed. This was clearly illustrated in shake-table tests on tubular bracing members (Elghazouli et al., 2005). High strains typically develop upon local buckling in the corner regions of the cross-section. Cracks eventually form in these regions, as shown in Figure 6.9 (Elghazouli et al., 2005), and gradually propagate through the cross-section under repeated cyclic loading.

The initiation of local buckling and fracture is influenced by the width-to-thickness ratio of the elements of the cross-section, as well as the applied loading history. Seismic codes rely on the limits imposed on the width-to-thickness ratios of the cross-section in order to delay or prevent local buckling and hence reduce the susceptibility to low cycle fatigue and fracture. There is also a dependence on the overall member slenderness of the brace (Elghazouli, 2003). Seismic codes also normally impose an upper limit on the member slenderness to limit sudden dynamic loading effects as well as the extent of post-buckling deformations.

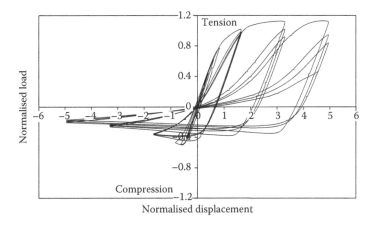

Figure 6.8 Typical response of a bracing member under cyclic axial loading.

Figure 6.9 Fracture of tubular steel bracing member during shake table testing. (a) Failure of tubular bracing member. (b) View of fracture during testing.

6.5.2 Design requirements

The provisions of EC8 for CBFs (provided mainly in Section 6.7 of EN1998-1) typically consider that the horizontal seismic forces are mainly resisted by the axially loaded members. Design should allow yielding of the diagonals in tension before yielding or buckling of the beams or columns and before failure of the connections. Due to buckling of the compression braces, tension braces are considered to be the main ductile members, except in V and inverted-V configurations.

In diagonal bracings of the types shown in Figure 6.7a and b, the cyclic horizontal forces can be assumed in EC8 to be resisted by the corresponding tension members only, with the contribution of the compression bracing members neglected. To avoid significant asymmetric response effects, the value of $A\cos\alpha$ (as shown in Figure 6.10) must not vary significantly between two opposite braces in the same storey, such that

$$\frac{(A^+ - A^-)}{(A^+ + A^-)} \leq 0.05 \tag{6.5}$$

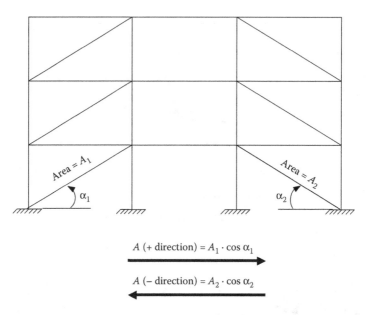

$$A \ (+ \text{direction}) = A_1 \cdot \cos \alpha_1$$

$$A \ (- \text{direction}) = A_2 \cdot \cos \alpha_2$$

Figure 6.10 Symmetry of lateral resistance in concentrically braced frames.

where A is the area of the cross-section of the tension diagonal and α is the slope of the diagonal to the horizontal.

In V-bracing, both tension and compression bracing members are needed to resist horizontal seismic forces effectively, hence both should be included in the elastic analysis of the frame. Also, the beams should be designed for gravity loading without considering the intermediate support of the diagonals, as well as account for the possibility of an unbalanced vertical action after brace buckling. In other frames, only the tension diagonals are considered. However, accounting for both braces is allowed in EC8 provided a non-linear static or time history analysis is used, both pre-buckling and post-buckling situations are considered and background studies justifying the models adopted are provided. It should be noted that ignoring the compression brace can have favourable or detrimental effects on the actual response, depending on the frame configuration and design situation (Elghazouli, 2003). On the other hand, K-bracing, such as that shown in Figure 6.7d where the diagonals meet at an intermediate point in the column, do not offer ductile behaviour due to the demand for a column yielding mechanism. Consequently, it is not appropriate for dissipative design and its use is not recommended in EC8.

In the design of the diagonal members, the non-dimensional slenderness $\bar{\lambda}$ used in EC3 plays an important role in the behaviour. This is discussed in detail elsewhere (Elghazouli, 2003). In earlier versions of EC8, an upper limit of 1.5 was proposed to prevent elastic buckling. However, further modifications have been made in the current version of EC8 and the upper limit has been revised to a value of 2.0, which results in a more efficient design. Moreover, no upper limit is needed for structures up to two storeys high. On the other hand, in frames with X-diagonal braces $\bar{\lambda}$ should be between 1.3 and 2.0. The lower limit is specified to avoid overloading columns in the pre-buckling stage of diagonals. Satisfying this lower limit can, however, result in some difficulties in practical design. It should also be noted that in frames with non-intersecting diagonal bracings (e.g. Figure 6.7a), the code stipulates that the design should account for forces that may develop in the columns due to loads from both the tension diagonals and pre-buckling forces in the compression diagonals.

Gravity loading ($N_{Ed,G}$)   Lateral loading ($N_{Ed,E}$) $N_{Ed,G} + N_{Ed,E}$

Figure 6.11 Axial forces due to gravity and lateral loading in the seismic design situation.

All columns and beams should be capacity designed for the seismic combination actions. In summary, the following relationship applies for the capacity design of non-diagonal members, where the design resistance of the beam or column under consideration $N_{Ed}(M_{Ed})$, with due account of the interaction with the bending moment M_{Ed} is determined as

$$N_{Ed}(M_{Ed}) \geq N_{Ed,G} + 1.1\gamma_{ov}\Omega N_{Ed,E} \tag{6.6}$$

where $N_{Ed,G}$ and $N_{Ed,E}$, are the axial load due to gravity and lateral actions, respectively, in the seismic design situation, as illustrated in Figure 6.11; Ω is the minimum value of axial brace over-strength over all the diagonals of the frame and γ_{ov} is the material over-strength. However, Ω of each diagonal should not differ from the minimum value by more than 25% in order to ensure reasonable distribution of ductility. It is worth noting that unlike in moment frames, gravity loading does not normally have an influence on the accuracy of Ω. It should also be noted that the 25% limit can result in difficulties in practical design; it can be shown (Elghazouli, 2007) that this limit can be relaxed or even removed if measures related to column continuity and stiffness are incorporated in design.

US provisions (ASCE, 2010; AISC, 2010) differ from those in EC8 in terms of the R factors recommended as well as cross-section limits for some section types. However, the most significant is related to the treatment of the brace buckling in compression, which may lead to notably different seismic behaviour depending mainly on the slenderness of the braces. This is discussed in more detail elsewhere (Elghazouli, 2003) and has significant implications on the frame over-strength as well as on the applied forces and ductility demand imposed on various frame components.

6.5.3 Bracing connections

Many of the failures reported in CBFs due to strong ground motion have been in the connections. In principle, bracing connections can be designed as rotationally restrained or unrestrained, provided that they can transfer the axial cyclic tension and compression effectively. The in- and out-of-plane behaviour of the connection, and their influence on the beam and column performance, should be carefully considered in all cases. For example, considering gusset plate connections (see Figure 6.12), satisfactory performance can be ensured by allowing the gusset plate to develop plastic rotations. This requires that the free length between the end of the brace and the assumed line of restraint for the gusset can be sufficiently long to permit plastic rotations, yet short enough to preclude the occurrence of plate buckling prior to member buckling (Astaneh et al., 1986). Alternatively, connections with stiffness in two directions, such as crossed gusset plates, can be detailed.

As in the case of moment frames, the design of connections between bracing members and the beams/columns in a CBF is only dealt with in a conceptual manner in EC8. Accordingly,

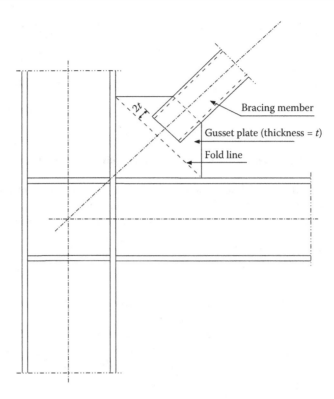

Figure 6.12 Brace-to-gusset plate connection in concentrically braced frames.

designers can adopt details available from existing literature, or based on prototype testing. The performance of bracing connections, such as those involving gusset plate components, has attracted significant research interest in recent years (e.g. Lehman et al., 2008; Yoo et al., 2008). Designing bracing connections in an efficient and practical manner can be complex and time-consuming, which has led to the development of various 'pre-engineered' proprietary solutions (Elghazouli and Packer, 2014). Supplementary European guidance, through complementary manuals, on the design and detailing of recommended bracing connections for seismic resistance is also underway.

6.6 ECCENTRICALLY BRACED FRAMES

6.6.1 Frame characteristics

In this type of structural system, as shown in Figure 6.13, the bracing members intersect the girder at an eccentricity 'e', and hence transmit forces by shear and bending. The length of the girder defined by 'e' is termed a 'link beam', which may behave predominantly in either shear or bending. While retaining the advantages of CBFs in terms of drift control, EBFs also represent an ideal configuration for failure mode control. Another important advantage is that by providing an eccentricity, a higher degree of flexibility in locating doors and windows in the structure is achieved. By careful design of the link beam, significant energy dissipation capacity can be obtained. Moreover, zones of excessive plastic deformations can be shifted away from beam–column connections, thus improving the overall integrity of the frame.

The length of the link zone has a direct influence on the frame stiffness. The relation between eccentricity ratio (e/L) and the lateral stiffness is illustrated in Figure 6.14. As e/L tends to

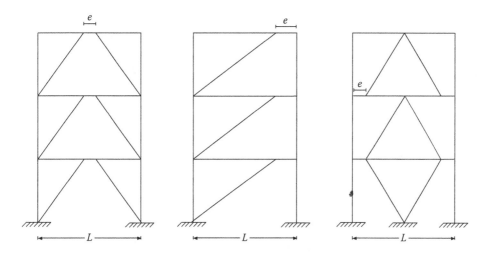

Figure 6.13 Possible configurations of eccentrically braced frames.

unity, the stiffness of the MRF is obtained, while the zero eccentricity ratio corresponds to the CBF stiffness. There is also a direct relationship between the frame drift angle (θ) and the rotational demand in the link (γ). Simple analysis of plastic collapse mechanisms of a single link in EBF, gives a relationship between frame and link deformations (see Figure 6.15) as

$$\theta L = \gamma e \qquad (6.7)$$

Since the span of the frame is significantly larger than the eccentricity '*e*', it follows that the ductility demand in the link is considerably higher than that for the frame. It is also evident that shorter links would have higher demand for the same level of frame drift.

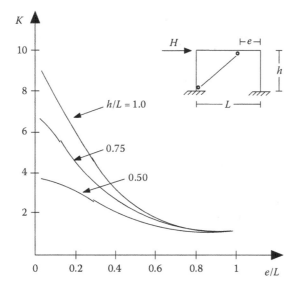

Figure 6.14 Relationship between link length and lateral stiffness of eccentrically braced frames.

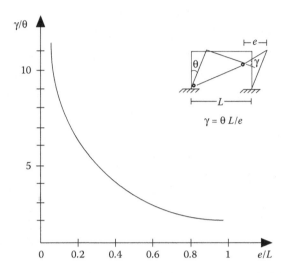

Figure 6.15 Relationship between link length and ductility demand in eccentrically braced frames.

As in other codes, EBFs are designed in EC8 so that beams are able to dissipate energy by formation of plastic bending or plastic shear mechanisms in the links. Specific rules are given to ensure that yielding in the bending or shear links of the beams will take place prior to yielding or failure in other members, which would therefore be capacity designed. The most recent version of EC8 incorporates detailed provisions (mainly in Section 6.8 of ENV1998-1), which are largely in accordance with North American design procedures.

6.6.2 Link beams

Whereas short links suffer from high-ductility demands, they yield primarily in shear. Experimental evidence (e.g. Hjelmstad and Popov, 1983; Kasai and Popov, 1986; Engelhardt and Popov, 1989) showed that shear link behaviour in steel is superior to that of flexural plastic hinges. However, other considerations such as architectural requirements may necessitate the use of long links. Assuming no strain-hardening or moment–shear interaction, the theoretical dividing length (e_c) between shear and flexural yielding is

$$e_c = \frac{2M_{p,link}}{V_{p,link}} \tag{6.8}$$

where $M_{p,link}$ and $V_{p,link}$ are the plastic moment and plastic shear capacities of the cross-section, respectively.

Experimental evidence, however, shows that strain-hardening is significant in link behaviour. The ultimate shear and bending strengths may be significantly higher than $V_{p,link}$ and $M_{p,link}$, with different ratios. Accordingly, in EN1998-1, for I-section links where equal moments occur at both ends, the links are defined as

- Short links:

$$e < e_s = \frac{1.6\, M_{p,link}}{V_{p,link}} \tag{6.9}$$

- Long links:

$$e > e_L = \frac{3.0\, M_{p,link}}{V_{p,link}} \tag{6.10}$$

- Intermediate links:

$$e_s < e < e_L \tag{6.11}$$

On the other hand, in designs in which only one plastic hinge forms at one end in I-sections:

- Short links:

$$e < e_s = 0.8\,(1+\alpha)\,\frac{M_{p,link}}{V_{p,link}} \tag{6.12}$$

- Long links:

$$e > e_L = 1.5\,(1+a)\,\frac{M_{p,link}}{V_{p,link}} \tag{6.13}$$

- Intermediate links:

$$e_s < e < e_L \tag{6.14}$$

where α is the ratio of the absolute value of the smaller-to-larger bending moments at the two ends of the link.

If the applied axial force exceeds 15% of the plastic axial capacity, reduced expressions for the moment and shear plastic capacities are provided in EC8 to account for the corresponding reductions in their values.

EC8 also provides limits on the rotation θ_p in accordance with the expected rotation capacity. This is given as 0.08 radians for short links and 0.02 radians for long links, whilst the limit for intermediate links can be determined by linear interpolation. The code also gives a number of rules for the provision of stiffeners in short, long and intermediate link zones.

6.6.3 Other frame members

Other members not containing seismic links, such as the columns and diagonals, should be capacity designed. These members should be verified considering the most unfavourable combination of axial force and bending moment with due account for shear forces, such that

$$N_{Rd}(M_{Ed}, V_{Ed}) \geq N_{Ed,G} + 1.1 y_{ov} \Omega N_{Ed,E} \tag{6.15}$$

where the actions are similar to those previously defined for CBFs. However, in this case, Ω is the minimum of the following: (i) min of $\Omega_i = 1.5\, V_{p,link,i}/V_{Ed,i}$ among all short links, and (ii) min of $\Omega_i = 1.5\, M_{p,link,i}/M_{Ed,i}$ among all intermediate and long links, where $V_{Ed,i}$ and

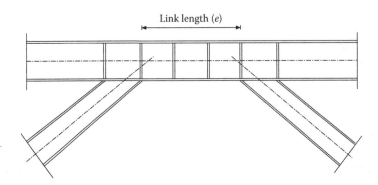

Figure 6.16 Full depth web stiffeners in link zones of eccentrically braced frames.

$M_{Ed,i}$ are the design values of the shear force and bending moment in link 'i' in the seismic design situation, whilst $V_{p,link,i}$ and $M_{p,link,i}$ are the shear and bending plastic design capacities of link 'i'. It should also be checked that the individual values of Ω_i do not differ from the minimum value by more than 25% in order to ensure reasonable distribution of ductility.

If the structure is designed to dissipate energy in the links, the connections of the links or of the elements containing the links should also be capacity designed with due account of the over-strength of the material and the links, as before. Semi-rigid and/or partial-strength connections are permitted with some conditions similar to those described previously for moment resisting frames.

Specific guidance is given for link stiffeners in EN1998-1. Full-depth stiffeners are required on both sides of the link web at the diagonal brace ends of the link as indicated in Figure 6.16. These stiffeners should have a combined width not less than $b_f - 2t_w$ and a thickness not less than $0.75t_w$ or 10 mm whichever is larger, where b_f and t_w are the link flange width and link web thickness, respectively.

Intermediate web stiffeners in shear links should be provided at intervals not exceeding $(30t_w - d/5)$ for a link rotation angle of 0.08 radians, or $(52t_w - d/5)$ for link rotation angles of 0.02 radians or less, with linear interpolation used in-between, where d is the section depth. Links of length greater than $2.6\,M_{p,link}/V_{p,link}$ and less than $5M_{p,link}/V_{p,link}$ should be provided with intermediate web stiffeners placed at a distance of 1.5 times b_f for each end of the link. Both requirements apply for links of length between 1.6 and $2.6M_{p,link}/V_{p,link}$, and no intermediate web stiffeners are required in links of lengths greater than $5\,M_{p,link}/V_{p,link}$. Intermediate link web stiffeners are required to be full depth. For links that are less than 600 mm in depth, stiffeners are required on only one side of the link web. Lateral supports are also required at both the top and bottom link flanges at the end of the link. End lateral supports of links should have design strength of 6% of the expected nominal strength of the link flange.

Design of link-to-column connections should be based upon cyclic test results that demonstrate inelastic rotation capability 20% greater than that calculated at the design storey drift. On the other hand, beam-to-column connections away from links are permitted to be designed as pinned in the plane of the web.

6.7 MATERIAL AND CONSTRUCTION CONSIDERATIONS

In addition to conforming to the requirements of EN1993 (Eurocode 3, 2005), EC8 incorporates specific rules dealing with the use of a realistic value of material strength in dissipative zones. In this respect, according to Section 6.2 of EN1998-1, the design should conform to one of the following conditions:

- The actual maximum yield strength $f_{y,max}$ of the steel of the dissipative zones satisfies the relationship: $f_{y,max} \leq 1.1 \, \gamma_{ov} \, f_y$, where f_y is the nominal yield strength and the recommended value of γ_{ov} is 1.25.
- The design of the structure is made on the basis of a single grade and nominal yield strength 'f_y' for the steels both in dissipative and non-dissipative zones, with an upper limit '$f_{y,max}$' specified for steel in dissipative zones, which is below the nominal value 'f_y' specified for non-dissipative zones and connections.
- The actual yield strength '$f_{y,act}$' of the steel of each dissipative zone is determined from measurements and the over-strength factor is assessed for each dissipative zone as $\gamma_{ov,act} = f_{y,act}/f_y$.

In addition to the above, steel sections, welds and bolts should satisfy other requirements in dissipative zones. In bolted connections, high-strength bolts (8.8 and 10.9) should be used in order to comply with the requirements of capacity design.

In terms of detailed design and construction requirements, in addition to the rules of EN1993, several specific provisions are given in Section 6.11 of EN1998-1. The details of connections, sizes and qualities of bolts and welds as well as the steel grades of the members and the maximum permissible yield strength $f_{y,max}$ in dissipative zones should be indicated on the fabrication and construction drawings.

Checks should be carried out to ensure that the specified maximum yield strength of steel does not exceed by more than 10%. It should also be ensured that the distribution of yield strength throughout the structure does not substantially differ from that assumed in design. If any of these conditions is not satisfied, new analysis of the structure and its details should be carried out to demonstrate compliance with the code.

6.8 DESIGN EXAMPLE: MOMENT FRAME

6.8.1 Introduction

The same eight-storey building considered in previous chapters is utilised in this example. The layout of the structure is reproduced in Figure 6.17. The main seismic design checks are carried out for a preliminary design according to EN 1998-1. For the purpose of illustrating the main seismic checks in a simple manner, consideration is only given to the lateral system in the X-direction of the plan, in which resistance is assumed to be provided by moment resisting frames spaced at 4 m. It is also assumed that an independent bracing system is provided in the transverse (Y) direction of the plan. Grade S275 is assumed for the structural steel used in the example.

6.8.2 Design loads

The gravity loads are adapted from those described in Chapter 3, and are depicted in Table 6.3. On the other hand, the seismic loads are evaluated based on the design response spectrum and on the fundamental period of the structure, which is estimated to be 1.06 s from the simplified expression in EC8 (Cl. 4.3.3.2.2). The total seismic mass, obtained from the self weight as well as an allowance of 30% of the imposed load, is found to be 8208 t. A behaviour factor of 4 is adopted assuming ductility class medium (DCM). The total

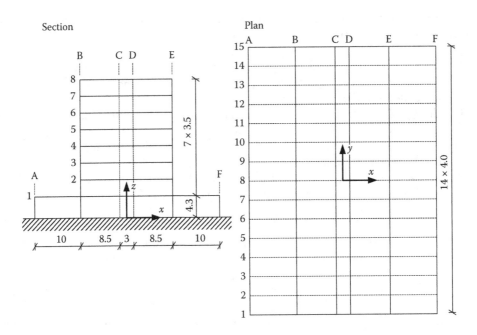

Figure 6.17 Frame layout.

design base shear for the whole structure is, therefore, estimated as 8,372 kN. The design base shear per frame is therefore considered as 558 kN.

The moment frame located on GL2 is selected for illustration in this example. Although the structure is symmetric in plan, an account should be made for torsional effects resulting from the accidental eccentricity. Using the simplified approach suggested in Cl. 4.3.3.2.4(1) of EC8, the design base shear for this frame is increased by a factor of about 1.26 to approximately 703 kN.

According to Cl. 4.3.3.2.3 of EC8, the design base shear should be applied in the form of equivalent lateral loads at the floor levels. These loads are obtained by distributing the base shear in proportion to the fundamental mode shape of the frame or, in cases where the mode follows a linear variation with height, by distributing the base shear in proportion to the mass and height of each floor. This simplified approach is adopted in this example and the floor loads and the values are given in Table 6.4.

Table 6.3 Summary of gravity loads

Type of load	Description	Value (kN/m²)
Dead load	150 mm thick solid slab	3.6
	Finishing	1.0
	External walls	3.25
	Internal walls	1.7
Imposed load	Roof	2.0
	Corridors	4.0
	Bedrooms	2.0
	Roof terrace	4.0

Table 6.4 Floor seismic loads (GL2 frame)

Floor	Seismic force (kN)
8	98.2
7	143.5
6	123.7
5	103.9
4	84.0
3	64.1
2	44.2
1	41.6

The frame on GL2 was first designed for the non-seismic/gravity loading combinations corresponding to both ultimate and serviceability limit states according to the provisions of EC3 (EN1993-1). On this basis, the initial column sections adopted were HEB450 for the four lower stories and HEB300 for the upper five stories, whilst IPE550 was selected for the beams.

6.8.3 Seismic design checks

6.8.3.1 General considerations

A preliminary elastic analysis was first carried out using the estimated seismic loads for the frame incorporating the initial member sizes. These initial member sizes were however found to be inadequate to fulfil both strength and damage limitation requirements. Accordingly, the columns were increased to HEA550 in the lower four storeys and to HEA500 in the upper five storeys. On the other hand, the initial size of the external (8.5 m) beams was retained, but the size of the internal (3 m) beams was reduced to IPE500 as this provided a more optimum solution in terms of the column sizes required to satisfy capacity criteria. It is also worth noting that controlling the lateral stiffness through the column sizes is often more optimal with respect to capacity design requirements.

The seismic design combination prescribed in Cl. 6.4.3.4 of Eurocode 0 (2002) is

$$\sum_{j \geq 1} G_{k,j} + \sum_{i \geq 1} \psi_{2,i} \cdot Q_{k,i} + A_{Ed}$$

where G_K, Q_K are the action effects due to the characteristic dead and imposed loads, respectively. The parameter Ψ_2 is the quasi-permanent combination factor, which, in this example, is taken as 0.3. In the same combination, A_{Ed} refers to the action effects due to the seismic loads.

A view of the frame model showing the element numbering is given in Figure 6.18. The results of elastic analysis for the seismic loading combination are initially used in the evaluation of the inter-storey drift sensitivity coefficient θ, as listed in Table 6.5. As shown in the table, θ does not exceed the limit of 0.1 according to Cl. 4.4.2.2(2) of EC8, and hence second-order effects do not have to be considered in the analysis.

The design checks for the beams and columns require the knowledge of the internal actions. As an example, the bending moment diagrams due to the vertical (i.e. $M_{Ed,G}$ due to $G_k + 0.3Q_k$) and earthquake (i.e. $M_{Ed,E}$ due to E) loads are presented in Figure 6.19. The final bending moment diagram for the seismic combination (i.e. M_{Ed}) is shown in Figure 6.20.

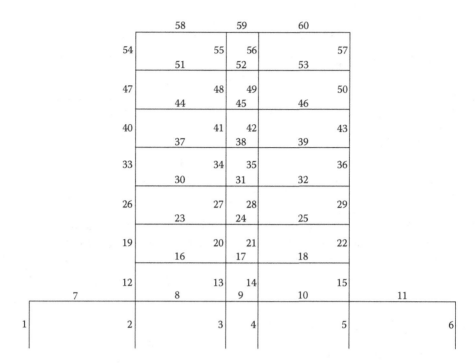

Figure 6.18 Frame model with element numbers.

Table 6.5 Calculation of inter-storey drift sensitivity coefficient

Level	d_e (mm)	d_s (mm)	d_r (mm)	P_{tot} (kN)	V_{tot} (kN)	h (mm)	θ
8	78.4	313.6	18.0	453	98.2	3,500	0.024
7	73.9	295.6	28.0	1,170	241.7	3,500	0.039
6	66.9	267.6	38.4	1,888	365.4	3,500	0.057
5	57.3	229.2	46.8	2,606	469.3	3,500	0.074
4	45.6	182.4	50.4	3,323	553.3	3,500	0.086
3	33.0	132.0	53.6	4,041	617.4	3,500	0.100
2	19.6	78.4	47.6	4,758	661.6	3,500	0.098
1	7.7	30.8	30.8	5,476	703.2	4,300	0.056

Figure 6.19 Bending moment diagrams due to (a) gravity and (b) earthquake loads.

Figure 6.20 Bending moment diagrams for the seismic combination.

6.8.3.2 Beam design checks

For illustration, the beam design checks are performed for a critical member, which is the 3 m internal beam located at the second floor (Element 17 in Figure 6.18). The internal forces at both ends of the member are listed in Table 6.6.

Based on the values from the table, the seismic demands on the beam are

$M_{Ed} = -33.2 + (-392.6) = -425.8 \text{ kN m}$
$N_{Ed} = 24.7 + 0 = 24.7 \text{ kN}$
$V_{Ed} = V_{Ed,G} + V_{Ed,M} = 45.1 + (603 + 603)/3.0 = 447.1 \text{ kN}$

According to Cl. 6.6.2(2) of EC8 and considering the properties of the beam section (which is Class 1 according to EC3):

$M_{Ed} \le M_{pl,Rd} \rightarrow 425.8 \le 2194.10^{-6} \times 275.10^3 \rightarrow 425.8 \text{ kN m} \le 603 \text{ kN m}$
$N_{Ed} \le 0.15.N_{pl,Rd} \rightarrow 24.7 \le 0.15 \times 116.4.10^{-4} \times 275.10^3 \rightarrow 24.7 \text{ kN} \le 480 \text{ kN}$
$V_{Ed} \le 0.5.V_{pl,Rd} \rightarrow 447.1 \le 0.5 \times 59.87.10^{-4} \times 275.10^3/\sqrt{3} \rightarrow 447.1 \text{ kN} \le 476 \text{ kN}$

6.8.3.3 Column design checks

The columns should be capacity designed based on the weak beam-strong column approach. According to Cl. 6.6.3 of EC8, the design forces are obtained using the following combination:

$E_d = E_{d,Gk+0.3Qk} + 1.1\gamma_{ov}\Omega E_{d,E}$

Table 6.6 Internal forces in element 17

	Left end		Right end	
	$G_k + 0.3Q_k$	E	$G_k + 0.3Q_k$	E
M	−33.2	392.6	−33.2	−392.6
V	45.1	−261.7	−45.1	−261.7
N	24.7	0	24.7	0

Table 6.7 Internal forces in element 4

	Bottom end		Top end	
	$G_k + 0.3Q_k$	E	$G_k + 0.3Q_k$	E
M	23.3	−383.4	−44.6	148.7
V	−15.8	123.7	−15.8	123.7
N	−1502.4	−755.0	−1495.4	−755.0

where

γ_{ov} is the over-strength factor assumed as 1.25.

$\Omega = \min (M_{pl,Rd,i}/M_{Ed,i}) = 603/425.8 = 1.42$.

The column design combination is therefore:

$$E_d = E_{d,Gk+0.3Qk} + 1.95E_{d,E}$$

The design forces for a critical column (Element 4 in Figure 6.18) are presented in Table 6.7 for illustration.

Based on these values, the seismic demands at the bottom end of the column are

$M_{Ed} = 23.3 + 1.95 \times (-383.4) = -724.3$ kN m
$N_{Ed} = -1502.4 + 1.95 \times (-755.0) = -2974.7$ kN
$V_{Ed} = -15.8 + 1.95 \times 123.7 = 225.4$ kN

The design checks are performed according to EC3. For brevity, only cross-section checks are presented, but clearly all EC3 resistance checks including those for member stability should also be satisfied. Considering the properties of HEA450 section (which is Class 1 according to EC3):

$M_{Ed}/M_{pl,Rd} + (N_{Ed}/N_{pl,Rd})^2 \leq 1.0$
$724.3/(5591.10^{-6} \times 275.10^3) + (2974.7/254.10^{-4} \times 275.10^3)^2 \leq 1.0$
$0.47 + 0.18 \leq 1.0$
$0.65 \leq 1.0$
$V_{Ed} \leq 0.5\ V_{pl,Rd}$
$225.4 \leq 0.5 \times 83.72 \times 10^{-4} \times 275 \times 10^3/\sqrt{3}$
225.4 kN ≤ 665 kN

In addition to the member checks, Cl. 4.4.2.3(4) of EC8 also requires that at every joint the following condition is satisfied:

$$\frac{\sum M_{Rc}}{\sum M_{Rb}} \geq 1.3$$

where $\sum M_{Rc}$ and $\sum M_{Rb}$ are the sum of the design moments of resistance of the columns and of the beams framing the joint, respectively. For illustration, this check is performed for an internal joint located at the first floor of the frame:

$\sum M_{Rc} = 2 \times 5591.10^{-6} \times 275.10^3 = 3{,}075 \text{ kN m}$

$\sum M_{Rb} = 2787.10^{-6} \times 275.10^3 + 2194.10^{-6} \times 275.10^3 = 1{,}370 \text{ kN m}$

$\sum M_{Rc}/\sum M_{Rb} \geq 1.3$

6.8.3.4 Joint design checks

According to Cl. 6.6.3(6) of EC8, the web panel zones at beam-to-column connections should be designed to resist the forces developed in the adjacent dissipative elements, which are the connected beams. For each panel zone, the following condition should be verified:

$$\frac{V_{wp,Ed}}{V_{wp,Rd}} \leq 1.0$$

where $V_{wp,Ed}$ is the design shear force in the web panel accounting for the plastic resistance of the adjacent beams/connections and $V_{wp,Rd}$ is the shear resistance of the panel zone according to EC3. For illustration, these checks are performed for an internal and an external panel.

External Panel Zone: (HEA 550 + IPE 550)

$$V_{wp,Ed} = \frac{M_{pl,Rd}}{(d_b - t_{bf})} = \frac{766}{(0.550 - 0.0172)} = 1{,}438 \text{ kN}$$

$$V_{wp,Rd} = \frac{0.9 \times f_{y,wc} \times A_{vc}}{\left(\sqrt{3} \times \gamma_{M0}\right)} + \frac{4 \times M_{pl,fc.Rd}}{(d_b - t_{bf})} \quad \text{(Cl. 6.2.6 of EC3 Part 1 - 8)}$$

$$= \frac{0.9 \times 275.10^3 \times 83.72.10^{-4}}{\sqrt{3} \times 1} + \frac{4 \times 0.300 \times 0.024^2 \times 275.10^3}{(4 \times (0.550 - 0.0172))}$$

$$= 1{,}196 + 89 = 1{,}285 \text{ kN}$$

$$\frac{V_{wp,Ed}}{V_{wp,Rd}} = \frac{1{,}438}{1{,}285} \geq 1.0 \rightarrow \text{Doubler plate required!}$$

Internal Panel Zone: (IPE 550 + HEA 550 + IPE 500)

$$\frac{V_{wp,Ed} = \sum M_{Rb}}{(d_b - t_{bf})} = \frac{1370}{(0.500 - 0.016)} = 2831 \text{ kN}$$

$$\frac{V_{wp,Ed}}{V_{wp,Rd}} = \frac{2{,}831}{1{,}285} \geq 1.0 \rightarrow \text{Doubler plates required}$$

Design of a single supplementary doubler plate with a length of 300 mm:

$V_{wp,Rd} \geq 1{,}438 \text{ kN}$

$1{,}285 + t_{dp} \times 0.300 \times 0.9 * 275.10^3/\sqrt{3} \geq 1438$

$t_{dp} \geq 3.6 \text{ mm} \rightarrow t_{dp} = 4 \text{ mm}$

6.8.4 Damage limitation

According to Cl. 4.4.3.2(1) for the damage limitation (serviceability) limit state:

$$d_r v \leq 0.01\,h$$

where d_r is the design inter-storey drift, v is a reduction factor that takes into account the lower return period of the frequent earthquake and is assumed as 0.5, and h is the storey height. The limit of 1% is applicable to cases where the non-structural components are fixed to the structure in a way that does not interfere with structural deformation. For cases with non-ductile or brittle non-structural elements, this limit is reduced to 0.75% and 0.5%, respectively.

Based on the results provided in Table 6.5, the maximum inter-storey drift occurs at the third floor:

$$d_r = 53.6\,\text{mm}$$
$$d_r\,v \leq 0.01\,h$$
$$53.6 \times 0.5 \leq 0.01 \times 3500$$
$$26.8\,\text{mm} < 35\,\text{mm}$$

6.9 DESIGN EXAMPLE: CONCENTRICALLY BRACED FRAME

6.9.1 Introduction

The same eight-storey building considered previously is utilised in this example. The main seismic design checks are carried out for a preliminary design according to EN 1998-1. For the purpose of illustrating the checks in a simple manner, consideration is only given to the lateral system in the X-direction of the plan, in which resistance is assumed to be provided by CBFs spaced at 8 m. With reference to the plan shown before in Figure 6.17, eight braced frames are considered at grid lines 1, 3, 5, 7, 9, 11, 13 and 15. It is also assumed that an independent bracing system is provided in the transverse (Y) direction of the plan. Grade S275 is considered for the structural steel used in the example.

6.9.2 Design loads

The gravity loads per unit area are the same as those adopted in the moment frame example as indicated in Table 6.3. The equivalent lateral seismic loads are evaluated based on an estimated fundamental period of 0.62 s using the simplified expression proposed in EC8 (Cl. 4.3.3.2.2). The behaviour factor considered is 4 and the total seismic mass is 8208 tons. Accordingly, the resulting base shear is estimated as 14,302 kN. The design base shear per frame is therefore considered as 1,788 kN.

The moment frame located on GL1 is selected for illustration in this example. Although the structure is symmetric in plan, an account should be made for torsional effects, resulting from the accidental eccentricity. Using the simplified approach suggested in Clause 4.3.3.2.4(1) of EC8, and for the purpose of preliminary design, the design base shear for this frame is increased by a factor of about 1.3 to approximately 2,324 kN. The base shear is applied to the frame in the form of floor loads distributed in proportion to the mass and height of each floor, as given in Table 6.8.

The frame on GL1 was firstly designed for the non-seismic/gravity loading combinations corresponding to both ultimate and serviceability limit states according to the provisions

Table 6.8 Floor seismic loads (frame on GL1)

Floor	Seismic force (kN)
8	324.4
7	474.3
6	408.8
5	343.3
4	277.5
3	212.0
2	146.2
1	137.6

of EC3 (EN1993-1). On this basis, the initial column sections adopted were HEB300 for the four lower stories and HEB220 for the upper five stories. For the beams, IPE450 was selected for the 3 and 8.5 m beams, whilst IPE 550 was necessary for the 10 m beams located on the first floor.

6.9.3 Seismic design checks

6.9.3.1 General considerations

A preliminary elastic analysis was first carried out using the estimated seismic loads for the frame incorporating the initial member sizes. Preliminary considerations indicated that a suitable arrangement consists of X-bracing over each two consecutive storeys on the 8.5 m bays. Due to the different height of the first storey, there is a change of brace angle at this level, which requires particular attention when examining the actions on the first floor beams. The initial column sizes were increased to HEM360 in the lower four storeys and to HEB320 in the upper five storeys, in order to satisfy strength and damage limitation requirements. The drifts and lateral shears related to the modified frame are given in Table 6.9, whilst the four different sizes selected for the braces are indicated in Table 6.10.

In the elastic analysis, the columns were assumed to be continuous along the height and pinned at the base. Beams and bracing members were also considered pinned at both ends. A view of the frame model indicating the element numbering is provided in Figure 6.21. The results of the elastic analysis for the seismic loading combination are initially used in the evaluation of the inter-storey drift sensitivity coefficients, which are listed in Table 6.9. As shown in the table, θ does not exceed the limit of 0.1 and hence second-order effects do not have to be considered in the analysis.

Table 6.9 Calculation of the inter-storey drift sensitivity coefficient

Level	d_e (mm)	d_s (mm)	d_r (mm)	P_{tot} (kN)	V_{tot} (kN)	h (mm)	θ
8	91.0	364.0	−5.6	453	324.4	3,500	0.004
7	92.4	369.6	68.4	1,171	798.7	3,500	0.043
6	75.3	301.2	37.6	1,888	1,207.5	3,500	0.021
5	65.9	263.6	56.8	2,606	1,550.8	3,500	0.029
4	51.7	206.8	57.6	3,323	1,828.3	3,500	0.028
3	37.3	149.2	36.8	4,041	2,040.3	3,500	0.016
2	28.1	112.4	30.4	4,758	2,186.5	3,500	0.013
1	20.5	82.0	82.0	5,476	2,324.1	4,300	0.027

Table 6.10 Axial forces in the braces for the seismic combination

Storeys	Element no.	Section	N_{Ed} (kN)
1	69	$200 \times 120 \times 12.5$	1,601
	70	$200 \times 120 \times 12.5$	1,336
2	71	$200 \times 120 \times 12.5$	1,957
	72	$200 \times 120 \times 12.5$	1,653
3	73	$200 \times 100 \times 10.0$	1,056
	74	$200 \times 100 \times 10.0$	1,125
4	75	$200 \times 100 \times 10.0$	1,291
	76	$200 \times 100 \times 10.0$	1,339
5	77	$200 \times 100 \times 8.0$	788
	78	$200 \times 100 \times 8.0$	868
6	79	$200 \times 100 \times 8.0$	975
	80	$200 \times 100 \times 8.0$	1,038
7	81	$200 \times 100 \times 5.0$	212
	82	$200 \times 100 \times 5.0$	274
8	83	$200 \times 100 \times 5.0$	335
	84	$200 \times 100 \times 5.0$	387

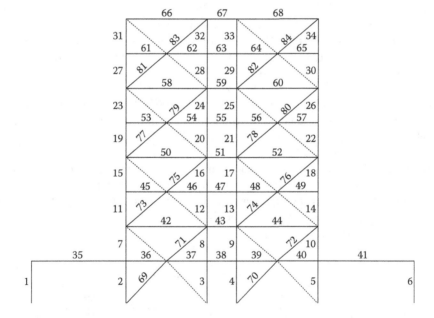

Figure 6.21 Frame model with element numbers.

6.9.3.2 Brace design checks

The design checks for the braces are conducted based on the axial forces, given in Table 6.10, from the structural analysis for the seismic design combination. Applying Cl. 6.7.3(5) (i.e. $N_{Ed} \leq N_{pl,Rd}$) and Cl. 6.7.3 (1) (i.e. $1.3 \leq \bar{\lambda} \leq 2.0$) of EC8 for a critical brace in the frame (Element 71 in Figure 6.21) as an illustration:

Table 6.11 Summary of design checks for the braces

Storeys	Element no.	Section	Lambda bar	$N_{pl,Rd}/N_{Ed} = \Omega$
1	69	200 × 120 × 12.5	1.50	1.24
	70	200 × 120 × 12.5	1.50	1.48
2	71	200 × 120 × 12.5	1.36	1.01
	72	200 × 120 × 12.5	1.36	1.20
3	73	200 × 100 × 10.0	1.59	1.43
	74	200 × 100 × 10.0	1.59	1.34
4	75	200 × 100 × 10.0	1.59	1.17
	76	200 × 100 × 10.0	1.59	1.13
5	77	200 × 100 × 8.0	1.56	1.56
	78	200 × 100 × 8.0	1.56	1.42
6	79	200 × 100 × 8.0	1.56	1.26
	80	200 × 100 × 8.0	1.56	1.19
7	81	200 × 100 × 5.0	1.52	3.72
	82	200 × 100 × 5.0	1.52	2.88
8	83	200 × 100 × 5.0	1.52	2.36
	84	200 × 100 × 5.0	1.52	2.04

$N_{Ed} \leq N_{pl,Rd} \rightarrow 1957 \leq 72.1 \times 10^{-4} \times 275 \times 10^{3} \rightarrow 1,957 \text{ kN} \leq 1,983 \text{ kN}$
$\bar{\lambda} = L_{cr}/i \times 1/\lambda_1$
$L_{cr} = 5.51 \text{ m}$
$i = 0.0466 \text{ m}$
$\lambda_1 = 93.9 \times \sqrt{(235/275)} = 86.8$
$\bar{\lambda} = (5.51/0.0466) \times (1/86.8) = 1.36 \leq 2.0$

The design checks for the remaining braces are summarised in Table 6.11.

In addition to the checks presented above, EC8 stipulates in Clause 6.7.3 (8) that the maximum brace over-strength (Ω) does not differ from the minimum value by more than 25%. As shown in Table 6.11, for this preliminary design, the over-strength in the braces exceeds this limit in several cases, with notable differences at the two upper storeys. As discussed before in Section 6.5, enforcing this limit can lead to impractical and inefficient design and may not be necessary if continuous and relatively stiff columns are adopted as is the case in this example. By increasing the brace sizes significantly throughout the frame, the code limit may be satisfied, yet this will be at the expense of the efficiency of the design; difficulties will also be encountered in satisfying the lower slenderness limit of 1.3, which is another limit that can be replaced by appropriate consideration of the post-buckling residual compressive capacity of the braces in the design of the frame.

6.9.3.3 Other frame members

Beams and columns, as well as connections, should be capacity designed to ensure that dissipative behaviour is provided primarily by the braces. According to Clause 6.7.4 of EC8, the design forces are obtained using the following combination:

$$E_d = E_{d,Gk+0.3Qk} + 1.1.\gamma_{ov}\Omega E_{d,E}$$

where

γ_{ov} is the over-strength factor assumed as 1.25
$\Omega = \min (N_{pl,Rd,i}/N_{Ed,i}) = 1983/1957 = 1.01$

The design combination is therefore:

$$E_d = E_{d,Gk+0.3Qk} + 1.39 \cdot E_{d,E}$$

The design forces for a critical beam (Element 42 in Figure 6.21) and a critical column (Element 10 in Figure 6.21) are presented in Tables 6.12 and 6.13, respectively, for illustration.

Based on these values, the seismic demands at mid-span of the beam are

$M_{Ed} = 248.5 + 1.39 \times 0 = 248.5 \text{ kN} \cdot \text{m}$
$N_{Ed} = -3.6 + 1.39 \times (-908.1) = -1,266 \text{ kN}$
$V_{Ed} = 0 \text{ kN}$

The design checks are performed according to EC3. For brevity, only cross-section checks are presented, but clearly all EC3 resistance checks including those for member stability should also be satisfied. Considering the properties of IPE450 section (which is Class 1 according to EC3):

$M_{Ed}/M_{pl,Rd} + (N_{Ed}/N_{pl,Rd})^2 \leq 1.0$
$248.5/(1,702 \times 10^{-6} \times 275 \times 10^3) + (1,266/98.8 \times 10^{-4} \times 275 \times 10^3)^2 \leq 1.0$
$0.53 + 0.22 \leq 1.0$
$0.75 \leq 1.0$

On the other hand, the seismic demands on the top end of the selected column are

$M_{Ed} = -95.6 + 1.39 \times (-321.9) = -543 \text{ kN m}$
$N_{Ed} = -2555.3 + 1.39 \times (-1600.5) = -4,780 \text{ kN}$
$V_{Ed} = 47.1 + 1.39 \times 139.4 = 241 \text{ kN}$

Table 6.12 Internal actions in the critical beam (element 42)

	Mid-span	
	$G_k + 0.3Q_k$	E
M	248.5	0.0
V	0.0	0.0
N	-3.6	-908.1

Table 6.13 Internal actions in element 10

	Bottom end		Top end	
	$G_k + 0.3Q_k$	E	$G_k + 0.3Q_k$	E
M	106.7	259.1	-95.6	-321.9
V	-57.8	-166.0	47.1	139.4
N	-1432.3	-2555.3	-1200.3	-1600.5

For brevity, only cross-section checks are presented, but clearly all EC3 resistance checks including those for member stability should also be satisfied. Considering the properties of HEM360 section (which is Class 1 according to EC3):

$M_{Ed}/M_{pl,Rd} + (N_{Ed}/N_{pl,Rd})^2 \leq 1.0$
$543/(4,989 \times 10^{-6} \times 275 \times 10^3) + (4,780/319.10^{-4} \times 275 \times 10^3)^2 \leq 1.0$
$0.40 + 0.30 \leq 1.0$
$0.70 \leq 1.0$
$V_{Ed} \leq 0.5 V_{pl,Rd}$
$241 \leq 0.5 \times 102.4 \times 10^{-4} \times 275 \times 10^3/\sqrt{3}$
$241 \text{ kN} \leq 813 \text{ kN}$

6.9.4 Damage limitation

According to Clause 4.4.3.2(1), for the damage limitation (serviceability) limit state:

$$d_r \cdot \nu \leq 0.01 \cdot h$$

where d_r is the design inter-storey drift, ν is a reduction factor that takes into account the lower return period of the frequent earthquake and is assumed 0.5, and h is the storey height. The limit of 1% is applicable to cases where the non-structural components are fixed to the structure in a way that do not interfere with structural deformation. For cases with non-ductile or brittle non-structural elements this limit is reduced to 0.75% and 0.5%, respectively.

Based on the results provided in Table 6.9, the maximum inter-storey drift occurs at the seventh storey:

$d_r = 68.4 \text{ mm}$
$d_r \nu \leq 0.01 \text{ h}$
$68.4 \times 0.5 \leq 0.01 \times 3,500$
$34.2 \text{ mm} < 35 \text{ mm}$

REFERENCES

AISC 2010a. *American Institute of Steel Construction Inc., Seismic Provisions for Structural Steel Buildings*, AISC, Chicago, IL.

AISC 2010b. *American Institute of Steel Construction Inc., Specification for Structural Steel Buildings*, ANSI-AISC-360-05, AISC, Chicago, IL.

AISC 2011. Prequalified connections for special and intermediate steel moment frames for seismic applications, ANSI/AISC 358-10 and ANSI/AISC 358s1-11, American Institute of Steel Construction, Chicago, IL.

ASCE/SEI 2010. *ASCE7-10 – Minimum Design Loads for Buildings and Other Structures*, American Society of Civil Engineers/Structural Engineering Institute, Reston, VA.

Astaneh, A., Goel, S. C. and Hanson, R. D. 1986. Earthquake-resistant design of double angle bracing, *Engineering Journal*, ASCE, 23(4), 133–147.

Bertero, V. V., Anderson, J. C. and Krawinkler, H. 1994. *Performance of Steel Building Structures during the Northridge Earthquake*, Report No. UCB/EERC-94/04, EERC, University of California, Berkeley.

Broderick, B. M., Goggins, J. M. and Elghazouli, A. Y. 2005. Cyclic performance of steel and composite bracing members, *Journal of Constructional Steel Research*, 61(4), 493–514.

Castro, J. M., Davila-Arbona, F. J. and Elghazouli, A. Y. 2008. Seismic design approaches for panel zones in steel moment frames, *Journal of Earthquake Engineering*, 12(S1), 34–51.

Castro, J. M., Elghazouli, A. Y. and Izzuddin, B. A. 2005. Modelling of the panel zone in steel and composite moment frames, *Engineering Structures*, 27(1), 129–144.

EERI 1995. *The Hyogo-ken Nanbu Earthquake Preliminary Reconnaissance Report*, Earthquake Engineering Research Institute Report no. 95-04, 116 pp.

Elghazouli, A. Y. 1996. Ductility of frames with semi-rigid connections, in: *11th World Conf. on Earthquake Engineering*, Acapulco, Mexico, Paper No. 1126.

Elghazouli, A. Y. 2003. Seismic design procedures for concentrically braced frames, *Proceedings of the Institution of Civil Engineers, Structures and Buildings*, 156, 381–394.

Elghazouli, A. Y. 2007. Seismic design of steel structures to eurocode 8, *The Structural Engineer*, 85(12), 26–31.

Elghazouli, A. Y. 2010. Assessment of European seismic design procedures for steel framed structures, submitted for publication, *Bulletin of Earthquake Engineering*, 8(1), 65–89.

Elghazouli, A. Y., Broderick, B. M., Goggins, J., Mouzakis, H., Carydis, P., Bouwkamp, J. and Plumier, A. 2005. Shake table testing of tubular steel bracing members, *Proc. Instn. Civ. Engrs., Structures and Buildings*, 158, 229–241.

Elghazouli, A. Y., Kumar, M. and Stafford, P. J. 2014. Prediction and optimisation of seismic drift demands incorporating strong motion frequency content, *Bulletin of Earthquake Engineering*, 12(1), 255–276.

Elghazouli, A. Y., Malaga-Chuquitaype, C., Castro, J. M. and Orton, A. 2009. Experimental monotonic and cyclic behaviour of blind-bolted angle connections, *Engineering Structures*, 31(11), 2540–2553.

Elghazouli, A. Y. and Packer, J. A. 2014. Seismic design solutions for connections to tubular members, *Journal of Steel Construction*, 7(2), 73–83.

Elnashai, A. S. and Elghazouli, A. Y. 1994. Seismic behaviour of semi-rigid steel frames: Experimental and analytical investigations, *Journal of Constructional Steel Research*, 29, 149–174.

EN 1990. 2002. *Eurocode 0: Basis of Structural Design*, European Committee for Standardization, CEN, Brussels.

EN 1993-1. 2005. *Eurocode 3: Design of Steel Structures – Part 1.1: General Rules and Rules for Buildings*, European Committee for Standardization, CEN, Brussels.

EN 1998-1. 2004. *Eurocode 8: Design Provisions for Earthquake Resistance of Structures, Part 1: General Rules, Seismic Actions and Rules for Buildings*, European Committee for Standardization, CEN, Brussels.

Engelhardt, M. D. and Popov, E. P. 1989. On the design of eccentrically braced frames, *Earthquake Spectra*, 5, 495–511.

Faella, C., Piluso, V. and Rizzano, G. 2000. *Structural Steel Semirigid Connections*, CRC Press, Boca Raton, FL.

FEMA 1995. *Federal Emergency Management Agency, FEMA 267 (SAC 96-02). Interim Guidelines: Evaluation, Repair, Modifications and Design of Steel Moment Frames*, FEMA, Washington, DC.

FEMA 1997. *Federal Emergency Management Agency, NEHRP (National Earthquake Hazards Reduction Program) Recommended Provisions for Seismic Regulations for New Buildings*, FEMA, Washington, DC.

FEMA 2000. *Federal Emergency Management Agency, Recommended Seismic Design Criteria for New Steel Moment-Frame Buildings, Program to Reduce Earthquake Hazards of Steel Moment-Frame Structures*, FEMA-350, FEMA, Washington DC.

Goel, S. C. and El-Tayem, A. 1986. Cyclic load behavior of angle X-bracing, *Journal of Structural Engineering*, ASCE, 112(11), 2528–2539.

Hjelmstad, K. D. and Popov, E. P. 1983. *Seismic Behavior of Active Beam Links in Eccentrically Braced Frames*, Report No. UCB/EERC–83/15, Earthquake Engineering Research Center, University of California, Berkeley, CA.

Kasai, K. and Popov, E. P. 1986. General behavior of wf steel shear link beams. *Journal of Structural Engineering*, ASCE, 112(2), 362–382.

Kato, B. and Akiyama, H. 1982. Seismic design of steel buildings, *ASCE, ST Division 108(ST8)*, 1705–1721.

Kato, B. 1989. Rotation capacity of H-section members as determined by local buckling, *Journal of Constructional Steel Research*, 13, 95–109.

Lay, M. G. and Galambos, T. V. 1967, Inelastic beams under moment gradient. *Journal of Structural Division, ASCE*, 93(1), 389–399.

Lehman, D. E., Roeder, C. W., Herman, D., Johnson, S. and Kotulka, B. 2008. Improved seismic performance of gusset plate connections, *Journal of Structural Engineering, ASCE*, 134(6), 890–889.

Maison, B. F. and Popov, E. P. 1980. Cyclic response prediction for braced steel frames, *Journal of Structural Engineering, ASCE*, 106(7), 1401–1416.

Mazzolani, F. M. and Piluso, V. 1996. *Theory and Design of Seismic Resistant Steel Frames*, E & FN Spon, London, UK.

Nader, M. N. and Astaneh, A. 1992. *Seismic Behaviour and Design of Semi-Rigid Steel Frames*, Report No. UCB/EERC-92/06, EERC, University of California, Berkeley.

PEER, 2000. *Cover-Plate and Flange-Plate Reinforced Steel Moment-Resisting Connections*, Report No. PEER2000/07, Pacific Earthquake Engineering Research Center, University of California, Berkeley.

Peres, R. and Castro, J. M. 2010. Comparison of European and American approaches for consideration of P-Delta effects in seismic design, in: *Proc. 14th European Conf. on Earthquake Engineering*, Ohrid, Macedonia, Paper No 1142.

Popov, E. P. and Black, G. R. 1981. Steel struts under severe cyclic loadings, *Journal of Structural Engineering, ASCE*, 107(9), 1857–1881.

SAC 1995. *Survey and Assessment of Damage to Buildings Affected by the Northridge Earthquake of January 17, 1994*, SAC95-06, SAC Joint Venture, Sacramento, CA.

SAC 1996a. *Experimental Investigations of Beam-Column Sub-assemblages*, SAC95-09, SAC Joint Venture, Sacramento, CA.

SAC 1996b. *SAC 95-09 Technical Report: Experimental Investigations of Beam-Column Subassemblages*, SAC Joint Venture, Sacramento, CA.

Sanchez-Ricart, L. and Plumier, A. 2008. Parametric study of ductile moment-resisting steel frames: A first step towards Eurocode 8 calibration, *Earthquake Engineering and Structural Dynamics*, 37, 1135–1155.

Villani, A., Castro, J. M. and Elghazouli, A. Y. 2009. Improved seismic design procedure for steel moment frames, in: *Proc. STESSA 2009: Behaviour of Steel Structures in Seismic Areas*, Philadelphia, PA, pp. 673–678.

Yoo, J., Roeder, C. W. and Lehman, D. E. 2008. Analytical performance simulation of special concentrically braced frames, *Journal of Structural Engineering, ASCE*, 134(6), 881–889.

Design of composite steel/concrete structures

Ahmed Y. Elghazouli and José Miguel Castro

CONTENTS

7.1 INTRODUCTION

The design of composite steel/concrete buildings in EC8, covered in Section 7 of EN1998-1 (2004), largely follows the general methodology adopted for steel structures (Section 6 of EN1998-1). Accordingly, most of the approaches and procedures discussed in the previous chapter also apply to composite steel/concrete structures, with some differences related mainly to ductility requirements and capacity design considerations. This chapter highlights these differences, discusses a number of key behavioural and design aspects, and concludes with an illustrative design example.

Three general 'design concepts' are stipulated in Section 7 of EN1998-1, namely:

1. *Concept a*: Low-dissipative structural behaviour, which refers to DCL in the same manner as in steel structures. In this case, a behaviour factor of 1.5–2 (recommended as 1.5) can be adopted based largely on the provisions of EC3 (EN1993) and EC4 (EN1994) for steel and composite components, respectively.
2. *Concept b*: Dissipative structural behaviour with composite dissipative zones. In this case, DCM and DCH design can be adopted with additional rules to satisfy ductility and capacity design requirements as discussed in subsequent sections of this chapter.
3. *Concept c*: Dissipative structural behaviour with steel dissipative zones. In this case, critical zones are designed as steel to Section 6 of EN1998-1 in the seismic situation, although other 'non-seismic' design situations may consider composite action to EC4 (EN1994). Therefore, specific measures are stipulated to prevent the contribution of concrete under seismic conditions.

This chapter deals primarily with concept b, in which composite dissipative zones are expected, but some discussion of concept c, which implies steel-only dissipation, is also included. After outlining the structural types and associated behaviour factors, as stipulated in Section 7 of EN1998-1, the main ductility and capacity design requirements are summarised. Emphasis is then given to discussing design procedures related to composite beam and column members within moment frames and other lateral-resisting structural configurations.

7.2 STRUCTURAL TYPES AND BEHAVIOUR FACTORS

The same upper limits of the reference behaviour factors specified for steel framed structures (Section 6 of EN1998-1) are also employed in Section 7 of EN1998-1 for composite structures. This applies to composite moment-resisting frames, composite concentrically braced frames and composite eccentrically braced frames. However, whilst in composite moment frames the dissipative beam and/or column zones may be steel or composite, the dissipative zones in braced frames are, in most cases, only allowed to be in steel. In other words, the diagonal braces in concentrically braced frames, and the bending/shear links in eccentrically braced frames, should in most cases be designed and detailed such that they behave as steel dissipative zones. This limitation is adopted in the code as a consequence of the uncertainty associated with determining the actual capacity and ductility properties of composite steel/concrete elements in these configurations. As a result, the design of composite braced frames follows very closely those specified for steel, and are therefore not discussed in detail herein. On the other hand, several specific criteria related to the dissipative behaviour of composite moment frames are addressed in subsequent sections of this chapter.

A number of additional composite structural systems are also referred to in Section 7 of EN1998-1, as indicated in Table 7.1, including

- Steel or composite frame with connected infill concrete panels (Type 1), or reinforced concrete walls with embedded vertical steel members acting as boundary/edge elements (Type 2).
- Steel or composite coupling beams in conjunction with reinforced concrete or composite steel/concrete walls (Type 3).
- Composite steel plate shear walls consisting of vertical continuous steel plates with concrete encasement on one or both sides of the plates and steel/composite boundary elements.

Table 7.1 Structural types and behaviour factors

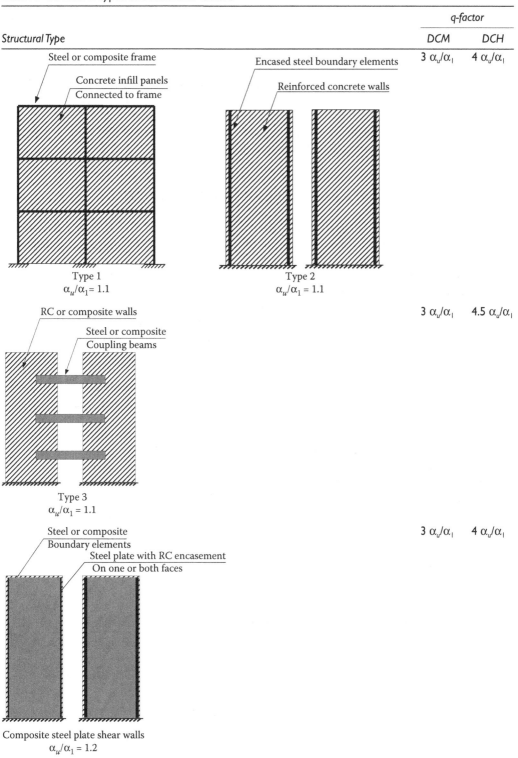

	q-factor	
Structural Type	DCM	DCH
Type 1 $\alpha_u/\alpha_1 = 1.1$ / Type 2 $\alpha_u/\alpha_1 = 1.1$	3 α_u/α_1	4 α_u/α_1
Type 3 $\alpha_u/\alpha_1 = 1.1$	3 α_u/α_1	4.5 α_u/α_1
Composite steel plate shear walls $\alpha_u/\alpha_1 = 1.2$	3 α_u/α_1	4 α_u/α_1

Steel or composite frame
Concrete infill panels
Connected to frame

Encased steel boundary elements
Reinforced concrete walls

Type 1
$\alpha_u/\alpha_1 = 1.1$

Type 2
$\alpha_u/\alpha_1 = 1.1$

RC or composite walls
Steel or composite
Coupling beams

Type 3
$\alpha_u/\alpha_1 = 1.1$

Steel or composite
Boundary elements
Steel plate with RC encasement
On one or both faces

Composite steel plate shear walls
$\alpha_u/\alpha_1 = 1.2$

Note: Additional to those in Section 6 of EN1998–1.

The upper limits of reference q for the above-listed systems are shown in Table 7.1 for DCM and DCH. As noted in previous chapters, these reference values should be reduced by 20% if the building is irregular in elevation. Also, an estimate for the multiplier α_u/α_1 may be determined from conventional non-linear 'push-over' analysis, but should not exceed 1.6. In the absence of detailed calculations, the default value of α_u/α_1 may be assumed as 1.1 for Types 1 through 3. For composite steel plate shear walls, the default value may be assumed as 1.2. It should be noted that for buildings that are irregular in plan, the default values of α_u/α_1 should be assumed as 1.05 and 1.1 for Types 1 through 3 and composite steel plate shear walls, respectively. In terms of dissipative zones, these can be located in the vertical steel sections and in the vertical reinforcement of the walls. The coupling beams in the case of Type 3 can also be considered as dissipative elements.

7.3 DUCTILITY CLASSES AND RULES FOR CROSS-SECTIONS

As in the case of dissipative steel zones, there is a direct relationship between the ductility of dissipative composite zones, consisting of concrete-encased or concrete-infilled steel members, and the cross-section slenderness. However, as expected, additional rules relating to the reinforcement detailing also apply in the case of composite members, as discussed in subsequent parts of this chapter.

If dissipative steel zones are ensured, the cross-section rules described in the previous chapter and in Section 6 of EN 1998-1 should be applied. For dissipative composite sections, the beneficial presence of the concrete parts in delaying local buckling of the steel components is accounted for by relaxing the width-to-thickness ratio as indicated in Table 7.2.

In Table 7.2 (which is adapted from Table 7.3 of EN1998-1), partially encased elements refer to sections in which concrete is placed between the flanges of I or H sections, whilst fully encased elements are those in which all the steel section is covered with concrete. The cross-section limit c/t_f refers to the slenderness of the flange outstand of length c and thickness t_f. The limits in hollow rectangular steel sections filled with concrete are represented in terms of h/t, which is the ratio between the maximum external dimension h and the tube thickness t. Similarly, for filled circular sections, d/t is the ratio between the external diameter d and the tube thickness t. The limits for partially encased sections may be relaxed even further if special additional details are provided to delay or inhibit local buckling. These aspects are discussed in subsequent sections of this chapter within the provisions related to the ductility and capacity design requirements in composite members and components.

Table 7.2 Cross-section requirements based on ductility classes and reference q factors

Ductility classes and reference q factors	Partially or fully encased H/I sections	Concrete-filled rectangular sections	Concrete-filled circular sections
DCM ($q \leq 1.5$–2.0)	$c/t_f \leq 20\sqrt{235/f_y}$	$h/t \leq 52\sqrt{235/f_y}$	$d/t \leq 90\,(235/f_y)$
DCM (1.5–$2.0 \leq q \leq 4.0$)	$c/t_f \leq 14\sqrt{235/f_y}$	$h/t \leq 38\sqrt{235/f_y}$	$d/t \leq 85\,(235/f_y)$
DCM ($q > 4.0$)	$c/t_f \leq 9\sqrt{235/f_y}$	$h/t \leq 24\sqrt{235/f_y}$	$d/t \leq 80\,(235/f_y)$

7.4 REQUIREMENTS FOR CRITICAL COMPOSITE ELEMENTS

7.4.1 Beams acting compositely with slabs

For beams attached with shear connectors to reinforced concrete- or composite-profiled slabs, a number of requirements are stipulated in Section 7.6.2 of EN1998-1 in order to ensure satisfactory performance as dissipative composite elements (concept b). These requirements comprise several criteria including those related to the degree of shear connection, ductility of the cross-section and effective width assumed for the slab.

Dissipative composite beams may be designed for full or partial shear connection according to EC4 (EN1994-1, 2004). However, the minimum degree of connection should not be lower than 80%. This is based on previous research studies (e.g. Bursi and Caldara, 2000; Bursi et al., 2005) which indicate that, at reduced connection levels, the connectors may be susceptible to low cycle fatigue under seismic loading. The total resistance of the shear connectors within hogging moment regions should also not be less than the plastic resistance of the reinforcement. In addition, EC8 requires the resistance of connectors (as determined from EC4) to be reduced by a factor of 75%. These two factors of 0.8 and 0.75, therefore, have the combined effect of imposing more than 100% in terms of degree of shear connection.

EC8 requirements also aim to ensure ductile behaviour in composite sections by limiting the maximum strain that can be imposed on concrete in the sagging moment regions of the dissipative zones. This is achieved by limiting the ratio x/d, as shown in Figure 7.1, where x is the distance from the neutral axis to the top concrete compression fibre and d is the overall depth of the composite section, such that

$$\frac{x}{d} < \frac{\varepsilon_{cu2}}{\varepsilon_{cu2} + \varepsilon_a} \tag{7.1}$$

in which ε_{cu2} is the ultimate compressive strain of concrete and ε_a is the total strain in steel at the ultimate limit state.

The code includes a table (Table 7.4 in EN1998-1), which proposes minimum values of x/d that are deemed to satisfy the ductility requirement depicted in Equation 7.1. The values are provided as a function of the ductility class (DCM or DCH) and yield strength of steel (f_y). Close observation of the limits stipulated in Table 7.4 of EN1998-1 suggests that they are derived based on assumed values for ε_{cu2} of 0.25% and ε_a of $q \times \varepsilon_y$, where ε_y is the yield strain of steel.

For dissipative zones of composite beams within moment frames, EN1998-1 requires the inclusion of 'seismic bars' in the slab at the beam-to-column connection region. The objective is to incorporate ductile reinforcement detailing to ensure favourable dissipative behaviour in the composite beams. The detailed rules are given in Annex C of EN1998-1 and

Figure 7.1 Ductility of dissipative composite beam section under sagging moment.

include reference to possible mechanisms of force transfer in the beam-to-column connection region of the slab. The provisions are largely based on background European research involving analytical and experimental studies (Plumier et al., 1998; Bouwkamp et al., 1998; Doneux and Plumier, 1999). It should be noted that Annex C of the code only applies to frames with rigid connections in which the plastic hinges form in the beams; the provisions in the annex are not intended, and have not been validated, for cases with partial strength beam-to-column connections.

Another important consideration related to composite beams is the extent of the effective width b_{eff} assumed for the slab, as indicated in Figure 7.1. EN1998-1 includes two tables (Tables 7.5 I and 7.5 II in the code) for determining the effective width. These values are based on the condition that the slab reinforcement is detailed according to the provisions of Annex C since the same background studies (Plumier et al., 1998; Bowkamp et al., 1998; Doneux and Plumier, 1999) were used for this purpose. The first table (7.5 I) gives values for negative (hogging) and positive (sagging) moments for use in establishing the second moment of area for elastic analysis. These values vary from zero to 10% of the beam span depending on the location (interior or exterior column), the direction of moment (negative or positive) and existence of transverse beams (present or not present). On the other hand, Table 7.5 II of the code provides values for use in the evaluation of the plastic moment resistance. The values in this case are as high as twice those suggested for elastic analysis. They vary from zero to 20% of the beam span depending on the location (interior or exterior column), the sign of moment (negative or positive), existence of transverse beams (present or not present), condition of seismic reinforcement, and, in some cases, on the width and depth of the column cross-section.

Clearly, design cases other than the seismic situation would require the adoption of the effective width values stipulated in EC4 (EN1994-1, 2004). Therefore, the designer may be faced with a number of values to consider for various scenarios. Nevertheless, since the sensitivity of the results to these variations may not be significant (depending on the design check at hand), some pragmatism in using these provisions appears to be warranted (Elghazouli, 2015). Previous research studies (Castro and Elghazouli, 2002; Amadio et al., 2004; Castro et al., 2007) indicate that the effective width is mostly related to the full slab width, although it also depends on a number of other parameters such as the slab thickness, beam span and boundary conditions.

7.4.2 Partially encased members

Partially encased members, in which concrete is placed between the flanges as shown in Figure 7.2a, are often used in beams and columns. This configuration offers several advantages in comparison with bare steel members particularly in terms of enhanced fire resistance

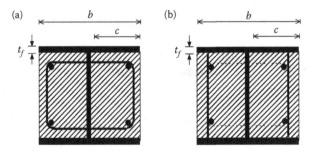

Figure 7.2 Partially encased composite sections. (a) Cross-section configuration and (b) straight bars welded to flanges.

(Schleich, 1988) as well as improved ductility due to the delay in local flange buckling (Ballio et al., 1987). In comparison with fully encased alternatives, this type of member enables the use of conventional steel connections to the flanges and reduces or eliminates the need for formwork. Several background studies on the inelastic behaviour of this type of member can be found elsewhere (Elghazouli and Dowling, 1992; Elghazouli and Elnashai, 1993; Broderick and Elnashai, 1994; Plumier et al., 1994; Elghazouli and Treadway, 2008).

Specific provisions for partially encased members are mainly included in Sections 7.6.1 and 7.6.5 of EN1998-1. In dissipative zones, the slenderness of the flange outstand should satisfy the limits given in Table 7.2 above. However, if straight links welded to the inside of the flanges (as shown in Figure 7.2b) are provided in the dissipative zones at a spacing s_1 (along the length of the member) which is less than the width of the flange outstand (i.e. $s_1/c < 1.0$), then the flange slenderness limits can be relaxed. For $s_1/c < 0.5$, the limits in Table 7.2 can be increased by 50%, and for $0.5 < s_1/c < 1.0$ linear interpolation can be employed. The weld of the straight bars should have a capacity of at least that of the tensile resistance of the bars. Also, a concrete cover of between 20 and 40 mm should be present, with the upper limit ensuring the effectiveness of the bar in delaying local flange buckling. The diameter d_{bw} of the straight welded bars should not be less than the larger of 6 mm or the value of

$$d_{bw} \geq \sqrt{\frac{t_f b}{8} \frac{f_{ydf}}{f_{ydw}}} \tag{7.2}$$

in which b is the overall width of the flange and t_f is the flange thickness, whilst f_{ydf} and f_{ydw} are the design yield strengths of the flange and straight welded bars, respectively.

Irrespective of whether straight-welded bars are employed or not, the longitudinal spacing of confining reinforcement within dissipative zones of partially encased members should be limited in order to ensure an adequate level of concrete integrity. This provision becomes particularly important if local buckling cannot be prevented at large inelastic deformation levels. The length l_{cr} of the critical dissipative zones, and the minimum longitudinal spacing s, need to be established for DCM and DCH; these requirements, which are also stipulated for fully encased members as noted in the following section, are largely based on the provisions for reinforced concrete members (Section 5 in EN1998-1) as discussed earlier in Chapter 5 of this book.

7.4.3 Fully encased columns

Composite members in which steel members are fully encased with concrete, as shown for example in Figure 7.3a, are often used as column elements in multi-storey buildings. These members clearly have inherent fire resistance properties and can provide relatively high axial and lateral loading capacity as well as significant ductility if properly designed and detailed.

A number of detailing requirements for fully encased composite columns are stipulated in Section 7.6.4 of EN1998-1. Although, in principle, the intended plastic mechanisms in frame systems may only imply the formation of column dissipative zones at the base and perhaps at the top storey, it is important that ductile detailing is provided in other critical column regions due to the adverse consequences of overstressing non-ductile concrete. This treatment is similar to that employed in the detailing of reinforced concrete columns (Chapter 5 of this book and Section 5 of EN1998-1) since the possibility of yielding in regions other than the intended dissipative zones exists due to factors such as higher dynamic modes, inelastic contra-flexure, bi-directional effects, amongst others.

Figure 7.3 Concrete encased open sections and concrete infilled tubular sections. (a) Fully encased section. (b) Infilled section.

As noted above, the detailing rules for critical regions are largely based on those for reinforced concrete columns. The length l_{cr} of the critical regions at the two ends of columns in moment frames depend on the length and depth of the column as well as on the ductility class (DCM or DCH). The code gives an expression (Equation 7.5 in EN1998-1) for the minimum volumetric ratio of hoop reinforcement. The spacing s of the confining hoops in the critical regions should also satisfy minimum values (Equations 7.7–7.9 in EN1998-1), which depend on dimensions of the concrete core, diameter of the longitudinal bars and the ductility class. The diameter d_{bw} of the hoops should satisfy minimum values (Equations 7.7–7.9 in EN1998-1) as a function of the ductility class and the maximum diameter of the longitudinal bars, as well as the yield strength of both the hoop and longitudinal reinforcement. Also, the minimum cross-section dimensions should not be less than 250 mm.

As indicated in Table 7.2, the code suggests the same flange slenderness limits for fully encased members as those for partially encased sections on the basis that the concrete cover is ineffective in providing additional restraint against local buckling. However, the presence of closely spaced confining hoops can clearly have a beneficial effect in delaying local flange buckling. Accordingly, this is treated in the code in the same manner as that of welded straight bars in partially encased members as discussed in the previous section. Therefore, if hoops are provided with spacing s, which is less than the width of the flange outstand (i.e. $s/c < 1.0$), then the flange slenderness limits can be relaxed. For $s/c < 0.5$, the limits in Table 7.2 can be increased by up to 50%, and for $0.5 < s/c < 1.0$ linear interpolation can be employed. Again, the diameter d_{bw} of the confining hoops used to delay local buckling should satisfy the minimum value resulting from Equation 7.2 above.

7.4.4 Filled composite columns

Tubular steel members of rectangular or circular/oval cross-sections can be filled with concrete to provide a highly effective solution for columns in buildings. Figure 7.3b shows an example of a rectangular hollow section filled with concrete. This type of member combines aesthetic appearance with favourable structural properties including stiffness, capacity and ductility, as well as enhanced fire resistance in comparison with bare steel configurations.

As for other types of composite member, the design should conform to the requirements of Eurocode 4 (EN1994, 2004). Additional specific criteria for filled columns are brief in EC8 and are given mainly in Section 7.6.6 of EN1998-1. For dissipative zones, the cross-section slenderness, represented by d/t or h/t, should satisfy the limits given in Table 7.2. Also, as for other types of composite member, the shear resistance in dissipative zones should be determined on the basis of the structural steel section only. However, it can also be based on the reinforced concrete section with the steel hollow section considered only as shear reinforcement.

In general, whether the member is encased or infilled, if the concrete is assumed to contribute to the axial and/or flexural resistance, complete shear transfer between the steel and reinforced concrete parts should be ensured. Due to the expected detrioration in shear strength under cyclic loading conditions, the design shear strength given in EC4 (EN1994-1, 2004) should be reduced by 50%. If shear transfer cannot be achieved through bond and friction, shear connectors should be provided to ensure full composite action. Also, in composite columns that are subjected to predominantly axial loads, sufficient shear transfer should be provided to ensure that the steel and concrete parts share the loads applied to the column at connections to beams and bracing members.

7.5 DESIGN OF STRUCTURAL SYSTEMS

7.5.1 Composite moment frames

Composite moment frames, consisting of steel (or composite) columns and steel (or encased/filled) beams acting compositely with reinforced concrete (or composite) slabs, can offer several behavioural and practical advantages over bare steel and other alternatives. The seismic behaviour of composite moment frames has been examined experimentally and analytically by several researchers (e.g. Plumier et al., 1998; Leon, 1998; Leon et al., 1998; Hajjar et al., 1998; Thermou et al., 2004; Spacone and El-Tawil, 2004; Bursi et al., 2004; Elghazouli et al., 2008). Several of these studies, among others, have dealt with modelling and design considerations including behaviour factors, slab effects, shear interaction, connections and capacity design, and have contributed to the development of design codes such as AISC (2010) and EC8 (EN1998-1, 2004).

Rules for the design and detailing of composite moment resisting frames are given in Section 7.7 of EN1998-1. With the exception of a number of specific criteria, this section of the code refers directly to the ductility and capacity design rules for steel moment frames (in Section 6 of EN1998-1 and Chapter 6 of this book), as well as the requirements for critical composite elements (Section 7.6 in EN1998-1) discussed above in Sections 7.3 and 7.4 of this chapter.

An important consideration is related to the flexural stiffness assumed in analysis. For composite beams, the code specifies two values EI_1 and EI_2 for positive bending (uncracked section) and negative bending (cracked section) regions, respectively. However, the code also allows the alternative use of an equivalent second moment of area EI_{eq}, which can be kept constant over the entire length of the beam, such that

$$EI_{eq} = 0.6 \, EI_1 + 0.4 \, EI_2 \tag{7.3}$$

The above equation clearly provides a more convenient representation of the composite beam for the purpose of analysis. On the other hand, if composite columns are used, the composite flexural stiffness of the column (EI_{comp}) can be represented as

$$EI_{comp} = 0.9 \, (E_s I_a + r E_{cm} I_c + E_s I_s) \tag{7.4}$$

where E_s and E_{cm} are the moduli of elasticity for steel and concrete, respectively, while I_a, I_c and I_s are the second moments of area for the steel section, concrete and reinforcement, respectively. The recommended value of r, which accounts for the influence of concrete cracking, is 0.5.

For composite columns, the code limits the applied axial load N_{Ed} to 30% of the plastic axial plastic capacity of the cross-section $N_{pl,Rd}$ to ensure that ductility is not significantly

Figure 7.4 View of column panel zone in a composite moment frame during testing.

reduced. The use of composite trusses as dissipative beams is also not permitted due to the uncertainty related to their performance under inelastic cyclic loading.

For steel panel zones in composite moment frames, as illustrated in Figure 7.4, the code refers to the rules for steel moment frames in Section 6 of EN1998-1. Previous studies (Castro et al., 2005) have, however, shown that the behaviour of panel zones in composite moment frames differs from that in steel moment frames due to the variation is stress distribution and distortional demand imposed on the panel. Accordingly, expressions used for the modelling and assessment for panel zones in steel frames may not be realistic for composite frames, and would need to be modified in order to account for the influence of beam–slab interaction.

For situations in which partially encased beams are utilised, the concrete encasement of the column web may be accounted for in determining the resistance of the panel zone. According to Section 7.5.4 of EN1998-1, the resistance can be evaluated as the sum of the contributions from the concrete and steel panels. However, the aspect ratio of the panel zone h_b/h_c has to be between 0.6 and 1.4, and the design shear force $V_{wp,Ed}$ derived from the plastic capacity of adjacent dissipative zones should be less than 80% of the shear resistance $V_{wp,Rd}$ of the composite steel–concrete web panel according to EN1994-1.

An important consideration is stipulated in Section 7.7.5 of EN1998-1 whereby the dissipative zones at the beam ends of composite moment frames can be considered as steel-only sections (i.e. following concept c). To achieve this, the slab needs to be totally disconnected from the steel members in a circular zone with a diameter of at least $2b_{eff}$ around the columns, with b_{eff} determined on the basis of the larger effective width of the connected beams. This 'total disconnection' also implies that there is no contact between the slab and the sides of any vertical element such as the columns, shear connectors, connecting plates, corrugated flange, etc. A similar approach has also been used in hybrid flat slab-tubular column connections (Eder et al., 2012), hence enabling the use of flat slabs in conjunction with steel-only dissipative members.

The above consideration, of disregarding the composite action and designing for steel-only dissipative zones, can be convenient in practical design (Elghazouli and Packer, 2014). Clearly, two *EI* values for the beams need to be accounted for in the analysis: composite in the middle and steel at the ends. The beams are composite in the middle, hence providing enhanced stiffness and capacity under gravity loading conditions. On the other hand, in the seismic situation, the use of steel dissipative zones avoids the need for detailed considerations in the slab, including those related to seismic rebars, effective width and ductility criteria associated with composite dissipative sections. This consideration also implies that the connections would be designed on the plastic capacity of the steel beams only. Also, the columns need to be capacity designed for the plastic resistance of steel instead of composite beam sections, which avoids over-sizing of the column members.

7.5.2 Composite – braced frames

As discussed before, in concentrically braced frames, the diagonal members, which are the main dissipative zones, should be in steel only according to the provisions of EC8. On the other hand, the beam and column members can be either steel composite. The seismic design rules are therefore directly based on those for steel concentrically braced frames in Section 6 of EN1998-1, since the ductility and capacity design requirements are largely related to the capacity of the diagonal braces. It should be noted, however, that buckling restrained braces (or unbonded braces) are not covered by the current version of EC8.

For composite eccentrically braced frames, the design rules again follow closely those stipulated for steel frames in Section 6 of EN1998-1. The code recommends that the link zones are steel sections, which should not be encased, although it is noted that they can be connected to the slab. The code stipulates that the links should be of short or intermediate length, and provides a number of additional requirements. Most importantly, if the link beam is connected to the slab, the concrete contribution should be ignored in determining the resistance of the link except when performing capacity design checks for members and components other than the dissipative zones.

7.5.3 Composite wall configurations

As discussed in Section 7.2, a number of composite wall systems are referred to in EN1998-1. Specific criteria related to the design and detailing of these systems are given Section 7.10 of the code. These include several useful figures outlining detailing requirements for partially encased and fully encased boundary elements for DCM and DCH, as well as details for coupling beams framing into walls. This part of the code offers guidance for the design and detailing of wall configurations including boundary elements, coupling beams and steel plates. For most aspects, it refers to the provisions of reinforced concrete design (Section 6 of EN1998-1 and Chapter 5 of this book) as well as other parts of Section 7 of the code that are related to rules for critical members.

7.6 OTHER DESIGN CONSIDERATIONS

In terms of material properties, apart from the requirements in the concrete and steel parts in EC8 (Sections 5 and 6 of EN1998-1), additional criteria are specified in Section 7.2 of the composite part (Sections 7 of EN1998-1). In dissipative zones, the concrete class should not be less than C20/25 and not higher than C40/50; the upper limit is imposed since the use of typical plastic capacity calculations for composite cross-sections may become unreliable when concrete of relatively high strength is employed. For dissipative composite zones, the reinforcement should be of Class B or C for DCM, and should be Class C for DCH. In addition, Class B or C reinforcement should be used in highly stressed regions of non-dissipative zones. Except for closed stirrups or cross ties, only ribbed bars are allowed as reinforcing steel in highly stressed regions. It is also important to note that non-ductile welded meshes are not recommended in composite dissipative zones. If they are used, ductile reinforcement duplicating the mesh should be placed and their resistance should be accounted for.

A number of general requirements related to the design and detailing of dissipative zones are also included in Sections 7.5 and 7.6 of EN1998-1. As in steel frames, it is stipulated that dissipative zones may be located in the structural members or in the connections; accordingly, capacity design checks of non-dissipative elements should be based on the plastic resistance of either the dissipative members or connections, respectively. In general, two plastic resistances for composite dissipative zones, reflecting the lower and upper bound estimates, should be determined. The former considers only the steel and reliably ductile concrete portions (for assessing the dependable resistance), whilst the latter accounts for the steel and concrete portions (for determining the over-strength necessary for capacity design checks).

7.7 DESIGN EXAMPLE – COMPOSITE MOMENT FRAME

7.7.1 Introduction

The same eight-storey building considered in previous chapters is utilised in this example. The layout of the structure is reproduced in Figure 7.5. The main seismic checks are carried out for a preliminary design according to EN1998-1. Consideration is only given to the lateral system in the X-direction of the plan, in which resistance is assumed to be provided my moment resisting frames spaced at 4 m. It is also assumed that an independent bracing system is provided in the transverse direction (Y) of the plan. Grades S275, S500 and C30/37 are assumed for structural steel, reinforcement and concrete, respectively.

The gravity and seismic loads are assumed to be the same as those adopted in the steel moment frame example presented in the previous chapter (Tables 6.3 and 6.4 of Chapter 6). As in the steel moment frame case, the example focuses on the design of the moment frame located on GL2, as indicated in Figure 7.5.

The frame on GL2 was firstly designed for the non-seismic/gravity loading combinations corresponding to both ultimate and serviceability limit states, according to the provisions of EC3 (EN1993-1) and EC4 (EN1994-1). On this basis, the initial sections adopted were partially encased HEA340 for all columns and IPE450 steel profiles (in conjunction with a 150 mm solid slab) for the beams. Composite action is achieved through the incorporation of shear studs in order to attain full interaction according to the provisions of EC4.

7.7.2 Seismic design checks

7.7.2.1 Initial considerations

A preliminary elastic analysis was first carried out using the estimated seismic loads for the frame incorporating the initial member sizes. These initial member sizes were, however, found to be inadequate to fulfil both strength and damage limitation requirements. Accordingly, the partially encased columns were increased to HEA500 in the lower four storeys and to HEA450 in the upper storeys.

Concept b, in which the contribution of concrete is accounted for in dissipative zones, is considered in this example. According to Cl. 7.6.3 of EN1998-1, the effective widths assumed in the seismic analysis and design of the frame, are presented in Table 7.3. It is assumed that seismic rebars can be anchored to a concrete cantilever edge strip or to a transverse beam.

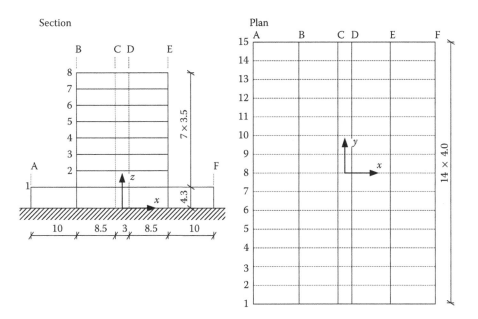

Figure 7.5 Frame layout.

Table 7.3 Effective widths according to EC8

	Analysis	Resistance
Positive moment	$2 \times 0.0375L = 0.075L$	$2 \times 0.075L = 0.15L$
Negative moment	$2 \times 0.05L = 0.1L$	$2 \times 0.1L = 0.2L$

According to Cl. 7.7.2(3) of EN1998-1, the structural analysis can be performed using equivalent properties for the entire beam instead of considering two flexural stiffnesses (for cracked and uncracked section). Therefore, EI_{eq} (as presented in Equation 7.3 above) was used within a linear elastic analysis of the frame. A view of the frame model showing the element numbers is shown in Figure 7.6. The results from the seismic loading combination are initially used in the evaluation of the inter-storey drift sensitivity coefficients θ as listed in Table 7.4. As shown in the table, θ does not exceed the lower limit of 0.1 and hence second-order effects do not have to be considered in the analysis.

7.7.2.2 Beam design checks

For illustration, the beam design checks are performed for one of the critical members, which is the 3 m composite beam located on the third floor (Element 17 in Figure 7.6). The internal forces at both ends of the member are listed in Table 7.5.

Based on the values from the table, the seismic demands on the beam are

$$M_{Ed,left} = -36.9 + 414.2 = -377.3 \text{ kN m}$$

$$M_{Ed,right} = -36.9 + (-414.2) = -451.1 \text{ kN m}$$

$$N_{Ed} = 27.8 + 0.0 = 27.8 \text{ kN}$$

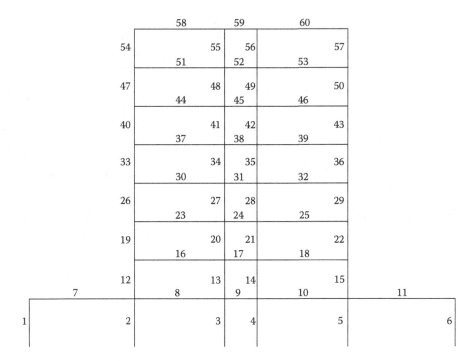

Figure 7.6 Frame model with element numbers.

Table 7.4 Calculation of the inter-storey drift sensitivity coefficients

Level	d_e (mm)	d_s (mm)	d_r (mm)	P_{tot} (kN)	V_{tot} (kN)	h (mm)	θ
8	77.7	310.8	16.4	453	98.2	3500	0.022
7	73.6	294.4	27.2	1170	241.7	3500	0.038
6	66.8	267.2	38.0	1888	365.4	3500	0.056
5	57.3	229.2	46.8	2606	469.3	3500	0.074
4	45.6	182.4	49.2	3323	553.3	3500	0.084
3	33.3	133.2	52.4	4041	617.4	3500	0.098
2	20.2	80.8	48.8	4758	661.6	3500	0.100
1	8.0	32.0	32.0	5476	703.2	4300	0.058

Table 7.5 Internal forces in Element 17

	Left end		Right end	
	$G_k + 0.3Q_k$	E	$G_k + 0.3Q_k$	E
M	−36.9	414.2	−36.9	−414.2
V	45.2	−276.2	−45.2	−276.2
N	27.8	0.0	27.8	0.0

According to Cl. 7.7.3(3) of EC8, which refers to Cl. 6.6.2(2), and considering the properties of the composite beam:

$$M_{Ed,left} \leq M_{pl,Rd} \rightarrow 377.3 \leq 704.0 \text{ kN m (considering 3 bars of 12 mm diameter within the }$$
$$0.45 \text{ m of effective width)}$$

$$M_{Ed,right} \leq M_{pl,Rd} \rightarrow 451.1 \leq 529.8 \text{ kN m (considering 4 bars of 12 mm diameter within the }$$
$$0.60 \text{ m of effective width)}$$

$$N_{Ed} \leq 0.15 N_{pl,Rd} \rightarrow 27.8 \leq 0.15 \times 98.8.10^{-4} \times 275.10^3 \rightarrow 27.5 \text{ kN} \leq 407.6 \text{ kN}$$

$$V_{Ed} \leq 0.5 V_{pl,Rd} \rightarrow V_{Ed} = V_{Ed,G} + V_{Ed,M} = 45.2 + (529.8 + 704.0)/3.0 = 456.5 \text{ kN}$$

$$\rightarrow 456.5 \leq 0.5 \times 50.85 . 10^{-4} \times 275.10^3/\sqrt{3}$$

$$\rightarrow 456.5 \text{ kN} \leq 403.7 \text{ kN not satisfied!}$$

From the above calculations, it is clear that the shear design check cannot be satisfied, which is largely a consequence of the short length of the beam. Increasing the beam strength would lead to higher moment capacity hence higher shears. Also, as a result of capacity design requirements, this would necessitate over-sizing of the columns. It is therefore suggested that specific measures are taken in order to increase the shear capacity through supplementary web plates within the critical short beams.

Cl. 7.6.2(7) of EC8 stipulates upper limits for the ratio x/d of the dissipative composite cross-section as indicated in Equation 7.1 and Figure 7.1 of this chapter. For the critical beam considered above, this ratio is found to be around 0.27 (i.e. 0.16/0.60), which is lower than the limit of 0.32 derived from the equation.

The adoption of concept b also requires that specific detailing rules are verified in order to ensure reliable dissipative behaviour in the concrete parts. According to Clauses C.3.2.2(2) and C.3.3.1(2) in Annex C of EC8, transverse reinforcement (or seismic rebars) should be positioned in the joint region in order to allow the mobilisation of Mechanism 2 (defined in Figure C.2 of EN1998-1, consisting of concrete diagonal compressive struts resisted by the internal region of the column). The area of reinforcement (A_T) is given by

$$A_T \geq F_{Rd2}/f_{yd,T} = 0.7 \times h_c \times d_{eff} \times f_{cd}/f_{yd,T}$$

where h_c is the depth of the column section, d_{eff} is the overall depth of slab, f_{cd} is the concrete design compressive strength and $f_{yd,T}$ is the design yield strength of the reinforcement.

The area of transverse reinforcement required at the joints within the lower four storeys is

$$A_T \geq 0.7 \times 0.490 \times 0.150 \times (30{,}000/1.5)/(500{,}000/1.15)$$

$$A_T \geq 23.67 \text{ cm}^2 \rightarrow 5 \text{ bars of 25 mm of diameter positioned over a beam length of 0.49 m.}$$

The area of seismic rebars to use at joints in the upper five storeys is given by

$$A_T \geq 0.7 \times 0.440 \times 0.150 \times (30{,}000/1.5)/(500{,}000/1.15)$$

$$A_T \geq 21.25 \text{ cm}^2 \rightarrow 5 \text{ bars of 25 mm of diameter positioned over a beam length of 0.44 m.}$$

7.7.2.3 Column design checks

Except at the base, the columns should be capacity designed according to the weak beam-strong column dissipative mechanism. According to Cl. 7.7.3(5) of EC8, which refers directly to Cl. 6.6.3, the design forces are obtained using the following combination:

$$E_d = E_{d,Gk+0.3Qk} + 1.1\gamma_{ov}\Omega E_{d,E}$$

where
  γ_{ov} is the overstrength factor assumed as 1.25
  $\Omega = \min (M_{pl,Rd,i}/M_{Ed,i}) = 529.8/476.5 = 1.11$
  (note that -476.5 kN m is obtained from Element 25)

The design combination for consideration in column checks is therefore given by

$$E_d = E_{d,Gk+0.3Qk} + 1.53E_{d,E}$$

The design forces for a critical column (Element 4 in Figure 7.6) are presented in Table 7.6 for illustration.

Based on these values, the seismic demands at the bottom end of the column are

$$M_{Ed} = 21.7 + 1.53 \times (-358.0) = -526.0 \text{ kN m}$$

$$N_{Ed} = -1529.5 + 1.53 \times (-828.3) = -2796.8 \text{ kN}$$

$$V_{Ed} = -14.8 + 1.53 \times 124.1 = 175.1 \text{ kN}$$

The design checks are performed according to EC4. For brevity, only cross-section checks are presented, but clearly all EC4 resistance checks including those for member stability should also be satisfied. Considering a partially encased HEA500 cross-section, for which the axial–bending interaction curve is depicted in Figure 7.7, it is evident that the composite column cross-section is able to satisfy the seismic demands. It is also necessary to check the level of shear applied on the cross-section:

$$V_{Ed} \leq 0.5V_{pl,Rd}$$

$$175.1 \leq 0.5 \times 74.72 \times 10^{-4} \times 275 \times 10^3/\sqrt{3}$$

$$175.1 \text{ kN} \leq 593.2 \text{ kN}$$

Table 7.6 Internal forces in Element 4

	Bottom end		Top end	
	$G_k + 0.3Q_k$	E	$G_k + 0.3Q_k$	E
M	21.7	−358.0	−41.8	175.7
V	−14.8	124.1	−14.8	124.1
N	−1529.5	−828.3	−1520.6	−828.3

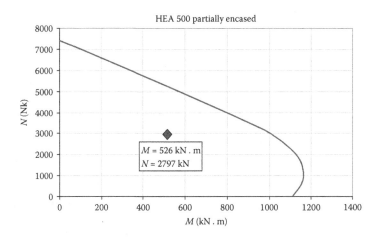

Figure 7.7 Interaction curve for the first storey composite column.

In addition to the member checks, Cl. 4.4.2.3(4) of EC8 also requires that at every joint the following condition is satisfied:

$$\frac{\sum M_{Rc}}{\sum M_{Rb}} \geq 1.3$$

where

$\sum M_{Rc}$ and $\sum M_{Rb}$ are the sum of the design moments of resistance of the columns and of the beams framing the joint, respectively. For illustration, this check is carried out for an internal joint located at the first floor of the frame, as follows:

$\sum M_{Rc} = 2 \times 1030 = 2060$ kN m (For a level of axial force of around 2800 kN)
$\sum M_{Rb} = 704.0 + 529.8 = 1233.8$ kN
$\sum M_{Rc}/\sum M_{Rb} \geq 1.3$

7.7.2.4 Joint design checks

According to Cl. 6.6.3(6) of EC8, the web panel zones at beam-to-column connections should be designed to resist the forces developed in the adjacent dissipative elements, which are the connected beams. For each panel zone the following expression should be verified:

$$\frac{V_{wp,Ed}}{V_{wp,Rd}} \leq 1.0$$

where $V_{wp,Ed}$ is the design shear force in the web panel accounting for the plastic resistance of the adjacent beams/connections and $V_{wp,Rd}$ is the shear resistance of the panel zone according to EC3. For illustration, these checks are performed for internal and external joint panel zones.

7.7.2.5 External panel zone

(HEA 500 + IPE 450):

$$V_{wp,Ed} = M_{pl,Rd}/(d_b-t_{bf}) = 704.0/(0.450-0.0146) = 1617 \text{ kN}$$

$$V_{wp,Rd} = 0.9\, f_{y,wc}\, A_{vc}/(\sqrt{3}\, \gamma_{M0}) + 4\, M_{pl,fc.Rd}/(d_b - t_{bf}) \text{ (Cl. 6.2.6 of EC3 Part 1-8)}$$

$$= 0.9 \times 275 \times 10^3 \times 74.72 \times 10^{-4}/\sqrt{3} + 4 \times 0.490 \times 0.023^2 \times 275$$
$$\times 10^3/(4 \times (0.450-0.0146)$$

$$= 1068 + 164 = 1232 \text{ kN}$$

$V_{wp,Ed}/V_{wp,Rd} = 1617/1232 \leq 1.0$ not staisfied! → A doubler plate is required

Design of a single supplementary doubler plate with a length of 300 mm:

$$1617 = 1232 + t_{dp} \times 0.300 \times 0.9 \times 275 \times 10^3/\sqrt{3}$$

$$t_{dp} = 8.98 \text{ mm} \rightarrow t_{dp} = 9 \text{ mm}$$

7.7.2.6 Internal panel zone

(IPE 450 + HEA 500 + IPE 450):

$$V_{wp,Ed} = \Sigma M_{pl,Rd}/(d_b - t_{bf}) = (704.0 + 529.8)/(0.450-0.0146) = 2834 \text{ kN}$$

$V_{wp,Ed}/V_{wp,Rd} = 2834/1232 \leq 1.0$ not satisfied! → A doubler plate is required.

Design of a single supplementary doubler plate with a length of 300 mm:

$$2834 = 1232 + t_{dp} \times 0.300 \times 0.9 \times 275 \times 10^3/\sqrt{3}$$

$$t_{dp} = 37.3 \text{ mm} \rightarrow t_{dp} = 38 \text{ mm}$$

7.7.3 Damage limitation

According to Cl. 4.4.3.2(1) for the damage limitation (serviceability) limit state:

$$d_r \nu \leq 0.01h$$

where d_r is the design inter-storey drift, ν is a reduction factor that takes into account the lower return period of the frequent earthquake and is assumed as 0.5, and h is the storey height. The limit of 1% is applicable to cases where the non-structural components are fixed to the structure in a way that does not interfere with structural deformation. For cases with non-ductile or brittle non-structural elements, this limit is reduced to 0.75% and 0.5%, respectively.

Based on the results provided in Table 7.4, the maximum inter-storey drift occurs at the third storey:

$d_r = 62.5$ mm

$d_r\, v \le 0.01$ h

$52.4 \times 0.5 \le 0.01 \times 3500$

26.2 mm < 35 mm (satisfies limit, provided that non-interfering elements are used).

REFERENCES

AISC 2005–2010. *American Institute of Steel Construction Inc., Seismic Provisions for Structural Steel Buildings*, AISC, Chicago, IL.

Amadio, C., Fedrigo, C., Fragiacomo, M. and Macorini, L. 2004. Experimental evaluation of effective width in steel-concrete composite beams. *Journal of Constructional Steel Research*, 60(2), 199–220.

Ballio, G., Calado, L., Iori, I. and Mirabella Roberti, G. 1987. *I Problemi Delle Grandi Costruzioni in Zona Sismica,* aicap, Roma, April, 1987, pp. 31–44.

Bouwkamp, J.G., Parung, H. and Plumier, A. 1998. Bi-directional cyclic response study of 3-D composite frame, *Proceedings of the 11th ECEE Conference*, Paris, France.

Broderick, B.M. and Elnashai, A.S. 1994. Seismic resistance of composite beam-columns in multi-story structures. *Journal of Constructional Steel Research*, 30, 231–258.

Bursi, O.S. and Caldara, R. 2000. Composite substructures with partial shear connection: Low cycle fatigue behaviour and analysis issues, *12th World Conference on Earthquake Engineering*, Auckland, New Zealand.

Bursi, O.S., Caramelli, S., Fabbrocino, G., Molina, J., Salvatore, W., Taucer, F. and Zandonini, R. 2004. *3D full-scale seismic testing of a steel-concrete composite building at ELSA*, Technical Report EUR 21299 EN, Joint Research Centre.

Bursi, O.S., Sun, F.F. and Postal, S. 2005. Non-linear analysis of steel-concrete composite frames with full and partial shear connection subjected to seismic loads. *Journal of Constructional Steel Research*, 61(1), 67–92.

Castro, J.M., Elghazouli, A.Y. and Izzuddin, B.A. 2005. Modelling of the panel zone in steel and composite moment frames. *Engineering Structures*, 27(1), 129–144.

Castro, J.M., Elghazouli, A.Y. and Izzuddin, B.A. 2007. Assessment of effective slab widths in composite beams. *Journal of Constructional Steel Research*, 63(10), 1317–1327.

Castro, J.M. and Elghazouli, A.Y. 2002. Behaviour of composite beams in moment-resisting frames. *12th European Conference on Earthquake Engineering*, London, Paper No. 521.

Doneux, C. and Plumier. 1999. Distribution of stresses in the slab of composite steel-concrete moment resistant frames submitted to earthquake action. Stahlbau, June.

Eder, M.A., Vollum, R.L. and Elghazouli, A.Y. 2012. Performance of ductile RC flat slab-to-steel column connections under cyclic loading. *Engineering Structures*, 36(1), 239–257.

Elghazouli, A.Y. 2015. Seismic Design Code Developments for Steel and Composite Structures, in *Perspectives on European Earthquake Engineering and Seismology*, Part 2, Chapter 5, Anal, A. (Editor). Geotechnical, Geological and Earthquake Engineering 39.

Elghazouli, A.Y. and Dowling, P.J. 1992. Behaviour of composite members subjected to earthquake loading. *10th World Conf. on Earthquake Engineering*, Madrid, Spain, 2621–2626.

Elghazouli, A.Y. and Elnashai, A.S. 1993. Performance of composite steel/concrete members under earthquake loading. *Earthquake Engineering and Structural Dynamics*, 22, 347–368.

Elghazouli, A.Y. and Packer, J.A. 2014. Seismic design solutions for connections to tubular members. *Journal of Steel Construction*, 7(2), 73–83.

Elghazouli, A.Y. and Treadway, J. 2008. Inelastic behaviour of composite members under combined bending and axial loading. *Journal of Constructional Steel Research*, 64(9), 2008, 1008–1019.

Elghazouli, A.Y., Castro, J.M. and Izzuddin, B.A. 2008. Seismic performance of composite moment frames. *Engineering Structures*, 30(7), 1802–1819.

EN 1993-1 2005. *Eurocode 3: Design of steel structures – Part 1.1: General rules and rules for buildings,* European Committee for Standardization, Brussels.

EN 1994-1 2004. *Eurocode 4: Design of composite steel and concrete structures – Part 1.1: General rules and rules for buildings.* European Committee for Standardization, Brussels.

EN 1998-1 2004. *Eurocode 8: Design provisions for earthquake resistance of structures, Part 1: General rules, seismic actions and rules for buildings.* European Committee for Standardization, Brussels.

Hajjar, J.F., Leon, R.T., Gustafson, M.A. and Shield, C.K. 1998. Seismic response of composite moment-resisting connections. II: Behaviour. *Journal of Structural Engineering,* 124(8), 877–885.

Leon, R.T. 1998. Analysis and design problems for PR composite frames subjected to seismic loads. *Engineering Structures*, 20(4–6), 364–371.

Leon, R.T., Hajjar, J.F. and Gustafson, M.A. 1998. Seismic response of composite moment-resisting connections. I: Performance. *Journal of Structural Engineering*, 124(8), 868–876.

Plumier, A., Abed, A., and Tilioune, B. 1994. Increase of buckling resistance and ductility of H-sections by encased concrete. In: Mazzolani, F.M. and Gioncu, V. (eds), *Behaviour of Steel Structures in Seismic Areas: STESSA'94*, E & FN Spon, London, 211–220.

Plumier, A., Doneux, C., Bouwkamp, J.G. and Plumier, C. 1998. Slab design in connection zones of composite frames. *Proceedings of the 11th ECEE Conference*, Paris, France.

Schleich, J.B. 1988. Fire Eng. Design of Steel Structures, *Steel Construction Today*, 2, 39–52.

Spacone, E. and El-Tawil, S. 2004. Nonlinear analysis of steel-concrete composite structures: State of the art. *Journal of Structural Engineering,* 130(2), 159–168.

Thermou, G.E., Elnashai, A.S., Plumier, A. and Doneux, C. 2004. Seismic design and performance of composite frames. *Journal of Constructional Steel Research*, 60(1), 31–57.

Design of timber structures

Christian Málaga-Chuquitaype and Ahmed Y. Elghazouli

CONTENTS

8.1 INTRODUCTION

Well-designed and well-constructed timber structures can have an excellent response under earthquake loading due primarily to the high strength to weight ratio of wood. Nevertheless, the seismic performance of timber buildings involves various inter-related factors that need to be properly understood. Many of the aspects related to the resistance of timber buildings spring from the atypical mechanical characteristics of wood as a construction material. In particular, there are significant differences in wood strength and stiffness depending on the orientation of the load with respect to the grain direction as depicted in Figure 8.1. It follows from the schematic strain–stress curves, indicated in Figure 8.1, that tension failures in wood are brittle and should be avoided while compressive behaviour (parallel to the grain) is a preferred mode of failure but should be limited. In fact, it is a typical approach of codes

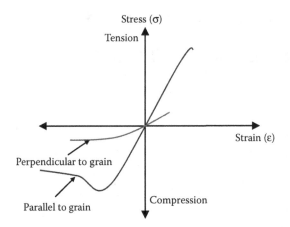

Figure 8.1 Schematic stress–strain relationship for wood.

of practice to ensure a ductile failure mechanism by inducing yielding in metallic connectors between timber members instead of the wood material itself in order to provide a sustained source of energy dissipation during seismic shaking.

In addition, like other materials, the building typology has a direct influence on the overall seismic behaviour of timber structures. For instance, the lightweight nature of typical wood-frame North American residential buildings, coupled with their high levels of structural redundancy, have been noted to favour a relatively limited earthquake damage in comparison with the heavy roof post-and-beam construction typically adopted in Japan (Karacabeyli and Popovski, 2003). Nonetheless, it has also been recognised that many of such wood-frame buildings (mainly constructed between the 1920s and 1990s in the West Coast of the United States) suffer from the lack of vertical continuity and are prone to soft-storey type of failure (Bahmani et al., 2014).

In earthquake-prone areas of Europe, whole-timber constructions are historically rare. As in many other regions of the world, wood has been traditionally employed in conjunction with other materials such as stone masonry to form masonry in-filled timber cross-walls. These structures have the benefit of an improved mechanical and thermal response that allows for the dissipation of seismic energy through friction and contact between the masonry units with the timber framing providing structural integrity (Meireles et al., 2012). Although the quality of construction in timber laced masonry buildings is variable, their remarkable seismic resilience has been demonstrated during numerous earthquakes worldwide (Langenbach, 2008). Indeed, recent full-scale experiments carried out at Imperial College, London (Elghazouli et al., 2013; Málaga-Chuquitaype et al., 2014a), have demonstrated the good cyclic behaviour of a new timber-frame design adapted from traditional timber cross-walls of Central American dwellings. This improved design uses timber frames with locally sourced cane to form a composite wall matrix, which is then plastered in cement to form shear-walls (Málaga-Chuquitaype et al., 2014b). A number of dwellings have been constructed using this timber-framed solution, which has proven to be inexpensive, sustainable and easy to maintain.

More recently, there has been a renewed interest in the use of timber as a main structural material largely due to a combination of the rising environmental concerns and the wider availability of newly developed high-performance timber-materials such as cross-laminated timber (CLT) and laminated veneer lumber (LVL). In response, timber design codes have begun to provide more information on the seismic design of timber buildings allowing for the use of wood in multi-storey applications. EC8 (EN1998-1, 2004) covers the design of Timber structures in Section 8. The scope of this section is limited in many respects, in particular with reference to timber construction systems of more generalised use today. In order to encompass aspects of more relevance to current practice, this chapter starts by reviewing fundamental concepts of the cyclic behaviour of timber structures with particular emphasis on current building practices. Subsequent sections present and discuss salient aspects of EC8 provisions. Finally, an illustrative example is presented for the preliminary design of a CLT shear-wall system for the eight-storey hotel building examined in previous chapters.

8.2 MODERN STRUCTURAL SYSTEMS

Three main timber structural systems prevail in current practice: (i) timber-framed buildings, (ii) solid wall timber buildings and (iii) moment resisting frames. The first is typically used in low-rise residential buildings. It consists of a timber frame built from studs, which are spanned by wooden or wood-based sheathing panels mechanically joined to the frame along their edges, typically by nailing. A timber-framed shear wall is usually formed by at least three risers (vertical timber studs) and an upper and a lower crossbeam while the sheathing panels are made from plywood, oriented strand board (OSB) or other wood-based sheathing material. Besides, timber-framed shear walls are normally connected at their base to the foundation by means of metal fasteners in order to resist uplift and overturning forces.

In addition to timber-framed construction, massive timber construction is becoming increasingly popular driven by the benefits of pre-fabrication it offers (e.g. increased quality control, on-site logistics, and speed of construction). In this sense, the most widely adopted solution is that of bearing timber panels constructed from cross-laminated timber (CLT). Much like timber-framed walls, these panels are employed as load-bearing elements carrying vertical and horizontal actions and are affected by in-plane and out-of-plane loads. They typically form box-like structures, which can have significant stiffness and resistance. Unlike timber-framed construction, CLT buildings make use of fewer fasteners, which are localised only in the contact between vertical panels and between vertical panels and their flooring system or foundation.

Finally, moment-resisting frames are mostly employed in single-storey sport facilities or industrial or commercial buildings where they take the form of portal frames. As generally expected, frames tend to be much more flexible than shear walls and it is important that such forms incorporate an adequate number of rigid connections in order to resist the lateral forces.

8.3 LATERAL DEFORMATION MODES OF TIMBER WALLS

The lateral force–displacement response of timber-framed shear walls is governed by the connections employed to join all their components. The two fundamental deformation modes of timber-framed shear walls are racking and rocking as depicted in Figure 8.2. More generally, a mixed deformation mode that combines the effects of these two mechanisms is also possible. It is evident from Figure 8.2 that while racking deformation tends to distribute plastic behaviour along the sheathing edges throughout the panel, rocking primarily affects the hold-down connectors. It is acknowledged that a large portion of the energy dissipation in timber-framed buildings takes place through yielding of the connectors (e.g. nails) between the sheathing panels and the timber frame which takes place due to differential deformations during racking (Filiatrault and Foschi, 1991; Filiatrault et al., 2002,

Figure 8.2 Fundamental inelastic deformation mechanisms for timber-framed shear walls. (a) Racking mode. (b) Rocking mode.

2003). Therefore, this deformation mode is usually targeted when applying the principles of capacity design discussed in previous chapters. On the other hand, when base connections have limited strength, a rocking or overturning failure mode can take place. However, this deformation mode is typically associated with limited ductility and energy dissipation, and should therefore be limited.

As with any other timber structures, the seismic deformation mechanisms observed in CLT panelised construction are also governed by the detailing of the connections. Previous experimental results (Acler et al., 2012; Popovski and Karacabeyli, 2012; Málaga-Chuquitaype et al., 2016) have proved that very different force–displacement responses can be achieved by varying the wall aspect ratio (length versus height ratio), level of vertical forces and the type of connecting fasteners employed. The connecting devices typically employed in CLT construction have the form of ribbed thin steel plates that are nailed or bolted to the panels and the floors such as those presented in Figure 8.3. A certain level of directionality is present in these connectors with shear brackets working primarily in shear and hold-downs used to resist vertical forces alone (Gavric and Popovski, 2014; Tomasi and Smith, 2015).

Two fundamental failure mechanisms have been observed in CLT shear walls at ultimate (Málaga-Chuquitaype et al., 2016), as shown in Figure 8.4: sliding at the base of the panel with shear angle bracket failure, and rocking of the panel with anchoring tie-down bracket failure. A mixed failure mode is also possible. In the case of multiple panels, the vertical joints between panels, which are usually nailed, provide a good source of energy dissipation and can significantly improve the ductility of the structure (Gavric et al., 2015). Similarly, the effect of vertical loads is to increase the strength of the wall, reduce the uplift forces on the base and increase the resisting friction forces at the base (Málaga-Chuquitaype et al., 2016). Nevertheless, due to the difficulties in estimating friction coefficients in real building configurations, these forces are usually ignored.

8.4 SEISMIC RESPONSE OF CONNECTIONS

The previous discussion has indicated that due to the inherent brittleness of timber, ductility in timber structures has to be provided by the design of ductile connections. In fact, the strength and stiffness of the modern shear wall systems, such as those discussed in the previous section can be estimated from the capacity of the anchoring metallic

Figure 8.3 Hold-down connector (a), shear bracket connector (b) and fully threaded screws (c) employed in CLT panelised construction. (Adapted from RothoBlaas. 2015a. *Wood Connectors and Timber Plates, Catalogue of Products*, RothoBlaas SRL, Cortaccia, Italy.)

connectors. Furthermore, the ductility available in any connection depends on its type and geometry.

Figures 8.5 and 8.6 present the effects of connection configuration on the force–displacement hysteresis of connections. A complete understanding of the behaviour of a joint can be illustrated with reference to these figures. During the first cycle, the joint develops the full stiffness and strength associated with the corresponding plastic mechanism adopted. In contrast, the following cycles will be associated with a systematic reduction in both connection strength and stiffness. As expected, the permanent embedment deformation observed in the wood brings about a sliding (or near sliding) phase in the response

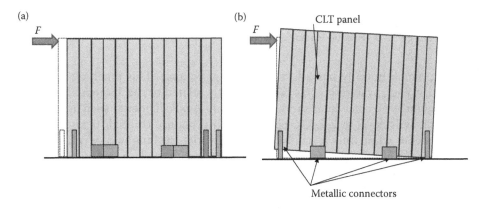

Figure 8.4 Fundamental inelastic deformation mechanisms for CLT shear walls. (a) Sliding mode, (b) rocking mode.

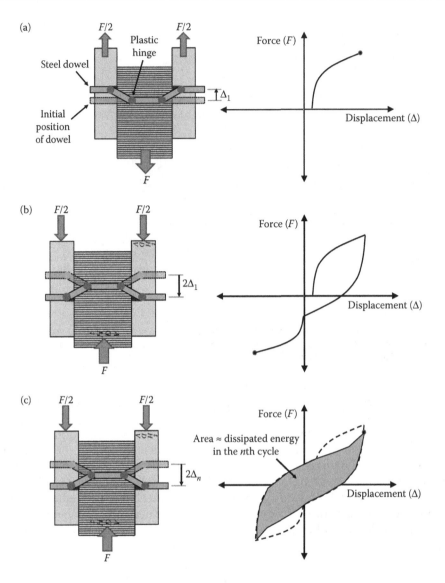

Figure 8.5 Moderately pinched hysteretic response of timber connections with dowel yielding. (a) First loading cycle. (b) First un-loading cycle. (c) Successive cycles.

of the joint, which initially has all its elements in contact. This pinching effect reduces the area within each hysteretic cycle that is directly related to the dissipated energy. It follows that a joint with steel connectors of relatively small diameter, such as that depicted in Figure 8.5, will develop good levels of ductility mainly through flexural yielding of the metal fastener while only a small amount of wood crushing will occur. On the other hand, when the joint incorporates a stiffer connector with a relatively large diameter as in the case of Figure 8.6, significant crushing and embedment of the wood fibres takes place (Karagiannis et al., 2016a) with only limited yielding of the steel. This latter behaviour will be associated with a more narrow hysteretic response and hence a reduced energy dissipation capacity. This is the reason behind the limits on dowel diameters imposed in Eurocode as discussed below.

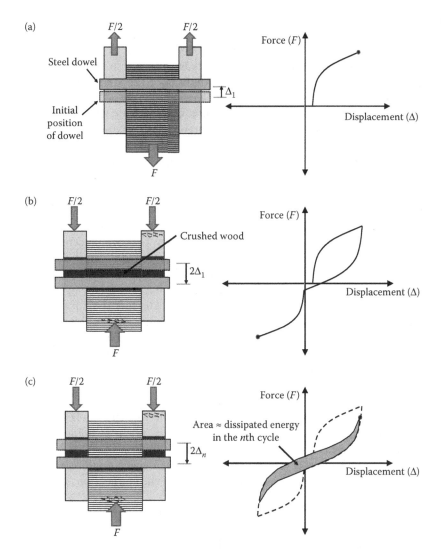

Figure 8.6 Severely pinched hysteretic response of timber connections with significant wood crushing. (a) First loading cycle. (b) First un-loading cycle. (c) Successive cycles.

8.5 EUROCODE 8 DEFINITIONS, DESIGN CONCEPTS AND MATERIAL PROPERTIES

The seismic design of timber structures is considered in Section 8 of Part 1 of EC8 (EN1998-1, 2004), which is concerned only with new buildings. Issues related to the reinforcement of existing timber structures are out of the scope of EC8. Moreover, Section 8 of the code starts with some fundamental definitions aimed at outlining the design philosophy and the element typologies on which the code relies on for the provision of adequate lateral seismic resistance.

The first definition presented in EC8 (EN 1998-1, 2004) is that of static ductility, taken as the ratio between the ultimate deformation and the deformation at the end of the elastic behaviour (first yield point) to be evaluated by means of quasi-static cyclic tests. This definition can be related to systems as well as connections. All the other definitions put forward by the code refer to timber joints. This reflects the paramount importance that timber connections have in the provision of ductility and strength in wood structures. In this context, semi-rigid joints are defined as those joints with significant flexibility, the influence of which hast to be taken into account in the analysis of the structures by following the guidance of Eurocode 5 (EN 1995-1-1, 2004). On the other hand, rigid joints are defined as joints with negligible flexibility (e.g. glued joints between solid timber elements). Dowel-type joints make use of metallic connectors like screws, nails or dowels that are loaded perpendicular to their axis. Finally, carpenter joints are also defined as those where loads are transferred directly from one element to the other by means of direct pressure and without the contribution of metallic connectors.

Similarly, EC8 recognises two approaches for the seismic design of timber structures aimed at achieving: low-dissipative or dissipative behaviour. The first is a design based on the elastic characteristics of the building with limited allowance for dissipation capacity. This design relies on the small levels of energy dissipation that are present in any structure and does not require any special consideration for the design and detailing of joints or elements beyond those already required by Eurocode 5 (EN 1995-1-1, 2004). Given the limited energy dissipation capacity aimed at in this design, a behaviour factor (q) not greater than 1.5 is specified by the code corresponding to a ductility class low (DCL).

On the other hand, when the design relies on the ability of the structure to dissipate energy, only the joints and connections can be considered as dissipative zones while the timber elements should remain elastic. This follows the general hierarchical seismic design philosophy adopted throughout EC8, where the zones in which inelastic deformations may be developed are selected beforehand with a view to developing, under ultimate design conditions, a pre-defined and controlled collapse mechanism. It is therefore necessary to guarantee that those zones allow for the development of the foreseen inelastic deformations. The behaviour factor, q, for dissipative structures may be taken as being greater than 1.5 depending on the ductility class adopted (i.e. medium or high).

EC8 gives specific advice on the use of material and their corresponding properties within the above-mentioned dissipative zones. In particular, for the connections it is stated that

1. Only materials and mechanical fasteners providing appropriate low-cycle fatigue behaviour may be used.
2. Glued joints shall be considered as non-dissipative zones.
3. Carpentry joints may only be used when they can provide sufficient energy dissipation capacity, without presenting risks of brittle failure in shear or tension perpendicular to the grain. Importantly, carpentry joints can only be used on the basis of experimental results illustrating adequate post-elastic capacity.

It is important to note that the code gives advice on a number of general connection typologies but does not consider any hybrid timber connection types (Andreolli et al., 2011; Karagiannis et al. 2016b). In this respect, it is worth recalling the Eurocode principles on the conception and design of timber joints whose configuration may not be directly addressed by the code. Aspects such as a clearly defined strength hierarchy should be incorporated ensuring that yielding is attained in the steel components well before the wood experiences any failure. The objective is to design a ductile joint response that can then be experimentally verified (EN 1998-1, 2004).

EC8 also provides criteria that shear walls and diaphragms should satisfy in order to guarantee their stiffness and strength, and to avoid potential instabilities that may arise when they are subjected to compression and shear. In particular, particle board panels should have a density of at least 650 kg/m³, plywood-sheathing should be at least 9 mm thick, and particleboard and fibreboard sheathing should be at least 13 mm thick. Similarly, the ductility properties of connections in trusses and between the sheathing material and the timber framing in Ductility Class M or H structures shall be tested for compliance by means of cyclic tests on the relevant combination of the connected parts and fastener.

8.6 DUCTILITY CLASSES AND BEHAVIOUR FACTORS

Table 8.1 of EC8 (reproduced below) indicates the maximum values of behaviour factors to be adopted for the design of structures of medium or high ductility as a function of their structural system. Since it is not always feasible to resort to experimental assessments for the evaluation of structural ductility, the code allows for ductility requirements to be considered as satisfied if the design is compliant with a number of geometric specifications. For instance, in timber-to-timber or steel-to-timber joints that use dowels, nails or screws, the minimum thickness of the elements to be connected should be of at least 10 times the connector diameter, which in turn should not be greater than 12 mm. Similarly, in shear walls and diaphragms where wood-based sheathing is employed, its thickness should be at least four times the nail diameter, which in turn cannot exceed 3.1 mm. These normative conditions have the purpose of ensuring the metal fastener is able to yield and dissipate energy through plastic deformations.

8.7 STRUCTURAL ANALYSIS

EC8 also includes a brief section on structural analysis, which recognises the importance of adequate structural modelling for timber buildings. For example, code requires that any joint idealisation should include slip deformations and that the instantaneous elastic modulus value (corresponding to 1.1 times the elastic modulus employed for short-term load calculations) should be used for parallel-to-the-fibre direction.

Regarding the structural representation, a rigid floor diaphragm assumption is permitted if a series of detailing rules are observed (as specified in Section 8.5.3 of the code) and the

Table 8.1 Structural types and behaviour factors

Energy dissipation capacity	Behaviour factor (q)	Examples
Low – DCL	1.5	Shelters, arches with two or three pinned joints, trusses joined with connectors.
Medium – DCM	2.0	Glued wall panels with glued diaphragms connected by nails and bolts, trusses with dowelled and bolted joints, mixed structures formed of timber framing and non-load bearing infills.
	2.5	Hyper-static portal frames with dowelled joints.
High – DCH	3.0	Nailed wall panels with glued diaphragms, connected with nails and bolts, trusses with nailed joints.
	4.0	Hyper-static portal frames with doweled and bolted joints.
	5.0	Nailed wall panels with nailed diaphragms connected with nails and bolts.

Source: Adapted from EN 1995-1-1 2004. *Eurocode 5: Design of Timber Structures – Part 1-1: General – Common Rules and Rules for Buildings*, European Committee for Standardization, CEN, Brussels.

size of the floor openings are such that they do not significantly affect its overall in-plane rigidity. The detailing rules considered include

1. Limits in the spacing between nails and connectors that aims at enabling adequate ductility.
2. Measures aimed at ensuring a certain degree of uniformity in the diaphragm that limits the possibility of unforeseeable loading conditions to occur and prevents fragile failure mechanisms. In this regard, specific requirements of beam continuity, diaphragm reinforcement around the openings, and limiting values of the depth-to-width ratios of un-restrained beams, are specified.

8.8 DETAILING RULES

In the case of joints, the code advice involves both the connections as well as the connected members. In the case of dowels or bolts, the connectors should be tight fitted and tightened. Also, the use of large diameters (greater than 16 mm) is discouraged. Similarly, connectors like dowels, smooth nails and staples should be avoided unless additional provisions are put in place against their withdrawal. These indications have the objective of ensuring a continuous energy dissipation capacity by limiting the possible degradation of the force–displacement hysteresis caused by excessive sliding or significant strength reduction in the connector.

For structural elements, the code explicitly stipulates additional provisions for avoiding splitting failure in tensioned timber elements. Likewise, compression members and connections (e.g. carpenter joints) should be prevented from separating under possible load reversals. These requirements are important also when evaluating the retrofit of structures, which is not directly addressed by the code.

8.9 SAFETY VERIFICATIONS AND CAPACITY DESIGN

The procedure specified in Eurocode 5 for the estimation of design material properties on the basis of characteristic ones should also be followed for seismic design calculations. In this case, the strength modification factor, k_{mod}, which accounts for the influence of the duration of the load and the environmental conditions on the strength of wood should be taken as that corresponding to instantaneous actions. Similarly, for ultimate limit state verifications, a value of partial safety factor for material properties, γ_M, as associated with fundamental load combinations is recommended for ductility class low (DCL) structures while a $\gamma_M = 1$ (corresponding to accidental load combinations) is permitted for medium and high ductility structures.

As with other structural materials, capacity design in timber requires that ductile yielding takes place only in the specially designated parts or elements and that all other structural parts are provided with adequate over-strength to prevent their failure. Predominantly, ductility in timber construction comes from the non-linear behaviour of steel components and connections and, as discussed previously, it is precisely these elements that can be designated as the dissipative elements according to EC8. That means that other parts of the structure should be capacity designed with adequate over-strength factors.

Unfortunately, EC8 is not explicit on the levels of over strength to be considered. Nevertheless, it should be noted that for nailed plywood connections in most timber structures, an over-strength factors of 2 can be adopted following past experimental research

(Buchanan, 2013). Likewise, an over-strength factor ranging between 1.3 and 1.6 can be recommended for CLT structures (Gavric et al., 2015). The over-strength factor represents the amount by which the actual strength may be higher than the design estimated value. The actual value of over strength depends on a number of factors such as the strength reduction factor used in the initial design, strain-hardening in the yielding steel components, the difference between the 5th percentile used for the design of yielding components (EN 1995-1-1, 2004) and the actual value present in the structure as well as other unintentional over-design capacities of the dissipative components.

8.10 DESIGN EXAMPLE

8.10.1 Introduction

The same eight-storey building considered in previous chapters is utilised herein for the design example. Figure 8.7 reproduces the building layout and elevation. A structural solution consisting of cross-laminated timber (CLT) shear-walls is provided in both x and y directions. Only the calculations related to the x-direction will be presented here. As it is typical in tall timber residential buildings where the wall density is maximized for redundancy purposes, CLT shear walls are provided in GLs 1 to 15 of Figure 8.7. Therefore, CLT shear walls spaced at 4 m between their in-plane centres are considered with two shear walls per GL as depicted in Figure 8.7.

Each CLT wall of 8.5 length is made of four 5-layered structural panels (of 2.125 m length each). For the first three storeys, 128 mm thickness (30-19-30-19-30) are used, while 95 mm thick panels (19-19-19-19-19) are employed for the upper storeys. Also, the floors and roof employ five-layer CLT panels of 200 mm. CLT panels of EN C24 timber class are

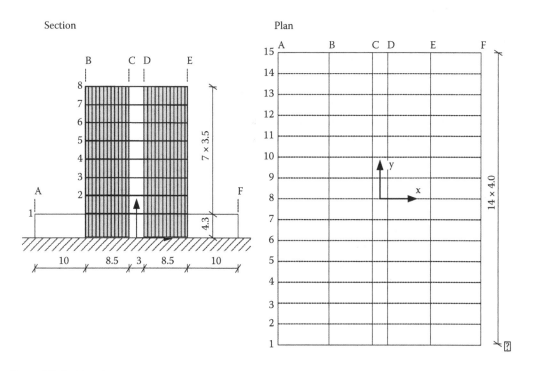

Figure 8.7 Building layout.

Table 8.2 Characteristic properties of EN C24 timber

Material property	Unit	Value
Strength in bending	$f_{m,k}$ (MPa)	24
Strength in tension in the fibre direction	$f_{t,0,k}$ (MPa)	14
Strength in tension perpendicular to the fibre direction	$f_{t,90,k}$ (MPa)	0.5
Strength in compression in the fibre direction	$f_{c,0,k}$ (MPa)	21
Strength in compression perpendicular to the fibre direction	$f_{c,90,k}$ (MPa)	2.5
Strength in shear	$f_{v,k}$ (MPa)	4
Rolling shear	$f_{T,k}$ (MPa)	1.75
Characteristic modulus of elasticity parallel to grain	$E_{0,k}$ (MPa)	7,400
Mean modulus of elasticity parallel to grain	E_0 (MPa)	11,000
Shear modulus	G_{mean} (MPa)	700
Characteristic density	ρ_k (kg/m³)	420
Mean density	ρ_{mean} (kg/m³)	350

considered in all cases. The characteristic material properties associated with this timber class are presented in Table 8.2. The design material properties are obtained from the characteristic values by multiplying with a $k_{mod} = 1.1$ and dividing by $\gamma_M = 1.0$ in accordance with Eurocode provisions for seismic scenarios. A rigid-diaphragm flooring systems is assumed.

8.10.2 Weight and mass calculation

An estimate of the total weight of the structure is necessary for the calculation of seismic forces. This is carried out by assuming the dimensions presented in Table 8.3 corresponding to floors and walls.

8.10.2.1 Dead load

1. Floor slabs
   Glued parquet floor: 10 mm × 8.00 kN/m³ = 0.08 kN/m²
   Final screed: 60 mm × 22.00 kN/m³ = 1.32 kN/m²
   Insulation of impact noise: 30 mm × 1.40 kN/m³ = 0.04 kN/m²
   Filling material: 60 mm × 20.00 kN/m³ = 1.20 kN/m²
   CLT: 200 mm × 5.00 kN/m³ = 1.00 kN/m²
   Suspended ceiling (gypsum plaster): 95 mm × 3.47 kN/m³ = 0.33 kN/m²
   Total = 3.97 kN/m²
2. Roof slab
   Mineral substrate and coating: 120 mm × 21.25 kN/m³ = 2.55 kN/m²
   Roof sheathing: 10 mm × 5.00 kN/m³ = 0.05 kN/m²
   Wooden formwork: 20 mm × 5.50 kN/m³ = 0.11 kN/m²
   Thermal insulation (with timber slats): 250 mm × 1.20 kN/m³ = 0.30 kN/m²
   CLT: 200 mm × 5.00 kN/m³ = 1.00 kN/m²
   Suspended ceiling (gypsum plaster): 95 mm × 3.47 kN/m³ = 0.33 kN/m²
   Total = 4.34 kN/m²
3. External CLT wall
   Façade plate: 15 mm × 8.00 kN/m³ = 0.12 kN/m²
   Counter battens/ventilation zone: 40 mm × 0.44 kN/m³ = 0.02 kN/m²
   Concealed façade: 160 mm × 1.9 kN/m³ = 0.31 kN/m²

Table 8.3 Geometrical characteristics of building elements

Element (total thickness)	Layer	Layer thickness (mm)
Floor slabs (402 mm)	Glued parquet floor	10
	Final screed	25
	Insulation of impact noise	12
	Filling material	60
	Cross laminated timber	200
	Suspended ceiling (gypsum plaster)	95
Roof slab (695 mm)	Mineral plant substrate (layer of vegetation)	90
	Protective coat with filtering features	30
	Roof sheathing	10
	Wooden formwork	20
	Thermal insulation with timber slats in-between	250
	Cross laminated timber	200
	Suspended ceiling (gypsum plaster)	95
External CLT shear wall (358 mm)	Façade plate	15
	Counter battens/ventilation zone	40
	Concealed façade insulating board with timber slats in between	160
	Cross laminated timber	128
	Gypsum plaster plate	15
Internal CLT shear wall (158 mm)	Gypsum plaster plate	15
	Cross laminated timber	128
	Gypsum plaster plate	15

CLT: 128 mm × 5.00 kN/m³ = 0.64 kN/m²
Gypsum plaster plate: 15 mm × 10.00 kN/m³ = 0.15 kN/m²
Total = 1.24 kN/m²
4. Internal CLT wall (load bearing)
Gypsum plaster plate: 15 mm × 10.00 kN/m³ = 0.15 kN/m²
CLT: 128 mm × 5.00 kN/m³ = 0.64 kN/m²
Gypsum plaster plate: 15 mm × 10.00 kN/m³ = 0.15 kN/m²
Total = 0.94 kN/m²

The dead load estimation is summarised in Table 8.4.

Table 8.4 Dead load calculation

Level		Calculation	Load (kN)	Total (kN)
8	Roof	(56 × 20) × 4.44	4,972.8	4,972.8
2–7	Slab	(56 × 20) × 3.97	4,446.4	
	External wall	((56 × 2) × 3.5 + (8.5 × 4) × 3.5) × 1.24	633.64	
	Internal wall	((56 × 2) × 3.5 + (8.5 × 26) × 3.5) × 0.94	1,095.57	6,175.6
1	Slab	(56 × 40) × 3.97	8,892.8	
	External wall	[(56 × 2) × 4.3 + (10 × 4) × 4.3] × 1.24	810.46	
	Internal wall	[(56 × 2) × 4.3 + (8.5 × 26) × 4.3] × 0.94	1,345.98	11,049.2
Total dead load, G				53,075.6

Table 8.5 Imposed load calculation

Level		Calculation	Load (kN)	Total (kN)
8	Roof	$(56 \times 220) \times 22.0$	2,240.0	2,240.0
2–7	Corridors	$((56 \times 23.0) + (8.5 \times 24) + (8.5 \times 28)) \times 24.0$	1,080.0	
	Bedrooms	$((56 \times 220) - 270) \times 22.0$	1,700.0	2,780.0
1	Tower area	As levels 2–7	2,780.0	
	Roof terrace	$(56 \times 220) \times 24.0$	4,480.0	7,260.0
Total imposed load, Q				26,180.0

8.10.2.2 Imposed load

The estimation of imposed load is summarised in Table 8.5.

8.10.2.3 Seismic mass

According to Cl. 3.2.4 the masses to be used in seismic analysis should be those associated with the load combination: $G + \psi_{E,i}{}^{*}Q$. Take $\psi_{E,i}$ to be 0.3.

The corresponding building weight is $53{,}075.6 + 0.3 \times 26{,}180 = 60{,}929.6$ kN
And the seismic mass can be calculated as: 6,211 tons.

It is important to recall that one of the aspects leading to a favourable earthquake performance of timber structures is their relatively lighter weight, which translates into lower active seismic masses. In this particular example, the timber building is 24% lighter than its steel or concrete options as presented in Chapter 3.

8.10.3 Seismic base shear

The Type 1 Spectrum for Soil Type C, used before, is also employed in this design. The spectral parameters associated with this (from EC8 Table 3.2) are

$$S = 1.15,\ T_B = 0.2\,\text{s},\ T_C = 0.6\,\text{s},\ T_D = 2.0\,\text{s}$$

As before, the reference peak ground acceleration is $\alpha_{gR} = 3.0$ m/s^2 and an importance factor $\gamma_I = 1.0$ is assumed, so the design ground acceleration is $\alpha_g = \gamma_I \times \alpha_{gR} = 3.0$ m/s^2.

Although EC8 does not address explicitly the definition of behaviour factors for CLT structures, recent research (Popovski and Karacabeyli, 2012; Pei et al., 2013) supports the use of a behaviour factor of at least $q = 3$ for CLT buildings designed in accordance with capacity design principles. Besides, considering that a maximum length of 2.5 m will be employed for each individual panel forming the shear walls, a $q = 3$ is considered adequate.

Estimate natural period, EC8 Equation 4.6

$$T_1 = C_T h^{3/4}$$

For timber buildings $C_T = 0.05$, hence

$$T_1 = 0.05 \times 28.8^{3/4}$$

$$T_1 = 0.62\ \text{s}$$

$T_C \le T \le T_D$ so EC8 Equation 3.15 applies

$$S_d = a_g S \frac{2.5}{q} \frac{T_C}{T_1}$$

Hence

$$S_d = 3 \times 1.15 \times \frac{2.5}{3} \frac{0.6}{0.62} = 2.78 \, \text{m/s}^2$$

And the base shear is

$$F_b = \lambda m S_d$$

In this case $T_1 < 2T_C$ so $= 0.85$, therefore

$$F_b = 0.85 \times 6211 \times 2.78$$

$$F_b = 14{,}676 \, \text{kN}$$

So the net horizontal force is $100 \times 14{,}676/60{,}930 = 24.1\%$ of the total building weight. Such high values of net horizontal forces are not unlikely in panel timber structures due to their stiff nature which places them in the short period range of the ground-motion response spectrum.

8.10.4 Shear load distribution and total moment calculation

Table 8.6 presents the lateral load distribution per storey assuming a linear mode shape approximation (Equation 4.11 in EC8).

The ratio of total base moment to the base shear gives the effective height of the resultant lateral force:

$$h_{eff} = \frac{287{,}732.5}{14{,}676} = 19.6 \, \text{m above the base, and} \quad \frac{h_{eff}}{h} = \frac{19.6}{28.8} = 0.68$$

Table 8.6 Lateral load distribution on the timber building using linear mode shape approximation

Level k	Height z_k (m)	Mass m_k (t)	$z_k m_k$ (mt)	Force F_k (kN)	Moment $= F_k z_k$ (kN m)
8	28.8	575.4	16,571.5	2,606.2	75,057.2
7	25.3	714.5	18,076.9	2,842.9	71,925.1
6	21.8	714.5	15,576.1	2,449.6	53,401.4
5	18.3	714.5	13,075.4	2,056.3	37,630.7
4	14.8	714.5	10,574.6	1,663.0	24,612.9
3	11.3	714.5	8,073.9	1,269.7	14,348.2
2	7.8	714.5	5,573.1	876.5	6,836.4
1	4.3	1,348.3	5,797.7	911.8	3,920.7
Totals		6,210.7	93,319.1	14,676.0	287,732.5

8.10.5 CLT shear wall actions

As explained above, the CLT shear walls are symmetrically distributed in plan. Assuming that a rigid diaphragm action can be ensured, the forces per storey should be distributed among the different walls in proportion to their relative stiffness. Besides, an accidental eccentricity as defined in EN 1998-1 Clause 4.3.2 should be considered. Table 8.7 presents the seismic design forces acting in the wall located on GL2 between GLs B and C.

The design forces (presented in Table 8.7 for the wall under consideration) are distributed among the connectors and CLT panels forming the shear wall in the way depicted in Figure 8.8. For simplicity, the shear brackets are only assumed to carry horizontal forces (N_h) while any lateral resistance of the tie-downs is neglected. The forces acting on the tie-downs at the extremes of each panel will depend on the storey shear (F_d), the number of panels and the lever arm from the point of rotation to the location of the tie-down, which can be assumed as the length of the individual panel under consideration. The vertical actions (w_v) and step-joint forces (N_s) can be conservatively assumed not to contribute towards the rocking resistance of the panel. Therefore, by establishing equilibrium, the final forces are

Table 8.7 Design forces in wall located on GL2 between GLs B and C

Level k	Shear force (kN)	Moment at wall's base (kN m)
8	87.3	28.0
7	182.6	80.5
6	264.7	154.0
5	333.6	245
4	389.4	350
3	431.9	465.5
2	461.3	588
1	491.8	742.8

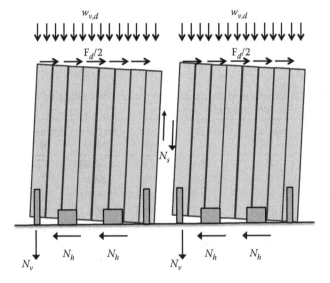

Figure 8.8 Distribution of forces in a CLT wall made of two panels.

Table 8.8 Final analysis and design of CLT wall laying in GL2 between GLs B and C

Level k	Number of panels	Shear per unit length (kN/m)	Uplift per panel (kN)	Number of AB shear brackets	Type of hold-down
8	4	10.9	3.5	1	WHT340
7	4	22.8	10.1	2	WHT340
6	4	33.1	19.3	2	WHT340
5	4	41.7	30.6	3	WHT340
4	4	48.7	43.8	3	WHT440
3	4	54.0	58.2	3	WHT540
2	4	57.7	73.5	4	WHT540
1	4	61.5	92.9	4	WHT620

obtained as summarised in Table 8.8. The total length of the shear wall between GLs B and C is 8 m. In order to ensure adequate levels of energy dissipation consistent with the behaviour factor $q = 3$ adopted, 4 CLT panels of 2 m long each are employed.

8.10.6 Seismic strength design checks

Two types of angle brackets made of galvanised mild steel are employed to connect the walls to the floor slabs. These mechanical connectors are taken from the catalogue of RothoBlaas (2015a) and their properties are summarised in Table 8.9. As before, the design resistances reported in Table 8.9 correspond to the characteristic strength of the connector multiplied by $k_{mod} = 1.1$ and divided by $\gamma_M = 1.0$. Angle brackets ABR105-R are used for transferring shear forces while hold-downs of different types (i.e. WHT340, WHT440, WHT540, WHT620) are employed at different storeys to resist uplift forces. These metallic connectors are pre-designed to fail in a ductile manner, provided the specifications for their installation are followed. All mechanical fasteners are installed with 4×60 anker nails as specified by the producer (RothoBlaas, 2015a).

8.10.7 Capacity design

To ensure that energy dissipation occurs through controlled yielding at the metallic connectors only, other joints need to be designed with sufficient over-strength to avoid premature failure. The connections to be over-designed are connections between floors and the walls underneath, connections between floor panels that form the rigid diaphragm and connections between perpendicular walls to ensure a box-like behaviour.

Table 8.9 Connectors employed from RothoBlaas (2015a)

Connector type	Main load type resisted	Characteristic load carrying capacity (N_k) (kN)	Design resistance (N_d) (kN)
ABR105-R	Shear ($N_{k,x}$)	19.1	21.01
WHT340	Uplift ($N_{k,y}$)	36.5	40.15
WHT440	Uplift ($N_{k,y}$)	57.9	63.69
WHT540	Uplift ($N_{k,y}$)	81.1	89.21
WHT620	Uplift ($N_{k,y}$)	100.4	110.44

Taking the second floor panel as an example, from Table 8.8, we can see that it needs to resist a design shear per unit length of $V_d = 57.7$ kN/m but with the current arrangement that was provided for 75 kN/m of shear capacity. If an over-strength factor of $\gamma_{ov} = 1.6$ is assumed, as explained above, the total force to be accommodated is $V_{\gamma,d} = 120$ kN/m. Therefore, self-tapping screws HBS 12 × 320 (RothoBlaas, 2015b), with a design shear strength for wood to wood of 9.44 kN (8.58 kN of characteristic shear strength) are required at 80 mm intervals.

Similarly, the in-plane shear capacity of the CLT panel can be verified to ensure no failure prior to base connector yielding. The same over-strength design load as before is used for this purpose $V_{\gamma,d} = 120$ kN/m. The representative volume sub-element (RVSE) method proposed by Bogensperger et al. (2010) is employed herein. This method is based on the definition of a series of i representative volumes of ideal thickness (t_i^*). In the case of the five-layer CLT panel used on the second level in this example, that thickness is always governed by the thinner 19 mm board. Hence

$$t_i^* = 19 \text{ mm}$$

$$t^* = \sum t_i^* = 4(19) = 76 \text{ mm}$$

therefore, assuming a unit length of 1 m and a constant shear stress distribution (Bogensperger et al., 2010) the volume shear force is

$$\tau_{0,RVSE}^* = \frac{n_{x,y}}{t^*} = \frac{120}{0.076} = 1{,}579 \text{ kN/m}^2$$

Two mechanisms need to be considered. The first involves a shear mechanism for which the following relationship needs to be verified:

$$\tau_{v,d} = 2\left(\tau_{0,RVSE}^*\right) \leq \frac{k_{mod} f_{v,k}}{\gamma_M}$$

$$\tau_{v,d} = 2(1.58) \text{ MPa} \leq \frac{1.1(4)}{1} \text{ MPa}$$

$$\tau_{v,d} = 3.16 \text{ MPa} \leq 4.4 \text{ MPa}$$

The second mechanism involves torsion failure on the glued interface between two boards for which the torsional stresses can be calculated as (Bogensperger et al., 2010):

$$\tau_{T,i}^* = 3\tau_{0,RVSE}^* \frac{t_i^*}{a}$$

where a is the width of the lamella in the CLT panel, therefore, by assuming a 1:4 thickness to width ratio of each lamella:

$$\tau_{T,i}^* = 3(1.58)\frac{1}{4} = 1.19 \text{ MPa}$$

which should be compared with the rolling shear strength of wood:

$$\tau_{T,d} \leq \frac{k_{mod}f_{T,k}}{\gamma_M}$$

$$1.19 \leq \frac{1.1(1.75)}{1} = 1.92\,\text{MPa}$$

A check should also be performed on the strength of the vertical joints that connect the four 2-metre panels to form the 8-metre long shear wall. The following inequality should be satisfied in order to ensure the individual panels rock separately as intended:

$$N_S \leq N_V + \sum N_{B,y,i}\frac{x_i}{l}$$

where N_v is the design factored resistance of the hold-down, $N_{B,y,I}$ is the vertical factored strength of the shear brackets that were conservatively assumed as zero before but can be obtained from the supplier, x_i are the horizontal distances between the corner of the panel and the centre of the shear brackets, and l is the total length of the panel. It is evident from this that if HBS 8 × 200 self-tapping screws with a design strength of 4.24 kN are employed in the vertical joints with a spacing of 200 mm centre to centre, the above equation will be satisfied.

8.10.8 Damage limitation and final design

Based on the preliminary design presented above, a structural model can be constructed with a view to refining the design and checking its damage limitation compliance (i.e. estimated storey displacements). The issue of an adequate numerical representation of the dynamic response of tall timber structures is complicated by the nature of wood construction and the high non-linear response of connections (Abeysekera and Málaga-Chuquitaype, 2015). A simple way to deal with the anisotropic nature of CLT buildings is by assuming an effective in-plane elastic stiffness as proposed by Blass and Fellmoser (2004):

$$k_{eff} = \frac{3EI_{eff}0.8GA_s}{3EI_{eff}H + 0.8GA_sH^3}$$

where E is the elastic modulus of timber, I_{eff} is the effective radius of gyration for a given cross-section, G is the shear modulus of timber, A_s the effective shear cross-section and H is the wall height. The application of this equation gives a k_{eff} = 65 kN/mm for the 128 mm panel and 59 kN/mm for the 95 mm panel.

A model was constructed in the finite element program Opensees (McKenna, 1997) employing four-noded quad elements with effective stiffness values to represent the CLT walls and zero-length elements in order to simulate the connectors at the base and between CLT panels. The numerical model had a fundamental period of 0.58 seconds, which is comparable to the 0.62 seconds estimated with the EC8 formula, as presented above. An elastic analysis disregarding the frictional forces was carried out from which a maximum deformation at the bottom storey of 25 mm was obtained, which is well below the maximum inter-storey drift of 1% specified by the code for the control of deformations in non-structural components that do not interfere with the deformation of structural elements:

$$d_r = 25 \text{ mm}$$
$$d_r v \leq 0.01 \text{ h}$$
$$25(0.5) \leq 0.01(4300)$$
$$12.5 \text{ mm} < 43 \text{ mm}$$

The final design should proceed to optimise the distribution of over-strengths along the height of the building so as to ensure a distribution of inelastic deformations as uniform in height as possible as uniform in height as possible and to ensure capacity design is implemented at each individual connection level.

REFERENCES

Abeysekera, I. and Málaga-Chuquitaype, C. 2015. Dynamic response of tall timber buildings, *SECED 2015 Conference: Earthquake Risk and Engineering towards a Resilient World*, Cambridge, UK.

Acler, E., Endrizzi, P. and Tomasi, R. 2012. Monotonic and cyclic in-plane behavior of CLT panels tested by using different types of metal devices, *in COST Action FP1004 – Early Stage Researchers Conference*, Bath, UK.

Andreolli, M., Piazza, M., Tomasi, R. and Zandonini, R. 2011. Ductile moment-resistant steel–timber connections, *Proceedings of the Institution of Civil Engineers: Structures and Buildings*, 164(SB2), 65–78.

Bahmani, P., van de Lindt, J., Gershfeld, M., Mochizuki, G., Pryor, S. and Rammer, D. 2014. Experimental seismic behavior of a full-scale four-story soft-story wood-frame building with retrofits. I: Building design, retrofit methodology, and numerical validation, *Journal of Structural Engineering*, Available Online, http://dx.doi.org/10.1061/(ASCE)ST.1943-541X.0001207.

Blass, H. and Fellmoser, P. 2004. Design of solid wood panels with cross layers, *World Conference on Timber Engineering, Volume II*, Lahti, Finland.

Bogensperger, T., Moosbrugger, T. and Silly, G. 2010. Verification of CLT-plates under loads in plane, *World Conference of Timber Engineering*, Riva del Garda, Italy.

Buchanan, A.H. 2013. *Timber Design Guide*, 3rd edition, New Zealand Timber Industry Federation, Wellington, New Zealand.

Elghazouli A.Y., Málaga-Chuquitaype, C., Lawrence, A., Kaminski, S. and Coates. K. 2013. *Seismic Testing of Sustainable Cane and Mortar Walls for Low-cost Housing in Developing Countries*, Institution of Civil Engineers, London, UK, Research and Development Enabling Fund Project Report.

EN 1995-1-1 2004. *Eurocode 5: Design of Timber Structures – Part 1-1: General – Common Rules and Rules for Buildings*, European Committee for Standardization, CEN, Brussels.

EN 1998-1 2004. *Eurocode 8: Design Provisions for Earthquake Resistance of Structures, Part 1: General Rules and Rules for Buildings*, European Committee for Standardization, CEN, Brussels.

Filiatrault, A., Fisher, D., Folz, B. and Uang, C.M. 2002. Seismic testing of two-story woodframe house: Influence of wall finish materials, *ASCE Journal of Structural Engineering*, 128(10), 1337–1346.

Filiatrault, A. and Foschi, R.O. 1991. Static and dynamic tests of timber shear walls fastened with nails and wood adhesive, *Canadian Journal of Civil Engineering*, 18, 749–755.

Filiatrault, A., Isoda, H. and Folz, B. 2003. Hysteretic damping of wood framed buildings, *Engineering Structures*, 25, 461–471.

Gavric, I., Fragiacomo, M. and Ceccotti, A. 2015. Cyclic behavior of CLT wall systems: Experimental tests and analytical prediction models, *ASCE Journal of Structural Engineering*, 141(11), 04015034.

Gavric, I. and Popovski, M. 2014. Design models for CLT shearwalls and assemblies based on connection properties, *International Network for Timber Engineering Research*, Paper: INTER/47-15-4, Bath, UK.

Karacabeyli, E., Popovski, M. 2003. Design for earthquake resistance, in Thelandersson, S. and Larsen H.J. (Eds) *Timber Engineering*, Wiley, Chichester, pp. 267–299.

Karagiannis, V., Málaga-Chuquitaype, C. and Elghazouli, A.Y. 2016a. Modified foundation modelling of dowel embedment in glulam connection, *Construction and Building Materials*, 102, 1168–1179.

Karagiannis, V., Málaga-Chuquitaype, C. and Elghazouli, A.Y. 2016b. Behaviour of hybrid timber beam-to-steel tubular column connections in bending, Submitted to *Engineering Structures*.

Langenbach, R. 2008. Learning from the past to protect the future, *Engineering Structures*, 30, 2096–2100.

Málaga-Chuquitaype, C., Kaminski, S., Elghazouli, A.Y. and Lawrence, A. 2014a. Seismic response of timber frames with cane and mortar walls, *Proceedings of the Institution of Civil Engineers – Structures and Buildings*, 167(12), 693–703.

Málaga-Chuquitaype, C., Kaminski, S., Elghazouli, A.Y. and Lawrence, A. 2014b. Seismic performance of low-cost timber frames with cane and mortar walls, *2nd European Conference on Earthquake Engineering and Seismology*, Istanbul, Turkey.

Málaga-Chuquitaype, C., Skinner, J., Dowdal, A. and Kernohan, J. 2016. Response of CLT shear walls under cyclic loads, *World Conference on Timber Engineering*, Vienna, Austria.

McKenna, F. 1997. Object Oriented Finite Element Programming: Frameworks for Analysis, Algorithms and Parallel Computing. PhD thesis, University of California, Berkeley, Berkeley, CA. http://opensees.berkeley.edu.

Meireles, H., Bento, R., Cattari, S. and Lagomarsino, S. 2012. A hysteretic model for "frontal" walls in Pombalino buildings, *Bulletin of Earthquake Engineering*, 10(5), 1481–1502.

Pei, S., van de Lindt, J. and Popovski, M. 2013. Approximate R-factor for cross-laminated timber walls in multistory buildings, *Journal of Architectural Engineering*, 117, 245–255.

Popovski, M. and Karacabeyli, E. 2012. Seismic behavior of cross-laminated timber structures, *15th World Conference on Earthquake Engineering*, Lisbon, Portugal.

RothoBlaas. 2015a. *Wood Connectors and Timber Plates, Catalogue of Products*, RothoBlaas SRL, Cortaccia, Italy.

RothoBlaas. 2015b. *Screws for Wood, Catalogue of Products*, RothoBlaas SRL, Cortaccia, Italy.

Tomasi, R. and Smith, I. 2015. Experimental characterization of monotonic and cyclic loading responses of CLT panel-to-foundation angle bracket connections, *Journal of Materials in Civil Engineering*, 27(6), 04014189

Chapter 9

Design of masonry structures

Matthew DeJong and Andrea Penna

CONTENTS

9.1 INTRODUCTION

A large proportion of buildings are masonry, and the material continues to be widely used in new construction. However, there is huge variability within materials classified as 'masonry', including block type (stone, clay brick, concrete, aerated concrete, hollow, rammed earth, etc.), mortar type (dry joints, cement mortar, lime mortar, etc.), bonding pattern, and function (load bearing, infill, cladding, partition wall, thermal barrier). In addition, each of these aspects may vary from country to country. This wide scope and variability makes the development of masonry building codes challenging. In the case of the Eurocode, this variability has forced the code to be more general than for other materials, leaving more responsibility with local (country) building officials, and more variation between practices within Eurocode countries.

Regardless of the wide variety within masonry structures, from the perspective of structural performance, and particularly seismic design, there is an obvious distinction that must be made between unreinforced and reinforced masonry. Generally, the provision of reinforcement, if properly detailed, can cause a more ductile response during an earthquake, and EC8 therefore allows more dissipative inelastic behaviour in the seismic design of reinforced masonry structures. For unreinforced masonry structures, the behaviour is typically less ductile, or at least the inelastic behaviour may be less consistent. Thus, only a very modest level of inelastic behaviour is currently allowed for in design. Further, EC8 imposes strict limits on unreinforced masonry building characteristics (building heights, wall slenderness, etc.), as will be highlighted in the following sections.

In this chapter, the primary focus is on unreinforced masonry structures, though reinforced and confined masonry (little detailed in EC8) are also discussed. The following section seeks to highlight and clarify important aspects of conceptual design, including general guidelines and simplified design limits that are generally applicable to all types of masonry structures. Subsequently, general structural behaviour and design concepts for the specific building typologies of reinforced, confined and unreinforced masonry are presented. The primary analysis methods available to design masonry structures are then discussed, followed by a design example that highlights the benefits and limitations of these analysis options.

Throughout this chapter, comments on the limitations of EC8 for masonry are included for engineers who seek to extend their designs beyond the code specifications. Additionally, examples of the variation between country annexes and codes are discussed, with particular emphasis on the Italian Building Code (NTC, 2008) that provides significantly more detail on seismic design of masonry than EC8 or many other country specifications. It also should be mentioned that this chapter focuses on new design as specified in EC8 Part 1 (EN1998-1, 2004), rather than assessment and retrofitting of existing structures, which is covered in EC8 Part 3 (EN1998-3, 2005), although some beneficial concepts from EC8 Part 3 are highlighted where appropriate.

9.2 CONCEPTUAL DESIGN PRINCIPLES

The general design goal, common to all Eurocodes, is structural robustness. For seismic design, this goal is stated in terms of performance requirements in Section 2 of EC8 Part 1. Essentially, for a major seismic event global or local collapse must not occur, and for a moderate seismic event, the cost of damage must not be disproportionate to the cost of the structure. To achieve these design goals, general fundamental principles of seismic design are outlined in Section 4 of EC8 Part 1, in addition to material specific design requirements in the following parts of the code. This section highlights and clarifies how these conceptual design principles are reflected in the masonry specifications of Section 9 of EC8 Part 1. While detailed design depends on the method of structural analysis, as will be discussed in Section 9.4, these fundamental concepts are independent of the analysis method employed. The concepts and requirements mentioned in this section are applicable to all masonry structures, whether classified as reinforced or unreinforced.

Shear walls typically provide the basic lateral resisting system for masonry structures. The design criteria provided in EC8 §9.5.1 provide general rules for the connectivity and geometry of the shear walls and the remaining structures, independent of detailing. These general rules address two aspects of seismic design: global and local resistance to seismic action.

First, to ensure that the global capacity of the structure is adequate, it is important that shear walls are provided in at least two orthogonal directions, and that building elements have some connectivity, even if the building is generally classified as 'unreinforced'. EC8 §9.5.1 and §9.5.2 specify that horizontal concrete beams or steel ties are required at every floor level, and intermediate horizontal ties are required if storey heights exceed 4 m. These horizontal ties extend around the entire building, providing what is sometimes referred to as 'box behaviour'. The ties have numerous purposes, including helping to ensure that all walls within a plane act together by providing a drag tie to connect their in-plane resistance and redistribute load if one element fails, helping to prevent out-of-plane collapse generally, and helping to prevent out-of-plane collapse from extending over several stories. Effective diaphragm action of the floors is also required, ensuring the safety of the floors themselves, but also ensuring the benefits of the horizontal reinforcement mentioned above.

Figure 9.1 shows some pictures of earthquake-induced out-of-plane collapse in poorly detailed existing masonry buildings. This type of damage constitutes the main source of seismic vulnerability and is primarily responsible for the bad reputation of the performance of masonry structures during earthquakes. Conversely, properly designed reinforced, confined and unreinforced masonry buildings (with some limitations) have proven capable of adequately withstanding earthquake shaking.

A second concept to ensure global stability is that of redundancy, introduced in EC8 §4.2 and also mentioned in Section 4.2.3 of this book. Redundancy is the idea that it is better for several elements to work together to resist the seismic load, so that if one element fails then others can still provide lateral resistance. This is obviously beneficial compared to having a single lateral resisting element, where the entire lateral capacity is compromised if the single element fails. Redundancy is not directly mentioned in Section 9 of EC8 Part 1, but the horizontal floor ties and diaphragms required ensure that load can be redistributed so redundancy can be exploited. The redistribution of loads can be determined directly using non-linear analysis (see Section 9.4), or standard redistribution limits are allowed for in EC8 §9.4 when linear analysis is used. Specifically, the design shear in any one wall may be reduced by up to 25% or increased by up to 33%, so long as the total design base shear is

Figure 9.1 Examples of collapse due to poor connectivity of perpendicular walls and lack of horizontal ties at floor levels to ensure box behaviour.

achieved. This provides a simple way of accounting for load redistribution due to non-linear behaviour without requiring a full non-linear analysis.

The general guidelines of EC8 §9.5.1 also provide rules which help to ensure local seismic resistance. Geometric height/thickness ratio limits help prevent local out-of-plane collapse of shear walls. Height/length ratios of shear walls help ensure that the in-plane stiffness of a single element is sufficient compared to adjacent elements so that load can be shared before an element experiences complete local failure. Note that the height used for the height/length ratio is that of the largest opening adjacent to the wall, as the aspect ratio of this pier within the wall generally governs the behaviour and capacity. Recommendations for these geometric limits are directly provided in EC8 Table 9.2, but values for each country are provided in their National Annex. Finally, the very general requirement of connections between all floors and walls obviously helps prevent local failure (particularly out-of-plane) that could then compromise global stability.

9.3 MASONRY TYPOLOGIES AND STRUCTURAL BEHAVIOUR

As mentioned in the introduction, the term 'masonry' includes several different types of units, mortar, assemblage and reinforcing. The materials and bonding of masonry are differentiated in some detail in EC6 (EN1996-1-1, 2005); EC8 Section 9 simply specifies that masonry units must be robust enough to avoid local brittle fracture, and specifies minimum unit and mortar strengths that must be provided. Though materials are given relatively little attention, the effect of reinforcing is distinguished in more detail.

EC8 §9.3 specifies three primary types of masonry structures: reinforced, confined and unreinforced. In effect, reinforced masonry behaves more like reinforced concrete than unreinforced masonry. Further, confined masonry, which must be distinguished from infill masonry, is effectively a partially reinforced system where much of the seismic resistance is provided by reinforced concrete tie beams and columns. These reinforced systems are not the primary focus of this chapter, but general discussion of all three systems follows.

9.3.1 Reinforced masonry

As discussed in Section 3.4 of this book, a level of ductility is allowed for in elastic design and is specified through a behaviour factor, or q-factor, which relates the yield strength required if the structure were to remain elastic to the yield strength required in the design. Table 9.1 of EC8 specifies the q-factor for reinforced masonry to range from 2.5–3.0, leaving specific limits to be specified by country annexes. To utilise this level of ductility in design, the reinforcing requirements in EC8 §9.5.4 must be followed.

The q-factor for reinforced masonry essentially corresponds to the q-factor of 3.0 specified in EC8 Table 5.1 for an uncoupled reinforced concrete wall system of ductility class medium (DCM), which is the lower bound q-value for all reinforced concrete wall systems. This comparison is logical, as reinforced masonry structures are not subdivided into ductility categories as is done for concrete structures. More specifically, the coupling of shear walls is not directly discussed for masonry, nor is there subdivision of reinforced masonry into medium (DCM) or high (DCH) ductility classes.

For concrete, these ductility classes are defined as 'enabling the structure to develop stable mechanisms associated with large dissipation of hysteretic energy under repeated reversed loading, without suffering brittle failures' (EC8, §5.2.1). Specifications to meet these ductility classes are then provided in EC8 §5.4–5.6. For masonry, the variability of both the constituent materials and reinforcement details make similar specifications more difficult to codify, and they currently do not exist. Instead, EC8 allows the use of a higher *q*-factor for masonry systems that have 'enhanced ductility' only through experimental verification. There is obvious justification for this approach, although there remains an opportunity to create more specific guidelines to allow for higher ductility classes of reinforced masonry structures. In particular, the reinforcing requirements for masonry in EC8 §9.5.4 are currently minimal and vague. More specific requirements would help engineers generate more ductile solutions, rather than requiring them to design and test more ductile solutions on their own.

Alternatively, enhanced ductility could also be incorporated through displacement-based design criteria. However, displacement capacities for reinforced masonry are not yet provided in EC8, as discussed in more detail for unreinforced masonry in Section 9.4.2.

9.3.2 Confined masonry

Confined masonry is defined by EC6 (EN1996-1-1, 2005) as 'masonry provided with reinforced concrete or reinforced masonry confining elements in the vertical and horizontal direction'. This type of construction is common in seismically active countries in the developing world, and can be an efficient construction method because masonry walls can be used as formwork for casting confining columns. It is essential that the confining elements are integrated with the masonry walls to prevent the masonry acting as infill in a concrete frame, rather than acting as an integrated lateral resisting system. Importantly, the confined masonry is load bearing, which distinguishes it from infill masonry (discussed below).

Confined masonry has performed relatively well in earthquakes, providing a simple and cost-effective reinforcing solution. Table 9.1 of EC8 specifies a similar *q*-factor for both reinforced and confined masonry, allowing a similar dissipation capacity to be exploited in design. The specifications for confined masonry are provided in EC8 §9.5.4, and essentially include spacing, connectivity and reinforcement requirements for confining elements. For a much more detailed understanding of confined masonry, the EERI (2011) publication on confined masonry is recommended.

9.3.3 Infill masonry

As infill masonry is not discussed in detail in other chapters of this book, EC8 treatment of infill masonry is briefly discussed here. Contrary to confined masonry, infill masonry is typically added to a concrete or steel frame after the frame is in place, and is therefore not load bearing and is often not effectively integrated in the frame system. If 'infill' masonry is utilised in design as part of the seismic resisting system, then EC8 §4.3.6 specifies that it must be designed according to the criteria for confined masonry in EC8 Section 9. However, even if its capacity is not relied upon in design, infill can drastically change the stiffness and strength of the structure, and therefore change the natural period and the dynamic response. If not properly accounted for, it can lead to a change in the governing failure mechanism. For example, infill can create soft stories if infill panels are not uniformly distributed through the height of the building, initiate shear failure prior to bending failure in columns where infill acts as a diagonal strut similar to a 'brace', or change the location of plastic hinges in concrete columns if infill has openings or is only of partial story height. Masonry infill was ignored in the design of many buildings that exist today, but these issues (and others) are now addressed in detail in EC8 §4.3.6.

The level of connectivity between the frame and the infill drives the seismic performance of these elements. Even if the in-plane effects of infill are accounted for, out-of-plane collapse in the form of falling masonry must be prevented. EC8 suggests providing slight reinforcement (e.g., wire mesh) to improve the in-plane and out-of-plane integrity and behaviour, but the effect of the infill must still be accounted for in analysis. Alternatively, infill panels can be isolated from the surrounding frame, reducing their effect on in-plane motion, decreasing damage to the infill panel and its negative interaction with the frame structure. However, out-of-plane anchorage is particularly important when such isolation is provided. Current research continues to investigate effective detailing for this purpose.

9.3.4 Unreinforced masonry

The classification of unreinforced masonry may be somewhat misleading, as new structures must still contain some reinforcing to provide connections between adjacent walls and floors.

Unreinforced masonry has relatively low tensile strengths and limited ductility, so it is typically designed to remain primarily elastic. If standard design provisions are satisfied, EC8 Table 9.1 specifies a q-factor of 1.5–2.5 for elastic analysis, demonstrating the limited ductility. Further, EC8 §9.3 recommends that for the specified minimum strength of masonry units, unreinforced masonry should not be used where the design ground acceleration exceeds 0.2 g, though individual countries may modify this value in their national annex. In the UK, this value is 0.25 g due to the short duration of UK earthquakes, while in Italy, this limitation is substituted by a limitation to the maximum number of storeys (2) for unreinforced masonry in higher seismicity regions. The combination of the low q-factor and the limit on ground accelerations is stringent. In fact, the low q-factor alone can make it difficult to meet strength requirements for design ground accelerations as low as 0.1 g (Magenes, 2010). The impact of these limits is discussed further in Section 9.4.3 and through the case study in Section 9.5 of this chapter.

In reality, the ductility and displacement capacity of masonry depends significantly on the failure mechanism experienced. For in-plane loading of masonry walls and piers, typical failure modes are well defined, and depend largely on the aspect ratio of the pier, the overburden stress (i.e., axial load), and the shear capacity of the pier (Magenes and Calvi, 1997). Figure 9.2 depicts the typical in-plane failure mechanisms. For relatively small axial loads, slender piers tend to overturn and exhibit rocking, while stocky piers tend to exhibit sliding due to a lower normal force along horizontal joints. On the contrary, high axial loads tend to provide sufficiently large resisting moments to resist overturning and a sufficiently high normal stress to prevent sliding, so shear failure tends to govern for most aspect ratios. For intermediate axial loads, rocking or sliding may still govern, and rocking may be accompanied by increased toe-crushing. In EC8, the q-factors mentioned above are independent of the possible failure mechanism, even though the mechanism significantly affects the inelastic response and energy dissipation capacity. The effects of different failure mechanisms are captured by non-linear static methods of analysis, which are discussed in Section 9.5.5.

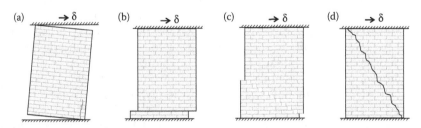

Figure 9.2 Typical in-plane failure mechanisms for unreinforced masonry walls or piers: (a) rocking, (b) sliding, (c) sliding-shear, (d) shear.

Figure 9.3 Typical out-of-plane failure mechanisms for unreinforced masonry structures.

For out-of-plane loading, typical failure mechanisms have been extensively observed in past earthquakes and are also well defined. Figure 9.3 provides some representative mechanisms, though numerous researchers have provided more extensive failure mechanism catalogues. Generally, the design concepts mentioned previously and specified in EC8 §9.5.1 help to prevent out-of-plane collapse. In particular, the requirements for connectivity between adjacent walls and floors and horizontal ties at floor levels tend to prevent out-of-plane mechanisms from spanning multiple stories, and minimum thickness and thickness/height ratios for shear walls tend to prevent individual walls from collapsing because they are excessively slender. However, the problem is complex, and these simple rules are not always sufficient. Regardless, EC8 does not directly provide guidance on how out-of-plane mechanisms should be checked.

The Italian building code (NTC, 2008) is more thorough in this respect, specifying a force-based method for checking out-of-plane capacity. The code essentially specifies that the elastic force demand should be calculated for the particular position of the wall within the structure, and a behaviour factor (q-factor) is specified to allow for some non-linear behaviour. Thus, linear analysis as discussed in Section 9.4.2 can be used for design.

Alternatively, the Italian code (NTC, 2008) more generally requires that all out-of-plane mechanisms must be checked for the assessment of existing structures. For these cases, an alternative displacement-based method specifies the calculation of an equivalent linear elastic oscillator to describe the rocking response of the out-of-plane mechanism. The capacity is then defined as an allowable displacement, which must exceed the displacement demand for the mechanism response taken from a linear elastic design spectrum. This approach is an alternative method of non-linear static analysis, which is discussed further in Section 9.4.3. The interested reader may wish to use these procedures recommended by the Italian code to more thoroughly design for out-of-plane collapse.

9.4 SIMPLE DESIGN RULES AND ANALYSIS METHODS

For the design of masonry structures, analysis must be conducted to verify the safety of every structural element as specified in EC8 §9.6, or the rules for 'simple masonry buildings' in EC8 §9.7 must be followed. These rules are first reviewed, followed by discussion of analysis methods available to achieve safety verification.

9.4.1 Rules for simple masonry structures

The rules for 'simple masonry buildings' in EC8 §9.7 essentially provide seismic safety by ensuring a conservatively large amount of lateral resistance in terms of total shear wall area, limiting the height as a function of the design ground acceleration, and eliminating irregularities (in plan, elevation and height) that are known to cause amplified seismic demands.

The height limits are quite restrictive for unreinforced masonry structures. For example, for ground acceleration levels greater than 0.1 g for building heights greater than two storeys, following these rules alone is not sufficient, and detailed analysis is required.

Regarding building irregularities, several specifications are made. Non-slender rectangular plan configurations are required to eliminate recessed corners, setbacks, and slender extensions that can 'flap' in an earthquake, causing excessive demands on floor diaphragms and shear walls. The specification of symmetric shear walls, separated by at least 75% of the building width, minimises potential torsional effects. The specifications that wall lengths be greater than 30% of the building length and that shear walls must carry 75% of the vertical loads ensure that walls are not too slender or lightly loaded, therefore enabling them to achieve their expected shear capacity. Shear walls must also be continuous through all stories, and the mass must be relatively uniformly distributed over the height of the building to minimise potential soft or weak storey effects.

One additional rule that is included in the Italian code (which also redefines the minimum shear wall area) but not in EC8 is the specification of a maximum allowable mean storey compressive stress. This rule prevents shear walls from being too highly loaded in compression due to dead load alone, prior to the additional compression and shear caused by seismic loads. Such a check helps prevent premature failure of the lateral resisting elements, and is recommended. Moreover, in the Italian Code the use of these 'simple' rules is not allowed in high seismicity areas.

9.4.2 Linear analysis

Two methods are specified for calculating the seismic demand using linear analysis: the lateral force method (EC8 §4.3.3.2), or the modal response spectrum method (EC8 §4.3.3.3). For masonry, the lateral force method is typically sufficient, as the natural period of most masonry structures is relatively short, rendering higher mode effects negligible. For example, the EC8 recommended approximate natural period of a four-storey unreinforced masonry structure with 4 m storey heights is $T_n = 0.05 \, h^{0.75} = 0.4$ s. This can be considered as a maximum expected period for a 'regular' unreinforced masonry structure, and is well below the $T_n = 2$ s limit, above which EC8 requires that higher mode effects be considered.

Using the natural period of the structure, either approximated as above or calculated directly using computational models, the elastic seismic demand can be read in the form of the spectral acceleration S_e from the relevant linear elastic design spectrum. The spectral acceleration can then be used to calculate the total design base shear. For regular structures, a straight line is used to approximate the first mode shape of the structure, that is, static horizontal forces that increase linearly with the storey height are applied (see Equation 3.32). Subsequently, the required force capacity of each structural element calculated using elastic analysis is then divided by the q-factor to account for ductility, and the design force capacity of each element is obtained, that is, $F_{design} = F_{elastic}/q$ (see Section 3.4.1).

In the stiffness calculation, all participating elements must be included. For masonry, this includes full storey-height shear walls as well as spandrel walls that connect shear walls. Where spandrel walls are regularly bonded to adjacent walls and connected to the floor tie beam above and the lintel below, the spandrel can be modelled as a coupling beam. When coupling beams are used to model spandrel walls, the shear walls can be modelled as columns, thus creating an 'equivalent frame' that represents the entire wall and spandrel system (see Figure 9.4). Alternatively, the walls and spandrels can be modelled directly without making an equivalent frame assumption, for example by using finite element modelling with shell elements.

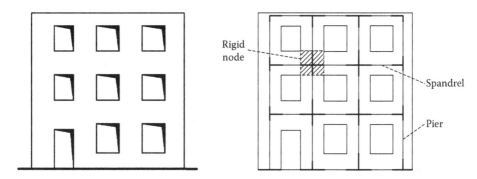

Figure 9.4 Example of a masonry façade and its 'equivalent' frame.

When simplifying piers and spandrel walls to an equivalent frame, the flexural, shear, and axial stiffness must be preserved. The strength of these elements must also be calculated, though the post-yield behaviour need only be considered when non-linear analysis is conducted (see Section 9.4.3). Either the uncracked or cracked stiffness of the elements may be used. In lieu of a more accurate evaluation, the cracked bending and shear stiffness may be taken as half of their uncracked values for the gross section.

Equivalent frame parameters can be computed using special software for modelling masonry structures, or these parameters can be calculated directly by hand. For example, consider the equivalent frame depicted in Figure 9.4. An example description of the calculation of strength and stiffness of masonry structural members is reported in the design example in Section 9.5.

Alternative to equivalent frame models are 'cantilever' models, that is, models in which multi-storey piers are modelled as cantilevers connected at the floor levels by links that only couple the horizontal displacements. Although cantilever models are useful for hand calculations, the use of such an approach typically provides over-conservative results.

It should be noted that, unlike for steel and concrete structures, EC8 does not specify use of an over-strength ratio (OSR) to adjust the q-factor for different building characteristics. The OSR essentially increases the q-factor (and therefore decreases the design base shear) when multiple connected lateral resisting elements exist that allow the load to be redistributed when an element reaches its capacity. The lack of an OSR for masonry structures, combined with the limit that redistribution of lateral loads is restricted to a reduction of 25% or an increase of 33% in any given wall, result in particularly strict design requirements when using linear analysis. To account for the actual ability of masonry to redistribute loads, the Italian Code (NTC, 2008) specifies a q-factor of 2.0, but recommends OSR values of 1.4 and 1.8 for single and multiple-storey buildings, respectively, resulting in an effective q-factor of up to 3.6.

9.4.3 Non-linear static analysis

Linear analysis is the most common way of meeting the requirements of EC8 for the design of unreinforced masonry structures. However, non-linear static analysis is becoming increasingly common because the low q-factors specified by EC8 significantly limit the possibility of using elastic analysis to meet code requirements, and because computational software tools are making non-linear static analysis more feasible. Moreover, the need for non-linear analysis procedures for masonry was acknowledged in the late 1970s, earlier than for concrete and steel structures (Tomažević, 1978; Braga and Dolce, 1982).

Non-linear static analysis, or pushover analysis, evaluates the response of a structure to gravity and a set of monotonically increasing horizontal loads (see Section 3.4.3 of this book for a more detailed explanation of non-linear static analysis). A minimum of two different vertical distributions of horizontal load must be checked. In one case, horizontal load is distributed in direct proportion to the storey mass. This assumption neglects the elastic mode shape and tends to concentrate damage at lower storeys. The second distribution attempts to account for the elastic modal response, effectively shifting mass up the building, and therefore increasing seismic demands at intermediate stories. Thus, for the two distributions, yielding may initiate at different heights of the building, which may therefore alter the subsequent non-linear response. In addition, EC8 still allows the use of separate storey models (Tomažević, 1987) for pushover analysis of low-rise masonry buildings, although these models can provide very approximate results.

The pushover curve, or capacity curve, is considered to represent the non-linear response of a single degree of freedom (SDOF) structure and can be used to identify displacement limit states. Using the pushover curve and the EC8 specified design spectrum, a target displacement, that is, a required design displacement, is then calculated. A limited ductility is considered for masonry members failing both in shear and in bending, thus they are only allowed to carry forces (which must not exceed their resistance capacity) until they reach their displacement/deformation capacity. After this point, their contribution is no longer accounted for in the global pushover curve.

For these reasons, unlike for concrete structures that may exhibit brittle failures (e.g., shear failures) not explicitly accounted for in the non-linear model, there is no need to determine the force and displacement demands placed on each individual element, and to check that they are less than the design resistance of the element. For masonry structures, the use of pushover analysis is thus limited to the comparison of the global displacement demand and the capacity threshold corresponding to the limit state of interest. This point is not entirely clear in the description in EC8 §4.4.2.2, as this section, similarly to other parts of EC8, was mainly written considering non-masonry structures.

The challenge for the design of masonry structures is that displacement limits, for example, the displacement capacity of a masonry pier, are not directly specified in EC8. The code only specifies drift limits related to non-structural components. For brittle non-structural components, EC8 §4.4.3.2 specifies a drift limit of $d_r{}^* \nu \leq 0.005\,h$, where d_r is the interstory drift, and ν is a reduction factor accounting for the importance class of the building and the return period of the earthquake for which the drift limit is set. So, in its current form, these non-structural drift limits serve as the only direct guideline for displacement limits, but they say nothing about the displacement capacity.

In future versions of EC8, displacement capacities for masonry may be directly specified. However, until then, other resources provide useful guidelines on displacement limits to be imposed. For existing masonry structures, Annex C of EC8 Part 3 (EN1998-3) recommends criteria to define both global and local displacement capacity limits.

The recommended near collapse (NC) global displacement capacity is defined by the magnitude of the roof displacement at the point when the lateral capacity has reduced to 80% of its previous maximum value. At this point, the stiffness of the global system has clearly become negative, and global collapse is a concern. The recommended significant damage (SD) global displacement capacity is then defined as 75% of the NC value. Meanwhile, for the damage limitation (DL) limit state, the displacement capacity is the displacement corresponding to the yield point of the idealised elasto-plastic pushover curve.

Regarding local displacement capacities, EC8 Part 3 (Annex C) directly recommends values for the SD limit state. When shear walls are governed by flexure, the recommended drift limits are 0.8% for primary shear walls and 1.2% for secondary shear walls (both multiplied by the ratio between the zero moment height and wall length). For shear walls governed by shear failure, these drift limits become 0.4% and 0.6%, respectively. For the NC limit state, the SD drift limits for both flexural and shear dominated response may be increased to 4/3 of the corresponding SD values.

It is also worth noting that the Italian Code (NTC, 2008), unlike EC8, does recommend local displacement capacities for the design of new structures. Specifically, the Italian Code recommends SD limit state values similar to those recommended for primary masonry shear walls by EC8 Part 3, Annex C (i.e. drift limits of 0.8% for walls governed by flexural failure and 0.4% for walls governed by shear failure).

9.4.4 Non-linear dynamic analysis

Finally, non-linear dynamic analysis can be conducted following the procedure in EC8 §4.3.3.4.3. However, the building response to seven earthquake time-histories must be simulated in order to use the average seismic demand for design. This requires extensive modelling and computation time, particularly for large structures with multiple elements that may exhibit non-linear behaviour, and is therefore less common. However, as computational times decrease and expertise increases, non-linear dynamic analysis may become more common for special masonry structures with unique design characteristics.

9.5 DESIGN EXAMPLE

9.5.1 Introduction

The eight-storey building considered in the previous chapters could not be realistically designed with a masonry load-bearing structure, particularly if we refer to the most common masonry typology, that is, unreinforced masonry. For this reason, a three-storey unreinforced masonry building is considered in this design example. The layout of the building is reproduced in Figure 9.5. The storey height of the ground storey was assumed to be

Figure 9.5 Plan (a) and 3-D view (b) of the design example masonry building (dimensions in cm).

3.75 m, whereas 3.5 m was considered for the upper storeys. The diaphragm structure (reinforced concrete slab) and imposed loads are the same as the frame structure design examples. The end use as a hotel is also maintained.

Similar to chapters that address other structural types, the seismic design checks reported in the following are performed for a preliminary design according to EN 1998-1. For the sake of simplicity, only one direction of analysis is considered. Additionally, this example focuses on in-plane seismic design. Out of plane checks should also be carried out, though they are not presented here.

The considered example is a mixed URM-RC structure, that is, a masonry building, which also includes internal reinforced concrete columns and beams that primarily carry vertical loads, allowing for a more open interior space. For the linear static analysis in Section 9.5.4, the reinforced concrete elements are assumed to only carry vertical loads, and thus the entire lateral seismic load is assumed to be taken by the masonry. On the contrary, in the non-linear analysis in Section 9.5.5 the resistance of concrete elements are included in the seismic design.

As mentioned in the previous sections, linear analysis can be extremely conservative in the case of masonry buildings. For this reason, the seismic safety checks presented in the next sub-sections include results of both linear and non-linear static analysis, clearly showing the advantages of the latter analysis method. In the case of linear static analysis, a significant limitation of the reference peak ground acceleration on rock ($a_{gR} = 0.5$ m/s^2) was hence necessary to satisfy the safety checks. Conversely, a_{gR} was set to 0.2 g in the example based on non-linear static analysis. It should be also noted that, according to clause 9.3(3) of EN 1998-1, the value of $a_g \cdot S$ should not exceed a certain limit $a_{g,urm}$, whose recommended value is 0.2 g.

9.5.2 Masonry mechanical properties

Masonry walls are considered to be constructed with perforated clay blocks and general-purpose mortar, used for both bed-joints and head-joints (entirely filled). The considered clay masonry units with vertical holes belong to Group 2 according to the classification proposed in Eurocode 6 (EN 1996-1-1, 2005) as the total volume of holes is assumed to be 45% of the gross volume of the block (this corresponds to the maximum allowed according to the Italian National Annex to EC8). The largest dimension of the block, corresponding to the wall thickness (single skin blockwork), was assumed to be 30 cm.

The normalised compressive strength of units in the direction parallel to the orientation of the holes (vertical), f_b, was assumed to be equal to 12 MPa and the compressive strength of mortar, f_m, equal to 10 MPa (M10). The characteristic vertical compressive strength of masonry, f_k, was evaluated according to Eurocode 6 as $f_k = K f_b^a f_m^\beta = 0.45 \cdot 12^{0.7} \cdot 10^{0.3} = 5.11$ MPa. The characteristic initial shear strength (cohesion), f_{vko}, was assumed to be equal to 0.3 MPa and the friction coefficient equal to 0.4. The masonry Young's modulus in compression, E, and the shear modulus, G, were then calculated as $E = 1000 f_k = 5,110$ MPa and $G = 0.4E = 2,044$ MPa, respectively.

9.5.3 Design loads

Dead and imposed loads are the same as those adopted in the design examples presented in Chapters 3 through 7 of this book. A self-weight of 10.0 kN/m^3, including plastering, was considered for the structural masonry walls. The presence of partition walls, with an equivalent distributed load of 1.0 kN/m^2 was accounted for in the room bays. Imposed loads consisted of 4.0 kN/m² on the roof terrace, 2.0 kN/m² in the rooms and 3.0 kN/m² in corridors, etc. The total seismic mass is 950.7 tons.

9.5.4 Seismic design checks: Linear static analysis

9.5.4.1 General considerations

A three-dimensional equivalent frame model was used in the analysis of the response of all elements. Members of the equivalent frame were masonry piers (vertical elements) and coupling beams (horizontal elements), consisting of reinforced concrete tie beams, as no masonry spandrels were assumed to be present in this example. The latter assumption corresponds to a common situation in modern masonry buildings, where the presence of heaters and rolling shutters results in a reduced thickness of the masonry below and above openings.

Structural members were modelled as Timoshenko beams, that is, accounting for both in-plane bending and shear deformability. Cracked bending and shear stiffness of the structural elements was accounted for by considering one half of the uncracked gross section elastic stiffness. The structural system resisting lateral loads was considered as constituted by the masonry wall systems oriented in the two perpendicular directions, whereas reinforced concrete columns were assumed to carry vertical loads only. Floor and roof diaphragms were modelled by means of membrane elements and they were rigid in their plane.

For linear analysis, a relatively small peak ground acceleration of 0.5 m/s² was considered. The same spectral shape (Type 1) and soil conditions (Soil C) considered in the design examples presented with other structural typologies were also assumed in this case. Considering the structural regularity, the number of storeys, and the redundancy of the example structure, a behaviour factor of $q = 2.5$ was specified, equalling the maximum of the range provided in EC8 Table 9.2. As noted previously, the Italian National Annex to EC8 increases this maximum value to 3.6, acknowledging the effect of global over-strength.

A simplified estimate of the first mode vibration period yields $T_1 = 0.05h^{0.75} \cong 0.3$ s. This period falls in the constant acceleration branch (0.2–0.6 s) of the acceleration design response spectrum for the assumed spectrum shape and soil conditions.

The ordinate of the design response spectrum was calculated as

$$S_d(T_1) = \frac{2.5a_gS}{q} = \frac{2.5 \cdot 0.5 \cdot 1.15}{2.5} = 0.575 \text{ m/s}^2$$

According to the EC8 lateral force method of analysis, the total base shear force is

$$F_b = S_d(T_1) \cdot m \cdot \lambda = 0.575 \cdot 950.7 \cdot 0.85 = 465 \text{ kN}$$

The analysis of the spatial building model was performed along the global X and Y directions. The analyses showed a lower capacity along X, although safety checks were satisfied with some margin for the considered low level of seismic action. Internal force distributions in the masonry members for one of the critical wall lines in the X direction (North façade) are depicted in Figure 9.6.

9.5.4.2 Member design checks

For unreinforced masonry, once the wall arrangement is specified by the architect, the degrees of freedom available for the structural engineer are limited, as constructive and

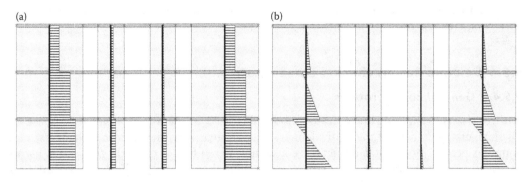

Figure 9.6 Distributions of shear forces (a) and bending moments (b) along the height of masonry piers in the North façade of the example structure.

structural members often coincide and no reinforcement design is carried out. Therefore, the structural design actually becomes to some extent a structural assessment process. Design checks were performed by comparing the values of the internal actions resulting from the analysis with strength capacity evaluated as described in the following. Design checks on reinforced concrete members are not presented.

Safety checks for unreinforced masonry members were based on the shear strength criterion proposed in Eurocode 6 (clause 6.2). The shear force calculated in each masonry pier was compared to the resisting shear force obtained from the following equation:

$$V_{Rd,s} = l_c t f_{vd}$$

where l_c is the compressed length of the masonry cross-section, t is the wall thickness and f_{vd} is the masonry shear strength obtained from

$$f_{vd} = \frac{f_{vko} + 0.4\sigma}{\gamma_M} \leq \frac{f_{vlt}}{\gamma_M}$$

where σ is the average stress acting on the compressed part of the cross section, $f_{vlt} = 0.065 f_b$ is the assumed limit shear force associated with cracking of units, and $\gamma_M = 2$ is the partial safety factor for masonry strength properties and is assumed equal to 2. This represents the minimum of the shear sliding strength along bed-joints (stepped cracks, Coulomb-type criterion) and the strength associated with cracks passing through masonry units (Mann and Mueller, 1982). In this case, the design masonry shear strength was equal to

$$f_{vd} = f_{vdo} + 0.2\sigma = 0.15 + 0.2\sigma \leq f_{vltd} = 0.39 \text{ MPa}$$

According to EC6 clause 6.2(5), the length of the compressed part of the wall should be verified for the vertical loading applied. This results in a minimum length of the compressed part of the masonry cross-section associated with the maximum possible eccentricity of the vertical force, and hence of the maximum in-plane bending moment and the associated shear force. Assuming a linear stress distribution on the compressed part of the masonry cross section, it has to be verified that

$$\sigma_{max} = 2\sigma = 2\frac{N_{Ed}}{t l_c} \leq f_d$$

where N_{Ed} is the design compression force acting on the cross-section, and $f_d = f_k/\gamma_M = 2.55$ MPa is the design masonry compressive strength.

As an example of safety checks against the ultimate limit state, a squat pier ($t = 0.3$ m, $l = 5.3$ m, $h = 3.7$ m) located at the ground storey of the South façade is considered. The following actions were calculated:

$$V_{Ed} = 52.7 \text{ kN}; \quad N_{Ed,top} = 212.7 \text{ kN}; \quad M_{Ed,top} = -202.9 \text{ kNm}; \quad N_{Ed,base} = 267.7 \text{ kN};$$
$$M_{Ed,base} = 397.8 \text{ kNm}$$

The eccentricity of the axial forces can be evaluated as $= |M_{Ed}|/N_{Ed}$, and if $e > l/6$ (cracked cross section) the length of the compressed part of the cross section is $l_c = 3(0.5l - e)$.

The compressed length at the top and bottom of the considered masonry pier are $l_{c,top} = 5.09$ m and $l_{c,base} = 3.49$ m. It was then verified that

$$\sigma_{max,top} = 0.28 < 2.55 = f_d; \quad \sigma_{max,base} = 0.51 < 2.55 = f_d$$

The shear strength values at the top and bottom cross sections were also evaluated as

$$f_{vd,top} = 0.18 \text{ MPa}; \quad f_{vd,base} = 0.20 \text{ MPa}$$

In the case of this structural member, the following values were calculated for the resisting shear force:

$$V_{Rd,top} = 271.5 \text{ kN}; \quad V_{Rd,base} = 210.7 \text{ kN}$$

The minimum calculated shear strength of the considered masonry pier is then $V_{Rd} = 210.7$ kN > 52.7 kN $= V_{Ed}$.

In this example, the low level of spectral acceleration allowed satisfactory safety checks in all structural members, so the shear force redistribution mentioned in Section 9.4.2 was not necessary. For higher levels of seismic action causing member shear forces to exceed capacity, redistribution could be employed, although the required procedure can be cumbersome and of limited benefit compared to non-linear analysis.

9.5.5 Seismic design checks: Non-linear static analysis

9.5.5.1 General considerations

For non-linear static (pushover) analysis, a higher value of peak ground acceleration of $0.2\,g$ was considered. A spatial equivalent frame model was again used. Structural members constituting the frame were modelled by means of non-linear frame elements (or macroelements). The simplest modelling approach of bilinear elastic perfectly-plastic elements was employed. The lateral stiffness (linear branch) was obtained, similarly to the case of linear models, by accounting for shear and flexural deformability and cracked conditions. The lateral strength was instead evaluated considering the minimum strength associated with the relevant shear criteria. Mean values of the mechanical properties were used in the computation of the lateral strength capacity.

Given the limiting bending moments at the top and bottom of the pier, a limit to the pier lateral strength associated with bending failure (also known as 'toe crushing') can be expressed by the following equation proposed in EC8 Part 3 (Annex C):

$$V_{Rd,f} = \frac{lN}{2h_0}\left(1 - 1.15\frac{N}{ltf_M}\right)$$

where N is the axial force, h_0 is the distance from the point where the flexural capacity is attained and the inflection point, and f_M is the mean masonry compressive strength. In this case f_M was assumed to be equal to $f_k/0.7 = 7.3$ MPa.

The following expression (Magenes and Calvi, 1997) directly provides the lateral force capacity associated with the shear failure modes considering the effect of cracking in the reduction of the compressed part of the cross-section, for the case of sliding failure along mortar joints:

$$V_{Rd,s-I} = N\frac{1.5f_{vo}lt + 0.4N}{N + 3f_{vo}h_ot}$$

where f_{vo} is the mean initial shear strength in the absence of compression (cohesion), assumed in this case to be equal to $f_{vko}/0.7 = 0.42$ MPa. Similarly, the shear capacity in the case of cracks through units can be expressed by the following equation:

$$V_{Rd,s-II} = N\frac{1.5f_{vlt}lt}{N + 3f_{vlt}h_ot}$$

As evident from these equations, the resisting shear forces evaluated using the different criteria are all strongly influenced by the contribution of vertical forces. For this reason, the axial forces used in the calculation of the lateral strength capacity need to be updated based on the effect of horizontal loading. The lateral capacity of each pier is hence given by

$$V_{Rd}(N) = \min\{V_{Rd,f}(N); V_{Rd,s-I}(N); V_{Rd,s-II}(N)\}$$

The relation between the resisting shear force obtained by the considered strength criteria and the axial compression can be represented graphically in a strength interaction diagram. In Figure 9.7a, an example interaction diagram is shown for a pier with the same dimensions

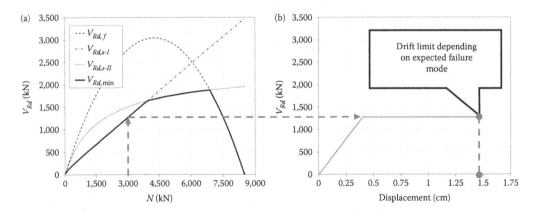

Figure 9.7 Resisting shear force – axial compression interaction diagram for the example masonry pier (a) and corresponding bilinear capacity curve for a given level of axial force (b).

as the one considered in the previous section (linear analysis), under the assumption of $h_0 = 0.5\ h$. Using this interaction diagram, the bilinear curve representing the lateral pier response can be directly derived (Figure 9.7b).

Moreover, the strength criteria identify different failure modes which are directly accounted for in the definition of the ultimate displacement capacity (i.e., the drift limit in Figure 9.7b). According to EC8 Part 3 (Annex C), in case of shear failure the ultimate lateral displacement can be assumed equal to $0.004\ h$, whereas for flexural failures, it can be taken as equal to $0.008\ h_0/l$. Recent studies (Petry and Beyer, 2014) support the use of a unified criterion for the definition of the displacement capacity, explicitly accounting for axial load ratio, boundary conditions and size effects.

As mentioned in the introduction, the use of non-linear analysis allowed consideration of the contribution of both the masonry and non-masonry elements (reinforced concrete frames) to the lateral resistance of the building model. An initial bilinear curve similar to Figure 9.7b was determined computationally for all elements of the structure given the computed axial forces. A non-linear static (pushover) analysis was then conducted computationally, which requires an iterative process that includes updating the bilinear curve according to the redistribution of loads (frame effect). The pushover analysis is carried out by imposing a predefined pattern of horizontal forces with constant relative ratios and controlling (increasing) the displacement of a control node usually selected at the top level. Loading patterns proportional to the first mode of vibration (or its linear approximation) and to the mass distribution are typically adopted. The computational procedure removes elements that exceed their ultimate drift capacity as the iteration progresses.

Pushover analysis on equivalent frame models provides capacity curves (top displacement vs. base shear). Importantly, the displacement controlled algorithm allows simulation of the post-peak (softening) branch on the pushover curve.

9.5.5.2 Pushover simulation results

For loading in the +X direction, Figure 9.8 depicts the capacity curve (total base shear vs. top displacement) obtained by means of the pushover analysis assuming a horizontal loading pattern proportional to the one adopted in the linear analysis. For the same loading case,

Figure 9.8 Capacity curve of the structure as obtained from pushover analysis (continuous line), with its bilinear idealisation (dashed line) and indication of the displacement demand (white circle).

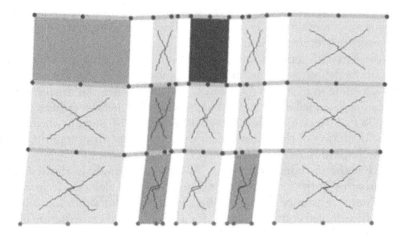

Figure 9.9 Qualitative depiction of shear and rocking cracks on the deformed shape of the North façade.

Figure 9.9 shows the qualitative expected damage pattern and failure modes of the North façade, as predicted by the analysis.

No accidental eccentricity effects were accounted for in this preliminary design. The capacity curve is plotted up to the attainment of the ultimate displacement capacity, that is, up to the roof displacement at which the total base shear dropped below 80% of the peak resistance of the structure, due to progressive failure of lateral load resisting elements, as suggested in EC8 Part 3 (Annex C). The ultimate displacement was $d_u = 2.4$ cm. In Figure 9.8, the significant over-strength with respect to the linear analysis limit is evident.

A bilinear approximation of the capacity curve was computed for the determination of the idealised elastic-perfectly plastic force–displacement relationship. As the procedure reported in Annex B of EC8 seems not to be fully applicable to masonry structures, the one suggested in the Italian building code was followed. This procedure consists of setting the linear branch secant at 70% of peak lateral strength and imposing that the areas below the two curves up to the ultimate displacement be equal. The lateral strength of the idealised elastic-perfectly plastic curve was $F_y = 1970$ kN and the yield displacement was $d_y = 0.42$ cm.

9.5.5.3 Displacement-based design checks

Individual member checks are automatically satisfied as they are embedded in the computational model by means of the adopted strength criteria. Therefore, in the case of non-linear static analysis, design checks move from local checks on single members to a global check on the overall building response. In particular, this design check consists of the comparison of the displacement capacity obtained from the analysis and the displacement demand.

In order to evaluate the displacement demand, it is necessary to define an equivalent SDOF system, whose capacity curve is obtained by dividing both force and displacement of the pushover curve by the participation factor Γ. In this case, the participation factor is $\Gamma = (\sum m_i \Phi_i)/(\sum m_i \Phi_i^2) = 1.31$ and the participating mass is $m^* = \sum m_i \Phi_i = 578$ tons. Hence, the parameters of the bilinear capacity curve for the equivalent SDOF system are

$$F_y^* = F_y/\Gamma = 1504 \text{ kN}; \quad d_y^* = d_y/\Gamma = 0.32 \text{ cm}; \quad d_u^* = d_u/\Gamma = 1.83 \text{ cm}$$

The period of vibration of the equivalent bilinear system was estimated as

$$T^* = 2\pi \sqrt{\frac{m^* d_y^*}{F_y^*}} = 0.22 \text{ s}$$

As expected, the period is short $(T^* < T_c)$ and falls in the constant acceleration branch of the response spectrum. The corresponding ordinate of the elastic acceleration response spectrum was then calculated:

$$S_e(T^*) = 2.5 a_g s = 2.5 \cdot 1.962 \cdot 1.15 = 5.64 \text{ m/s}^2$$

and the corresponding elastic displacement demand resulted in

$$d_{et}^* = S_e(T^*) \left[\frac{T^*}{2\pi} \right]^2 = 0.70 \text{ cm}$$

The ratio between the requested elastic force and the available lateral strength of the bilinear SDOF system was computed as

$$q_u = S_e(T^*) \frac{m^*}{F_y^*} = \frac{5.64 \cdot 578}{1504} = 2.17$$

The displacement demand (target displacement) of the equivalent SDOF system was then calculated as proposed in Annex B of EC8 (Fajfar, 2000):

$$d_t^* = \frac{d_{et}^*}{q_u} \left(1 + (q_u - 1)\frac{T_c}{T^*} \right) = \frac{0.70}{2.17} \left(1 + (2.17 - 1)\frac{0.6}{0.22} \right) = 1.34 \text{ cm}$$

Hence, the target displacement in the MDOF system (white circle in Figure 9.8) is

$$d_t = \Gamma d_t^* = 1.76 \text{ cm} < 2.4 \text{ cm} = d_u$$

Therefore, the in-plane capacity is proven satisfactory for the direction of loading shown. Similar analysis would need to be conducted for loading in the Y direction. Additionally, as mentioned previously, out-of-plane checks should also be conducted.

9.6 SUMMARY

Though masonry structures have less ductility than many other structures, they can still be safely designed to resist seismic action, and their non-linear behaviour can still be relied upon in design. As with other building typologies, proper detailing is essential. As demonstrated in this chapter, EC8 does provide considerable guidance on seismic design of masonry, both in terms of general concepts and guidelines, and more specific requirements. In particular, the design example has demonstrated how new structures classed as unreinforced masonry can be properly designed using both linear and non-linear methods.

However, EC8 currently contains less detail for masonry than for other building materials, perhaps due to the variability of masonry as a building material and its reputation for limited seismic capacity. There is significant opportunity to add detail to future revisions of the code. Both the general discussion and the design example in this chapter have mentioned additional guidelines specified by the Italian building code, which are useful as an additional resource for design, but also must be further extended.

REFERENCES

Braga, F. and Dolce, M. 1982. A method for the analysis of antiseismic multi-storey masonry buildings, *Proceedings of the 6th International Brick Masonry Conference*, Rome, Italy [in Italian].

EERI. 2011. *Seismic Design Guide for Low-Rise Confined Masonry Buildings*, Earthquake Engineering Research Institute, Oakland, California.

EN 1996-1-1. 2005. *Eurocode: Design of Masonry Structures, Part 1-1: A General Rules and for Reinforced and Unreinforced Masonry Structures*, European Committee for Standardization, CEN, Brussels.

EN 1998-1. 2004. *Eurocode 8: Design of Structures for Earthquake Resistance, Part 1: General Rules, Seismic Actions and Rules for Buildings*, European Committee for Standardization, CEN, Brussels.

EN 1998-3. 2005. *Eurocode 8: Design of Structures for Earthquake Resistance, Part 3: Assessment and Retrofitting of Buildings*, European Committee for Standardization, CEN, Brussels.

Fajfar, P. 2000. A non-linear analysis method for performance-based seismic design. *Earthquake Spectra*, 16(3), 573–592.

G.U. no. 47, 26/02/2009 (suppl. ord. no. 27). Circular 2/02/2009, no. 617. Instructions for the application of the new technical code for the design of constructions, issued by D.M. 14/01/2008 [in Italian].

Magenes, G. 2010. Earthquake resistant design of masonry structures: Rules, backgrounds, latest findings, *Proceedings of the 8th International Masonry Conference*, Dresden, Germany.

Magenes, G. and Calvi, G.M. 1997. In-plane seismic response of brick masonry walls, *Earthquake Engineering and Structural Dynamics*, 26(11), 1091–1112.

Mann, W. and Mueller, T. 1982. Failure of Shear-Stressed Masonry: An enlarged theory, tests and application to shear walls, *Proceedings of the British Ceramical Society*, 30, 223–235.

NTC. 2008. (Norme Tecniche per le Costruzioni) 2008. D.M.14/1/2008, S.O. n°30, G.U. n.29 del 4/02/2008

Petry, S. and Beyer, K. 2014. Influence of boundary conditions and size effect on the drift capacity of URM walls, *Engineering Structures*, 65, 76–88.

Tomaževič, M. 1978. *The computer program POR*, Report ZRMK, Ljubljana [in Slovenian].

Tomaževič, M. 1987. Dynamic modelling of masonry buildings: Storey mechanism model as a simple alternative, *Earthquake Engineering and Structural Dynamics*, 15(6), 731–749.

Seismic isolation and supplemental damping

Damian Grant

CONTENTS

10.1 INTRODUCTION

Seismic isolation and supplemental damping are passive protection strategies for reducing the seismic demand on structural elements. They are commonly used in retrofit applications, where the existing structure cannot be relied upon to resist a design earthquake ground motion in a ductile manner, and other interventions may be difficult or infeasible to carry out. Isolation is also very common in bridge applications, where the connection between pier and deck presents a natural and effective interface to allow relative movement.

For new buildings, either approach is often used in conjunction with a desire for improved seismic performance – whether in terms of reducing expected repair costs, or allowing uninterrupted functionality, following a large earthquake. This strategy lends itself naturally to buildings whose post-earthquake performance is essential, such as civil defence buildings and hospitals. Even valuable works of art, such as heavy statues, are protected by seismic isolation – in this case preventing damage due to toppling rather than through yielding of structural components.

Chapter 10 in EC8-1 covers seismically isolated structures in general, as well as specific rules for base isolation of buildings. In this chapter, background concepts on seismic isolation, types of devices available, numerical results and design criteria from EC8 are discussed in Sections 10.2 to 10.5, respectively, and a design example is provided in Section 10.6.

Chapter 10 of EC8-1 does not cover passive energy dissipation systems (supplemental damping) that are distributed over several storeys or levels. Nevertheless, in this chapter, a brief introduction is provided to the topic in Section 10.7, a summary of available devices is given in Section 10.8 and simplified design criteria for viscous and viscoelastic damping systems are outlined in Section 10.9.

Other relevant European standards which are not discussed in any detail are the following:

- EC8-2, for seismically isolated bridges.
- EC8-3, for assessment of existing buildings and isolation design for retrofit applications.
- EN 15129 Anti-seismic Devices, for much more detailed guidance on general design rules, compliance criteria, testing and verification, maintenance and installation, for isolators and damping devices.

These references should also be consulted, where relevant for a particular application.

10.2 CONCEPTS OF SEISMIC ISOLATION

In seismic isolation, flexible isolation bearings are placed between the primary mass of the structure and the source motion, effectively using inertia and a lengthened vibration period to limit deformation in structural components. In this manner, a building is isolated from its foundations, and the superstructure of a bridge is isolated from its piers.

The main benefits of seismically isolating a structure, either new or existing, are as follows:

1. The introduction of flexibility to the system increases the fundamental period, moving it outside the predominant frequency range of earthquake input. In short period structures, this can decrease the design forces substantially, although for longer period structures, this effect may be negligible. This is also a benefit for acceleration-sensitive non-structural components, such as sensitive equipment. This effect is illustrated with reference to a typical EC8 spectrum in the left half of Figure 10.1.

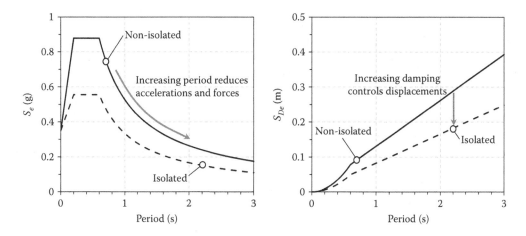

Figure 10.1 **Effect of seismic isolation: increasing the period reduces the seismic forces transmitted to the superstructure, while increasing damping controls the displacements.**

2. Although increased flexibility can lead to larger system displacements, inelastic deformation is confined to the isolation system, allowing elastic design of the remainder of the structure. Bearings are relatively easy to maintain and, if necessary, replace at least compared to beams or columns in the superstructure.
3. Significant seismic energy may be dissipated in the isolation system. Some systems provide damping inherently, through inherent viscous or hysteretic damping in their components. Hybrid systems, where flexibility and damping are provided in separate devices, take advantage of the fact that large movements will occur across the isolation interface, providing a convenient location to place damping devices. Higher damping has the effect of further decreasing the seismic forces and limiting the maximum displacement demand on the bearings. This effect is illustrated with reference to a typical EC8 spectrum in the right half of Figure 10.1.
4. The shear forces transmitted to the superstructure are limited by the amount of force that can be transmitted across the isolation interface, which, in some cases, allows the isolation device to act as a fuse for the structure. In the context of capacity design (see Section 4.6), a reliable ductile fuse allows effective control of the earthquake response, and gives some confidence that unwelcome brittle mechanisms are unlikely to form.

At first glance, Figure 10.1 presents an apparent contradiction, as we know that forces in a structure are approximately equal to both the mass times the acceleration and the stiffness times the displacement. Decreasing the acceleration while increasing the displacements does not seem compatible with this explanation. The apparent contradiction is resolved by realising that the isolated structure can be represented by two springs in series (as analysed in Section 10.4), one for each of the isolator and the superstructure. We are interested in the effect of introducing the isolator spring to the response of the superstructure spring. It is shown in Section 10.4 that the overall force is reduced and overall displacement is increased, as shown in Figure 10.1. However, the deformation in the spring representing the superstructure is reduced significantly from the non-isolated case, as almost all the displacement is carried in the isolator. Therefore, the increasing displacement shown in Figure 10.1 is not a particular concern for the superstructure, although the added damping is important for controlling the demands on the device itself.

Seismic isolation is widely used in bridges, in which most of the mass is concentrated in the deck, and pot bearings are typically provided anyway to transfer the gravity loads from the deck while allowing thermal and other movements. Isolation is also commonly adopted in the rehabilitation of existing buildings, especially heritage buildings, where it is technically difficult or architecturally unacceptable to make modifications to the existing structural elements.

For the design of new buildings, seismic isolation is often adopted as part of an enhanced performance objective, where better than 'no collapse' performance is desired after a strong earthquake. Typically, isolated buildings will be designed to remain elastic during a design level earthquake ground motion. In fact, non-linear response of the superstructure has the effect of lengthening the vibration period which can interfere with the effect of the isolation. Energy dissipation and ductility are expected to develop in the isolation system only. As discussed in Section 10.5, in EC8, this is reflected in the use of a low behaviour factor ($q = 1.5$) for the design of the superstructure in an isolated building – consistent with low ductility design elsewhere in the code – and the use of $q = 1.0$ in calculating the forces transmitted through the isolators to the substructure. In fact, EC8 design rules are not intended for use with a 'partial isolation' strategy, where the superstructure does not remain within the elastic range (Cl. 10.4(5)).

As with any other design decision, the benefits of seismic isolation should be weighed against the costs. In many cases, the material savings that can be realised due to both the reduction in seismic demand, and also in the relaxation of ductile detailing requirements, can outweigh the additional cost of the isolation system. In calculating the latter, the designer should take into account not just the unit cost of the devices, but also requirements for their testing, installation and maintenance, and the need to provide sufficient separation around the structure for it to move freely with respect to the ground (Cl. 10.5.4(1)). Building services crossing the interface will also need to be designed to accommodate this movement. Furthermore, separating the building from the ground means that gravity loads at the ground floor level will need to be transferred to the ground through the bearings, requiring additional structure.

An analysis based solely on monetary costs does not take into account the enhanced performance expected of an isolated structure in an earthquake. For critical facilities, such as hospitals or nuclear plants, seismic isolation may be adopted to minimise earthquake damage, and to allow the facility to be used immediately after the event. Even for non-critical facilities, in highly seismic areas, the expected post-earthquake repair costs over the life of the structure may be reduced by adopting seismic isolation, and this could be taken into account in the initial cost–benefit analysis. The REDi™ Rating System (Almufti and Willford, 2014) provides additional incentive for adopting enhanced performance objectives in seismic design, and the highest level rating all but requires the use of seismic isolation to achieve, at least in areas of high seismic hazard.

Although isolated structures have performed effectively in recent earthquakes, there is a lingering feeling in the community that extra conservatism should be incorporated into their design. The ground motion return period of 2475 years, adopted as the basis of seismic design of all buildings in the United States and replacing the previous 475-year return period, was originally introduced for seismic isolation, to ensure that the margin against collapse was sufficient in this relatively new technology (A. Whittaker, 2011, pers. comm.). EC8 maintains the same 475-year return period as the basis of design (Section 2.5.1), but introduces a magnification factor, γ_x, with a recommended value of 1.2 (1.5 for bridges), to increase the reliability required of isolation systems (Cl. 10.3(2)).

Furthermore, recognising that the behaviour of isolating devices can vary with rate of loading, vertical load, horizontal load in the transverse direction, temperature, and a change of properties over the projected service life, EC8 requires a bounding approach to be adopted for the modelling of isolators (Cl. 10.8(1)). Upper bound stiffness values (and minimum damping and friction) are used to estimate accelerations and inertia forces in the superstructure, and lower bound stiffness, damping and friction values are adopted for the estimation of displacements (Cl. 10.8(1) and (2)). The exception is for importance class I or II buildings with maximum and minimum values within 15% of the mean values, for which the mean values of physical and mechanical properties can be used (Cl. 10.8(4)).

10.3 ISOLATION SYSTEMS

There are a large number of isolation devices available on the market, each combining the three main requirements for an effective isolator: vertical stiffness, lateral flexibility and energy dissipation. The optimum device for a given project will depend not just on availability and price, but also on the technical requirements of the project. The reader is referred to a number of references which describe and compare various available products: Skinner et al. (1993), Kelly (1990), Christopoulos and Filiatrault (2006) and Priestley et al. (1996).

A brief description of some of the more common devices is provided in the following.

10.3.1 Elastomeric and lead-rubber bearings

The vertical stiffness and strength of a large block of rubber is limited by its tendency to bulge in the middle, which is a function of the shape of the block and the properties of the rubber. For use in seismic isolation, elastomeric bearings comprise alternating layers of elastomeric material and steel plate, which are vulcanised together under high pressure and temperature. The steel restrains the rubber layers from bulging, without significantly increasing the shear stiffness. A typical laminated rubber bearing is illustrated in Figure 10.2a.

The force–displacement response of a laminated rubber bearing is close to linear, with very low inherent damping. Therefore, they are typically relied upon to carry vertical load and provide lateral flexibility only, and modifications must be made to provide energy dissipation. The three main strategies for elastomeric bearings in seismic isolation are

1. Provide a supplementary damping device in parallel with the bearings that dissipates energy without adding significantly to the lateral stiffness. Viscous dampers, for example, discussed in Section 10.8.2, could be appropriate.
2. Make use of a high-damping rubber (HDR) in the bearing, which can provide 10%–15% viscous damping inherently. High-damping rubber bearing (HDRB) response departs somewhat from linear, tending to stiffen under large deformations (see Figure 10.3).

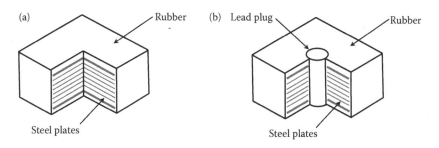

Figure 10.2 Cut-away view of (a) laminated rubber bearing and (b) lead-rubber bearing.

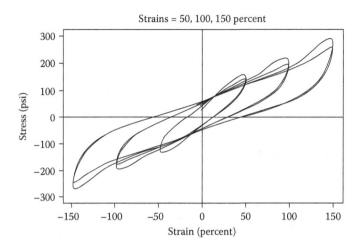

Figure 10.3 Schematic force–displacement response for an HDR bearing, showing stiffening response at large strains and scragging behaviour under cyclic loading.

This is often cited as an advantage, as it can limit bearing deformations under extreme ground motions, although this can interfere with the effectiveness of the isolation (Grant et al., 2005b). Furthermore, HDRBs exhibit scragging behaviour, which means that the virgin response (i.e. before subjecting the bearing to lateral load) is stiffer and stronger than subsequent cycles of load. This variability needs to be taken into account in design (Thompson et al., 2000).

3. In 'lead-rubber bearings' (LRB), the central portion of the rubber is replaced with a lead plug (see Figure 10.2b). Lead is an ideal material for this purpose, as it exhibits essentially elastic-perfectly-plastic response which is stable under cyclic loading, and is hot worked at room temperature, meaning that fatigue is not an issue. The lead plug adds to the initial lateral stiffness of the bearing, but when its yield capacity is reached, the remaining lateral stiffness is the shear stiffness of the rubber alone. This allows flexibility in design, as the relative dimensions of the rubber and lead can be determined to provide practically any combination of vertical and shear stiffness and strength, and significant energy dissipation.

Skinner et al. (1993) and Kelly (1990) are excellent resources for the design and analysis of elastomeric isolation bearings.

10.3.2 Friction-based isolation bearings

The basic concept of friction-based isolation bearings is for the building to slide on a controlled surface, which can transfer the weight of the building to the foundations while dissipating energy through friction. A flat sliding surface is not usually appropriate, as the lateral sliding of the building is not controlled, and the total movement over the duration of an earthquake ground motion could be very large. Flat sliders with low friction coefficients are sometimes used in conjunction with other energy dissipation devices, essentially just to transfer the gravity load and provide very little lateral resistance or damping of their own.

Friction pendulum system (FPS) bearings make use of an articulated slider on a spherical sliding surface and gravity to provide a restoring force, thereby limiting the amount of movement over an earthquake without significantly reducing the effectiveness of the isolation (see Figure 10.4a). The motion of an FPS bearing is analogous to that of a pendulum, from which their name is derived.

The force–displacement response of an FPS bearing (Figure 10.5) is essentially rigid until the static friction is exceeded, and with a post-slip stiffness (k_b) governed solely by the supported weight (W) and the radius of curvature (r) of the bearing:

$$k_b = \frac{W}{r}.$$

(10.1)

The vibration period (T_b) when sliding is independent of the weight, and depends solely on the radius of curvature:

$$T_b = 2\pi\sqrt{\frac{r}{g}}.$$

(10.2)

Note that this is not the same as the effective period defined based on the secant stiffness to the design displacement in EC8 (see Section 10.5.2.1).

The design of FPS bearings is extremely flexible, in that the two parameters defining the bilinear response in Figure 10.5 (slip force and post-slip stiffness) can be determined independently of one another, by modifying the friction coefficient of the sliding surfaces and the radius of curvature of the bearing. The near-rigid initial response allows wind and other loads to be resisted without movement. The post-slip response can be tuned to provide

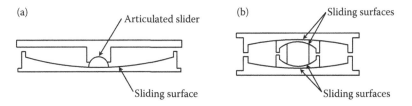

Figure 10.4 Section view of (a) an FPS bearing, and (b) a Triple Pendulum™ bearing.

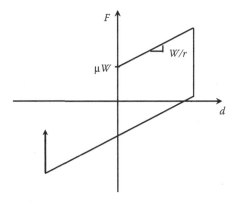

Figure 10.5 Schematic force–displacement response for an FPS bearing.

enough flexibility to limit forces transmitted to the structure, while limiting overall movement. Typically in US practice, stoppers are built into FPS bearings to prevent unseating under very large movements; note, however, that this is explicitly forbidden in EN 15129 Cl. 8.3.1.2.3, which notes that impact at stoppers can cause damage to the device. The alternative – complete unseating – should perhaps be of more concern.

The relatively recently introduced Triple Pendulum™ bearing (Figure 10.4b) allows even more flexibility in the design. It uses multiple nested articulated sliding surfaces, to control the small, moderate and large displacement response independently. In particular, this allows a stiffening behaviour to be introduced for large displacements, to limit the chance of unseating or impact with stoppers. The Triple Pendulum bearing also has the advantage over the FPS bearing of being significantly smaller in diameter for the same design bearing displacement.

10.3.3 Hybrid isolation/damping devices

As mentioned in the previous sub-sections, often multiple devices can be used in parallel to obtain desirable response and to satisfy the objectives of vertical stiffness and strength, lateral flexibility and energy dissipation. Any combination of the following could be used:

- Flat sliding surfaces and/or low-damping rubber bearings to transmit vertical loads.
- Viscous/viscoelastic, steel hysteretic or lead-extrusion dampers to provide damping without significantly increasing the lateral resistance.

Damping devices are discussed separately in Section 10.8.

10.3.4 Rocking

Isolation typically works by reducing the horizontal restraint at the base of a building (or the top of a bridge pier). An alternative strategy, which may also be thought of as seismic isolation, is to release or reduce the rotational and/or vertical restraint at the base, allowing the building or bridge to rock on its foundations. As with conventional isolation, this increases the effective vibration period, reducing the seismic energy that is transmitted to the structure. Furthermore, damping devices can be used to dissipate energy between the building and the ground, in addition to the energy that is dissipated in radiation into the ground when impacts occur.

The South Rangitikei viaduct in New Zealand is an example of a railway bridge that was designed to rock at the base of the piers (see Figure 10.6). Torsional-beam dampers (an example of steel hysteretic dampers) provide energy dissipation when the piers uplift. Gravity loads are transmitted to the pile cap through elastomeric bearings. The viaduct has been in operation since 1981.

Renzo Piano's Maison Hermès in Tokyo, Japan, is a building application of the rocking concept (Figure 10.7). The building site was limited in size, and the architect wanted the structural elements to be hidden at the rear of the building. The resulting slenderness made traditional seismic design strategies prohibitively expensive. A line of columns at the rear of the building were therefore allowed to uplift from their foundations when they were subjected to net tensile forces. Viscoelastic dampers were provided to dissipate energy and to prevent excessive uplifting. See Field et al. (2005) for more details of the analyses performed to verify the design.

Analysis and design approaches for rocking structures are described in Priestley et al. (1996). For more complex applications, like in Maison Hermès, non-linear time history

Figure 10.6 (a) South Rangitikei viaduct. (Photo by Alan O'Brien, Flickr user Raurimu_Spiral; https://www.flickr.com/photos/raurimu_spiral/. With permission) (b) Detail of rocking foundation. (Priestley, M.J.N., Seible, F. and Calvi, G.M.: *Seismic Design and Retrofit of Bridges*. 1996. Wiley, New York. Copyright Wiley-VCH Verlag GmbH & Co. KGaA. Reproduced with permission.)

Figure 10.7 (a) Maison Hermès building. (Photo by Michael Denance. With permission.) (b) Rendering of rocking foundation detail. (Courtesy of Arup Rendering.)

analysis is essential to ensure that the effects of impacts and non-linear geometry are properly accounted for.

10.4 THEORY OF SEISMIC ISOLATION

A well-designed isolation system can regularise the response of a building, and permit simplified analysis methods to be used.

Figure 10.8 Spring-damper model for analysis of seismically isolated building.

Christopoulos and Filiatrault (2006) present a theoretical analysis of linear seismically isolated systems, originally due to Kelly (1990), key results of which are summarised and extended in this section. The theory is based on the two degree of freedom system (2DOF) system illustrated in Figure 10.8. The top spring-damper-mass system (k_s, c_s and m_s, respectively) represents the superstructure, without isolation; the bottom spring-damper-mass represents the isolation system (stiffness, k_b, and damping ratio, c_b) and the base of the structure (mass, m_b).

The results can be conveniently expressed in terms of the following parameters:

$$\text{Superstructure natural frequency, } \omega_s = \frac{k_s}{m_s}, \tag{10.3}$$

$$\text{Isolation system natural frequency, } \omega_b = \frac{k_b}{m_s + m_b}, \tag{10.4}$$

$$\text{Squared frequency ratio, } \varepsilon = \left(\frac{\omega_b}{\omega_s}\right)^2, \tag{10.5}$$

$$\text{Mass ratio, } \gamma = \frac{m_s}{m_s + m_b}. \tag{10.6}$$

The damping ratios of the structure and the bearings, ξ_s and ξ_b, are defined with respect to the total supported mass; therefore

$$\xi_s = \frac{c_s}{2m_s\omega_s}, \quad \text{and} \quad \xi_b = \frac{c_b}{2(m_s + m_b)\omega_b}. \tag{10.7}$$

For small values of ε (of the order 10^{-2}) and $m_b < m_s$, Christopoulos and Filiatrault (2006) show that the two natural frequencies of the isolated system are approximately

$$\omega_1 = \omega_b\sqrt{1 - \gamma\varepsilon} \approx \omega_b, \tag{10.8}$$

$$\omega_2 = \frac{\omega_s}{\sqrt{1 - \gamma}}\sqrt{1 + \frac{\gamma\omega_b^2}{\omega_s^2}} \approx \frac{\omega_s}{\sqrt{1 - \gamma}}. \tag{10.9}$$

That is, the fundamental frequency of the system is approximately equal to the natural frequency of the isolation system (taking into account the total supported mass, $m_s + m_b$), and is not affected by the flexibility of the structure. The second natural frequency is larger than the superstructure natural frequency and increases with increasing mass ratio.

The damping ratios for the two modes of the system can be shown to be approximately

$$\xi_1 \approx \xi_b \left(1 - \frac{3}{2}\gamma\varepsilon\right),$$

(10.10)

$$\xi_2 \approx \frac{\xi_s}{\sqrt{1-\gamma}} + \frac{\gamma\xi_b\sqrt{\varepsilon}}{\sqrt{1-\gamma}}.$$

(10.11)

The first result shows that the damping ratio of the fundamental mode tends to the damping ratio of the isolation system for very small values of the squared frequency ratio; for example, for a period ratio of 3 ($\varepsilon = 1/3^2 = 0.11$) and a mass ratio of 0.9, the first mode damping ratio is 85% of that of the isolation system by itself. The second mode damping ratio can be significantly greater than the damping ratio of the bare structure alone, particularly since the damping ratio of the isolators will generally be much higher than that of the superstructure alone.

For design of isolation systems, typically the two main response quantities of interest are the deformations in the two springs in Figure 10.8 – the deformation in the top spring is related directly to the forces in the superstructure by the stiffness (k_s) and the deformation in the bottom spring is the isolation system displacement. The peak deformations in the superstructure and bearings are, respectively:

$$|v_s|_{max} \approx \varepsilon\sqrt{[S_{De}(\omega_1,\xi_1)]^2 + [S_{De}(\omega_2,\xi_2)]^2},$$

(10.12)

$$|v_b|_{max} \approx (1-\gamma\varepsilon)S_{De}(\omega_1,\xi_1).$$

(10.13)

where $S_{De}(\omega,\xi)$ is the spectral displacement for natural frequency (ω) and damping ratio (ξ). This derivation assumes that the spectral displacement at ω_2 is much smaller than at ω_1.

The expressions presented here (and derived fully in Christopoulos and Filiatrault, 2006) give a mathematical explanation of why seismic isolation works, and what parameters influence its effectiveness. As the squared frequency ratio approaches very small values, the following limits hold:

$$\lim_{\varepsilon \to 0} |v_s|_{max} = 0,$$

(10.14)

$$\lim_{\varepsilon \to 0} |v_b|_{max} = S_D(\omega_b,\xi_b).$$

(10.15)

That is, the deformation in the superstructure tends to zero, and the behaviour of the isolated system approaches that of the isolators alone supporting the total mass of the superstructure and base.

10.5 DESIGN CRITERIA FOR SEISMIC ISOLATION SYSTEMS

10.5.1 Design philosophy

The design philosophy for isolated buildings in Chapter 10 of EC8 has been introduced in previous sections. In summary:

- Only passive energy dissipation arranged on a single interface is considered, and not distributed damping systems (Cl. 10.1(4); see Sections 10.7 to 10.9 for more information on supplemental damping).
- Only full isolation is considered; i.e. the superstructure is assumed to remain elastic (Cl. 10.4(5)). A behaviour factor, q, of up to 1.5 is allowed for the design of the superstructure, which is consistent with a nominal amount of over-strength and little or no yielding. This also means that capacity design requirements do not need to be followed for the design of the superstructure, as it does not need to form a ductile mechanism to behave in a satisfactory manner. Ductility class DCL may be adopted.
- No q factor is adopted for the design of the isolators (i.e. $q = 1$), as their response is modelled explicitly, whether through an equivalent linear model or non-linear hysteresis model.
- Increased reliability is targeted for isolated systems by multiplying seismic displacements determined from analysis by a multiplication factor, γ_x, with a recommended value of 1.2 (Cl. 10.3(2)) (or 1.5 for bridges).

10.5.2 Analysis methods and assumptions

As has been discussed previously, seismic isolation is a relatively new approach to reducing seismic vulnerability, and therefore analysis and modelling requirements can be more onerous than for non-isolated structures. On the other hand, seismic isolation effectively regularises structural response – both in plan, by giving a uniform distribution of stiffness that can be tuned to the mass distribution of the structure to minimise torsion, and in elevation, by giving some confidence in the vertical distribution of seismic demand. EC8 recognises four analysis approaches that may be used for design of isolated buildings:

1. Simplified linear analysis
2. Modal simplified linear analysis
3. Conventional modal analysis
4. Time history analysis.

The requirements for each analysis method, and other considerations for design, are discussed in the following sub-sections.

10.5.2.1 Modelling of isolators

10.5.2.1.1 Bounding analysis

As discussed in Section 10.2, separate analyses should be carried out with upper- and lower-bound properties for the isolators, except for buildings of importance classes I and II, and where maximum and minimum values of isolator properties are within 15% of the mean where it is permissible to use the mean values (Cl. 10.8). Note that this requirement applies regardless of the analysis method adopted.

European Norm EN 15129 gives additional guidance on the determination of upper and lower bound properties for isolation devices.

10.5.2.1.2 Linear modelling

Cl. 10.9.2 gives a set of criteria that must be met if isolators are to be modelled as linear viscoelastic. Interestingly (and in contrast to other international design regulations), Cl. 10.9.2(1) only allows equivalent linear models to be used if the isolation system 'consists of devices such as laminated elastomeric bearings', and requires a bilinear hysteretic model to be used for 'elasto-plastic types of devices'. The latter would presumably include hysteretic and friction-based systems, and even lead-rubber bearings whose behaviour is characterised by yielding of the lead core. Therefore, it would not be possible to use the simplified linear or modal simplified linear analysis methods (described in the following subsections) for these systems, and time history analysis would be required. Chapter 10's primary author confirmed that this interpretation of Cl. 10.9.2(1) is correct, and that friction-based or hysteretic isolators should be verified using non-linear analysis, although linear methods could still be used at a preliminary design stage (P. Bisch, 2015, pers. comm.).

Equivalent linearisation of non-linear response is a commonly used simplification in earthquake engineering (see Grant et al., 2005a, for a discussion), and is especially common in modelling seismic isolation bearings. Figure 10.9 shows a bilinear hysteresis rule and a linear visco-elastic equivalent. Since visco-elastic response is frequency-dependent, the properties of the equivalent linear representation should be determined at the range of frequencies expected in the design ground shaking (Cl. 10.9.2(3)).

The objective of linearisation should be to obtain an accurate (or at least conservatively biased) approximation of the true non-linear response, while simplifying the analysis required. In general, we have two properties to specify – an effective stiffness (K_{eff}) and an equivalent viscous damping (ξ_{eff}) – and therefore it is possible to determine an optimal pair of values as a function of non-linear response parameters and peak displacement (Iwan and Gates, 1979).

In EC8, the effective stiffness is taken as the secant stiffness at the total design displacement of the isolation system, d_{db} (Cl. 10.2 and 10.9.2(2)). This is generally much smaller (i.e. effective period much longer) than the optimal value determined using the method of Iwan and Gates (1979), but with correspondingly higher effective damping values, adequate estimates of peak displacement response can be obtained (e.g. Grant et al., 2005a). Using the secant stiffness has the advantage that it has some physical significance, and relates the peak displacement and peak force transmitted into the superstructure directly.

For equivalent viscous damping, EC8 Cl. 10.9.2(3) states that 'the energy dissipation in bearings should be expressed from the measured energy dissipated in cycles with frequency

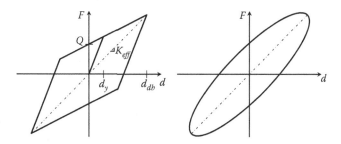

Figure 10.9 Bilinear hysteresis and equivalent linear approximation.

in the range of the natural frequencies of the modes considered', although it does not give specific advice on how to calculate it. An approach based on equating the area within a single hysteresis loop of the non-linear response is often used. Assuming the bearing is vibrating at its resonance frequency, this approach gives the following expression (e.g. Chopra, 2011):

$$\xi_{eff} = \frac{1}{4\pi} \frac{E_D}{E_{So}}, \tag{10.16}$$

where E_D is the area contained in one hysteresis loop (e.g. Figure 10.9), and E_{So} is the strain energy at the peak displacement and effective stiffness, $E_{So} = K_{eff} d_{db}^2/2$.

For the bilinear response shown in Figure 10.9, this gives the following expression:

$$\xi_{eff} = \frac{2Q(d_{db} - d_y)}{\pi K_{eff} d_{db}^2}. \tag{10.17}$$

where the parameters are identified in Figure 10.9.

Equations 10.16 and 10.17, or variants thereof, are commonly used in earthquake engineering, in conjunction with an effective period calculated from the secant stiffness. However, it has long been recognised (e.g. Iwan and Gates, 1979) that this significantly over-damps the response compared to the non-linear system (i.e. the peak non-linear response is underestimated). Many authors (e.g. Grant et al., 2005a) have determined alternative damping relationships that provide more accurate estimates of non-linear response, but the designer should recognise that there is often a large degree of uncertainty attached to their use. Note that some manufacturer's catalogues make use of Equation 10.17 in reporting equivalent viscous damping values, therefore exaggerating the effectiveness of their products. In lieu of specific advice from EC8 on calculating equivalent viscous damping, it is recommended that literature relationships for various hysteresis rules are adopted, rather than Equations 10.16 and 10.17. Of course, non-linear time history analysis is a more direct way of taking into account non-linear isolator response, and is always an option if there is doubt.

Since the secant stiffness and equivalent viscous damping are functions of displacement for non-linear systems, EC8 requires an iterative procedure to be adopted, based on assumed values for the isolator unit design displacement, d_{db} (Cl. 10.9.2(4)). Note that in general this is not equal to the design displacement for the system overall (d_{dc}); even for regularly distributed isolators, accidental eccentricity effects (required to be taken into account by Cl. 10.9.1(2)) will lead to higher displacements in devices with a higher torsional radius. However, the designer may consider that the effort involved in simultaneously iterating on displacements in every isolator unit is not warranted by the simplifications inherent in the equivalent linearisation method.

EC8 seeks to minimise the inaccuracy in using equivalent linearisation by limiting its use to isolators that meet the following additional criteria (Cl. 10.9.2(5)):

- The effective (secant) stiffness at system design displacement, d_{dc}, is not less than 50% of the effective stiffness at displacement $0.2d_{dc}$.
- The effective damping ratio at the design displacement does not exceed 30%. Note that this is less likely to be a problem using more realistic damping levels from the literature, rather than Equation 10.16 or 10.17. Note also that EN 15129 strongly recommends time history analysis when damping exceeds 15%.

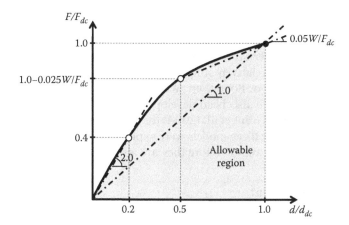

Figure 10.10 Allowable normalised force–displacement response to use equivalent linear modelling of isolators (response is only limited at the circles, but the line connecting them is shown for illustration).

- The force–displacement characteristics do not vary more than 10% due to rate of loading or variation in vertical loads.
- The increase in force in the system between $0.5d_{dc}$ and d_{dc} is not less than 2.5% of the total gravity load.

Given a desired performance point (d_{dc}, F_{dc}) for an isolation system (and neglecting the difference between the isolator unit displacement, d_{db}, and the isolation system displacement, d_{dc}) the first and fourth requirements both impose upper bounds on the forces at displacements less than d_{dc}. Figure 10.10 shows how these requirements translate into an upper bound on isolator response, normalised with respect to the performance point. All possible force–displacement backbone curves under the solid line (or at least under the two open circles at $0.2d_{dc}$ and $0.5d_{dc}$, where the rules are checked) meet the first and fourth requirements, including those with low initial stiffness that display hardening behaviour leading up to the design displacement. Other force–displacement response is of course permissible, but an equivalent linear approach cannot be used.

As discussed in Section 10.3, the force–displacement response of friction pendulum bearings depends only on the supported weight, the radius of curvature, and the friction coefficient. It can be shown that the first and fourth requirements are satisfied if the following inequality holds:

$$\frac{d_{dc}}{R} \geq \min(0.05, 3\,\mu). \tag{10.18}$$

Since the response of friction devices is heavily sensitive to vertical load, the third requirement above would only be satisfied if it could be shown that load variation was small. Note, however, the previous comments about use of equivalent linearisation for friction-based isolators.

10.5.2.1.3 Non-linear modelling

If the four requirements discussed above are not met, a non-linear model must be used for the isolators. Since the simplified linear analysis method and modal analysis require linear (or equivalent linear) response, this leads to the requirement to carry out time history analysis, discussed further in Section 10.5.2.5. Note that this does not obviate the need to use upper and lower bound properties, except where this is explicitly not required by EC8 Cl. 10.8.

A number of non-linear hysteresis models are available for modelling the response of isolators. For 2-D analysis models, relatively simple non-linear models used for other applications, such as bilinear (elasto-plastic), or Bouc-Wen (Wen, 1976) models are appropriate. More complicated non-linear models have also been developed for specific devices, such as Mosqueda et al. (2004) for FPS bearings, Grant et al. (2004) for HDR bearings, and Huang et al. (2000) for lead-rubber bearings. These three models were explicitly developed for modelling the bidirectional response of bearings, and therefore they are also appropriate for 3-D analysis.

10.5.2.2 Ground motion input

Generally, the ground motion input for isolated buildings is the same as for non-isolated structures, as discussed in detail in Chapter 2. The following additional considerations apply:

- The vertical component of seismic action should be considered for base-isolated structures when the peak vertical ground acceleration, a_{vg}, exceeds 0.25 g (Cl. 4.3.3.5.2). Arguably (P. Bisch, 2015, pers. comm.), it should always be considered for isolated structures, to ensure that uplift in devices is prevented (or at least is taken into account in isolator specification).
- Isolated buildings will usually exhibit viscous damping (or equivalent viscous damping) greater than the 5% of critical assumed for normal buildings. For the simplified linear analysis and modal analysis methods, this is taken into account by the damping correction of Cl. 3.2.2.2(3). This damping reduction only applies to the fundamental (isolated) mode of response in each principal direction, as higher mode damping is relatively unaffected by the isolators (see Equation 10.9). When analysis is carried out by hand, this is relatively straightforward to implement. When carrying out modal analysis in structural analysis software, this can be taken into account by inputting a hybrid spectrum with increased damping in the range of periods encompassing the fundamental period(s), and 5% damping associated with bare structural response for lower periods (see Figure 10.11). Also note the dependence of

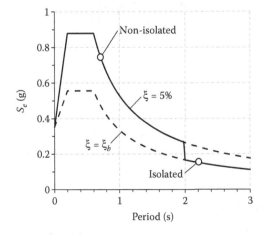

Figure 10.11 Hybrid spectrum for response spectrum analysis of isolated buildings in analysis software.

damping reduction on the expected duration of ground motion, not currently taken into account in EC8.

- Since isolated structures rely on long fundamental periods to reduce seismic response, their design can be particularly sensitive to the corner period, T_D, that marks the transition from linearly increasing spectral displacement with period, to constant spectral displacement. In EC8 this parameter is set at 2.0 s and 1.5 s for Type 1 and Type 2 spectra, respectively, whereas evidence suggests that higher values are appropriate for magnitudes greater than 6.0. This artificially favours seismic isolation as a design strategy, as it provides an upper limit on the displacement that the isolation system needs to accommodate, regardless of period. The fundamental period of isolated buildings will typically be in the 2–3 s range. In lieu of any additional information about magnitude and corner period appropriate for a project, it is recommended that the corner period is assumed to be greater than the isolated period for the design of isolated structures (i.e. spectral displacement continues to increase linearly).

- Isolation systems are usually designed such that the stiffness and therefore fundamental period in each principal direction is the same. There is presently some debate in the earthquake engineering community about the use of the maximum rotated ground motion (i.e. evaluating the spectral response at the orientation that produces the highest demand on the structure) in design, rather than the geometric mean measure adopted in EC8. However, for axisymmetric systems such as typical isolated buildings, the maximum rotated ground motion is certainly appropriate (Grant et al., 2011). Huang et al. (2008) have determined factors by which to multiply the geometric mean demand to obtain the maximum rotated demand in the near field. This will typically give more onerous input than using 100% of the demand in one direction and 30% in the other direction, as given in EC8 (Cl. 4.3.3.5.1(3)), which for axisymmetric structures gives a total demand of $\sqrt{(1^2 + 0.3^2)} = 1.04$ times the geometric mean.

10.5.2.3 Simplified linear analysis

Linear analysis has already been discussed in Section 10.5.2.1 from the point of view of the modelling of the isolators. The simplified linear analysis method is analogous to the lateral force method for non-isolated buildings, but explicitly incorporates displacement-based design ideas. It assumes rigid superstructure behaviour, giving an effective period of

$$T_{eff} = 2\pi \sqrt{\frac{M}{K_{eff}}}, \tag{10.19}$$

where M is the mass of the superstructure. The system effective damping (ξ_{eff}) is the effective damping of the isolators, as discussed in Section 10.5.2.1.

With rigid superstructure response, the expected displacement at the stiffness centre of the isolation system is simply the design spectral displacement at the effective period (using lower-bound stiffness for the isolators, if required by Cl. 10.8) and lower bound effective damping (Cl. 10.9.3(5)):

$$d_{dc} = S_{De}(T_{eff}, \xi_{eff}) = \frac{MS_e(T_{eff}, \xi_{eff})}{K_{eff}}. \tag{10.20}$$

Rigid superstructural response implies that the inertia forces will be uniformly distributed over the height of the superstructure (assuming a uniform distribution of mass), instead of the triangular distribution used for non-isolated buildings. The storey forces are therefore given by (Cl. 10.9.3.5(6)):

$$f_j = \frac{m_j}{M} F_b = \frac{m_j}{M} M S_e(T_{eff}, \xi_{eff}) = m_j S_e(T_{eff}, \xi_{eff}). \tag{10.21}$$

where m_j is the mass of level j, M is the total mass of the superstructure and F_b is the total base shear. Note that the correction factor, λ, from EC8 Cl. 4.3.3.2.2(1) is effectively taken as 1.0, regardless of the number of storeys for isolated buildings. This is because the effective modal mass of the isolated fundamental mode is assumed to be equal to the full superstructure mass, consistent with the assumption of rigid superstructural response.

Equations 10.20 and 10.21 give design displacements and forces for isolated buildings at a system level. For individual isolators, the design displacement must be multiplied by the reliability factor, γ_x, discussed in Section 10.5.1, as well as an amplification factor to take into account torsion about the vertical axis (Cl. 10.9.3(8)):

$$\delta_{xi} = 1 + \frac{e_{tot,y}}{r_y^2} y_i, \tag{10.22}$$

where r_y is the torsional radius of the isolation system, given by

$$r_y^2 = \sum \left(x_i^2 K_{yi} + y_i^2 K_{xi} \right) \Big/ \sum K_{xi}, \tag{10.23}$$

and x_i and y_i are the coordinates of the isolator with respect to the effective stiffness centre, $e_{tot,y}$ is the total eccentricity in the y direction (including the accidental eccentricity, e_{ai}), and K_{xi} and K_{yi} are the effective stiffness of unit i in the x and y directions. Equation 10.22 applies to response in the x direction, while an analogous expression holds for response in the y direction.

Design forces in the superstructure are also increased to take into account torsion, using the expression for non-isolated buildings in Cl. 4.3.3.2.4. They are then reduced by a behaviour factor, q, up to an allowable value of 1.5.

For buildings, the simplified linear analysis method may be used where equivalent linear isolation behaviour is allowed, and where the following conditions are met (Cl. 10.9.3):

- Maximum horizontal eccentricity between the centre of mass and centre of stiffness of the isolation system, including 5% accidental eccentricity from Cl. 4.3.2, should not exceed 7.5% of the plan dimension, in the two principal horizontal directions. Note that this clause (10.9.3(2)) is somewhat misleading, as it suggests that torsion may be 'neglected' if this requirement is satisfied, but Cl. 10.9.3(8) clarifies that torsion is taken into account by amplifying both the isolator displacements and superstructure forces by an amplification factor.
- The distance from site to nearest potentially active fault with $M_s \geq 6.5$ is greater than 15 km. At closer distances, there is a concern that the presence of velocity pulses and near-field 'fling' in the earthquake ground motion may be especially damaging

to isolated structures. This should be taken into account in the design by time history analysis with an appropriate selection of ground motions that incorporate pulses (Almufti et al., 2015) and/or fling (Kamai et al., 2014). On the other hand, it should also be noted that pulse-like motions are not only recorded within 15 km of a fault, particularly for very large magnitude events (Iervolino and Cornell, 2008).

- The largest dimension of the superstructure on plan is not greater than 50 m.
- The substructure (i.e. the foundations for buildings) is sufficiently rigid to minimise the effects of differential displacements of the ground, and all isolation devices bear directly on vertical supporting elements.
- The effective period from Equation 10.7 is between $3T_f$ and 3 seconds, where T_f is the fundamental period of the superstructure assuming a fixed base, using one of the simplified expressions from Cl. 4.3.3.2.2. This effectively implies that the simplified method cannot be used on buildings whose non-isolated period exceeds 1.0 seconds.
- The lateral-load resisting system of the superstructure is regularly and symmetrically arranged in plan.
- Rocking at the base of the substructure is negligible.
- The isolation system is sufficiently stiff vertically to satisfy the following two expressions:

$$\frac{K_v}{K_{eff}} \geq 150 \quad \text{and} \quad T_v = 2\pi\sqrt{\frac{M}{K_v}} \geq 0.1. \tag{10.24}$$

10.5.2.4 Modal simplified linear analysis

As for simplified linear analysis, modal analysis may be adopted for design if the isolation system may be modelled as equivalent linear (see Section 10.5.2.1; Cl. 10.9.4). When using analysis software to carry out the eigenvalue and response spectrum analysis, the approach illustrated in Figure 10.11 can be used to provide an appropriately damped input motion.

If all the conditions for simplified linear analysis discussed in the previous section are also met, except for the maximum horizontal eccentricity of 7.5%, a simplified modal analysis assuming rigid superstructure and substructure may be adopted (Cl. 10.9.4(2)). Total eccentricity of mass, including accidental eccentricity, must be taken into account in the evaluation of eigenmodes, which means that the rotational and translational displacements in the isolators will be accounted for without requiring amplification factors. Iteration will be required to obtain an effective stiffness for each bearing that is compatible with its total displacement.

Conventional modal analysis may be used when equivalent linear behaviour is allowed, but the other conditions are not met.

10.5.2.5 Time history analysis

If an equivalent linear model may not be used for the isolators, non-linear time history analysis is required. Generally, only the isolators will be modelled as non-linear, taking into account the considerations in Section 10.5.2.1. If superstructure or substructure non-linearity is acceptable for the design, this could also be considered explicitly in the analysis, although this would be outside the scope of EC8 Chapter 10, which explicitly does not consider partial isolation.

10.6 DESIGN EXAMPLE: BASE-ISOLATED CONCENTRICALLY BRACED FRAME

10.6.1 Introduction

The same eight-storey building considered previously is used in this example. A steel concentrically braced frame is provided for the lateral system of the superstructure, arranged as in Section 6.9. As before, the main seismic design checks are carried out for a preliminary design according to EN 1998-1, and only the lateral system in the x direction in plan is considered. This example outlines the main considerations in the design of the isolation system; indicative sizes are given for members in the superstructure to illustrate the potential savings that can be realised in seismically isolated structures, but the reader should refer to Chapter 6 for more detailed calculations for steel concentrically braced frame structures.

10.6.2 Design loads

The gravity loads per unit area are the same as those adopted in the steel moment frame and concentrically braced frame examples as indicated in Table 6.3. The estimated fundamental period of the non-isolated superstructure is 0.62 s, calculated as previously using the simplified expression proposed in EC8 (Cl. 4.3.3.2.2). Note that this is less than the 1.0 s limit identified in Section 10.5.2.3, below which the simplified base isolation design could be applied (provided the other conditions are met).

The design seismic forces for the superstructure depend on the characteristics (equivalent period and viscous damping) of the isolation system adopted, and these are calculated in the following sub-sections. Note that EC8 (Cl. 10.10(5)) allows a behaviour factor of 1.5 to be adopted for the design of the superstructure; seismic forces and spectral displacements for the design of the isolation system, however, are not reduced by a behaviour factor (EC8 Cl. 10.7).

The total seismic mass of the superstructure above ground level is 8,208 t, as previously. However, since a diaphragm and transfer structure must be provided at the ground level to distribute gravity loads to the foundations via the isolators and to transfer lateral loads (Cl. 10.5.3), additional seismic mass participates in the seismic response of the building. Assuming the same dead and imposed loading for the ground level slab as for level 1 (see Tables 3.1 and 3.2), except that the floor-to-floor height is increased to 4.3 m and there is no load due to external glazing (assumed to be supported from the floor above), the additional dead load is calculated to be 14,860 kN and the imposed load is 7,260 kN. With a $\psi_{E,i}$ factor of 0.3, this implies an additional seismic mass of 1,737 t, for a total of 9,945 t.

According to EC8, the simplified linear analysis method of Cl. 10.9.3 may not be used to validate the design, as the largest dimension of the superstructure in plan is 56 m, which is greater than the upper limit of 50 m allowed by Cl. 10.9.3(3)(b). Nevertheless, the simplified method is followed here for two reasons:

- It provides a useful preliminary design even for structures outside its scope, and therefore allows a rapid comparison of several different isolation design options before carrying out modal analysis, in accordance with Cl. 10.9.4(1) and 4.3.3.3, or time history analysis, following Cl. 10.9.5.
- To illustrate for the reader the steps involved in the simplified linear analysis method, which will be relevant in many practical applications.

Aside from the limit on plan dimensions, the simplified linear analysis method is appropriate provided that all the conditions discussed in Section 10.5.2.3 are satisfied.

The braced frame located on GL 1 is again selected for illustration in this example. As discussed in Section 6.9.2, the non-seismic/gravity design to EC3 (EN 1993-1) resulted in initial column sections of HEB300 for the four lower storeys and HEB220 for the upper five storeys, initial beam sizes of IPE450 for the 3 and 8.5 m length beams and IPE550 for the 10 m beams at the first floor level. A desirable base isolation design would sufficiently reduce the imposed seismic demand on the superstructure to allow these sizes to be retained also for the seismic design.

Brace sizes were determined for the non-isolated superstructure in Section 6.9, assuming that seismic demand governed over wind demand. For the base-isolated building, seismic demand may be reduced sufficiently such that wind loads govern. In any case, the brace sizes required to resist the reduced seismic forces are calculated in the following sub-sections, which would need to be checked also for wind loads in a complete design.

10.6.3 Seismic design checks

10.6.3.1 Design of base isolation system

Any of the devices introduced in Section 10.3 could be used for the seismic isolation of the example building. In this example, lead-rubber bearings are considered.

One advantage of seismic isolation is that a structure with eccentric centres of stiffness and mass can be regularised by a judicious arrangement of isolation devices. In this example, the superstructure is regular, and isolators can be conveniently placed at the base of some or all columns to transfer gravity loads directly to the foundation. We consider isolators on the same gridlines as the braced frames in the x-direction – that is, 1, 3, 5, 7, 9, 11, 13 and 15 – and at all gridlines in the y direction, for a total of $8 \times 6 = 48$ isolators. Gravity loads from other column locations would need to be transferred to the isolators and foundations through a transfer structure at ground level. Computer analysis of the building suggests that isolators should be designed for a peak vertical reaction of 5,220 kN, for a factored gravity load combination from EC8. In addition, the load combination $G + \psi_{E,i}Q$ gives a peak reaction of 3,220 kN, which must be resisted simultaneously with the design isolator displacement (these values apply to frames on the internal gridlines 3, 5,..., 13, which have higher gravity loads, but since we are assuming identical isolators at all locations, this capacity requirement holds for GL1 also).

Since the building is of importance class II, the mean values of effective stiffness and viscous damping may be used, provided that the upper and lower bound values are within ±15% of the mean values (Cl. 10.8(4)). Note that this is not allowed for buildings of importance classes III and IV, in which case upper and lower bound properties should be evaluated separately, with the former controlling acceleration and inertia forces in the superstructure, and the latter controlling maximum isolator displacements.

In order to apply the simplified linear analysis method in EC8 Cl. 10.9.3 (notwithstanding the plan dimension limit of 50 m), the effective period, T_{eff}, should be greater than three times the period assuming a fixed base, but less than or equal to 3 seconds, to ensure that the first two modes of the isolated structure are sufficiently decoupled to assume rigid superstructure response (see Section 10.5.2.3). For this building, the lower limit is 3×0.62 s = 1.86 s. Since this is almost equal to the value of $T_D = 2$ s for the Type 1 design spectrum in EC8, it is likely that an effective isolation design will be at or near the displacement plateau of the EC8 design displacement spectrum (i.e. in the $1/T^2$ portion of the design acceleration spectrum). As noted in Section 10.5.2.3, however, such a low value of T_D is not supported by current understanding of long period response spectral ordinates.

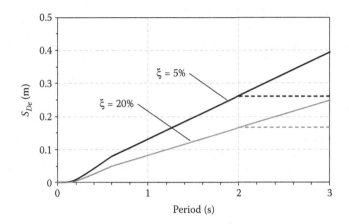

Figure 10.12 Displacement design spectrum for 5% and 20% damping; dashed lines for $T_D = 2$ s, and solid lines for $T_D > 3$ s.

The elastic displacement design spectrum obtained from the acceleration design spectrum in Section 3.6.3 is shown in Figure 10.12. Isolator displacements are one of the key design parameters to check in a seismic isolation design. As discussed in Section 10.5.2.3, the isolator displacement check in the simplified linear analysis procedure is essentially a direct displacement-based design (Priestley et al., 2007), and therefore the displacement design spectrum is the most effective way of demonstrating the effect of design variables on response of the seismic isolation system. The spectrum is shown for damping values of 5% and 20%, the latter corresponding to a typical effective damping value for lead-rubber bearings. For damping values greater than 28%, the lower limit of 0.55 for the damping reduction factor, η, applies (Cl. 3.2.2.2(3)). The solid lines show the displacement spectrum for $T_D > 3$ s, and the dashed lines show the displacement plateau for $T_D > 2$ s.

10.6.3.1.1 Design assuming corner period (T_D) as in EC8

If we retain the value of $T_D = 2$ s from EC8, and assume a damping value of 20% is attainable from a lead-rubber bearing, then the design displacement for effective periods greater than 2 s may be calculated from EC8, Equations 3.5 and 3.7:

$$S_{De}(T) = 3.0 \text{ m/s}^2 \times 1.15 \times \sqrt{\frac{10}{5+20}} \times 2.5 \times \left[\frac{0.6 \text{ s} \times 2.0 \text{ s}}{T^2}\right] \times \left[\frac{T}{2\pi}\right]^2 = 0.166 \text{ m}.$$

The spectral displacement is equivalent to Equation 10.5 from EC8, and gives the design displacement of the stiffness centre of the isolation system due to seismic action (in each horizontal direction), d_{dc}.

The simplified linear analysis method may only be applied when the total eccentricity (including accidental eccentricity) between the stiffness centre of the isolation system and the centre of mass of the superstructure is less than 7.5% (Cl. 10.9.3(2)). In this case, the centre of stiffness and centre of mass are concentric, and therefore only the accidental eccentricity of 5% of plan floor dimension (Cl. 4.3.2(1)) applies.

Assuming $K_{xi} = K_{yi}$ is constant for all isolators, the torsional radius from Equation 10.23 is calculated as $r_y = 22.4$ m. Isolators on gridline 1 have $y_i = 28$ m. The accidental eccentricity from EC8 Cl. 4.3.2(1) is 5% of the plan dimension perpendicular to the direction of load application, or 0.05×58 m $= 2.9$ m. Therefore, the highest value of the amplification factor from Equation 10.22 is

$$\delta_x \left(\text{gridlines 1A,1F}\right) = 1 + \frac{2.9}{22.4^2} \times 28 = 1.16.$$

The design isolator displacement must also be multiplied by the magnification factor, γ_x, for which the value is found in the National Annex, but which has a recommended value of 1.2 (Cl. 10.3(2)).

Therefore, the design isolator displacement for each isolator is given by

$$d_{db} = 1.2 \times 1.16 \times 0.166 \, \text{m} = 0.231 \, \text{m} = 231 \, \text{mm}.$$

Isolation design is typically a trade-off between displacement across the isolation interface and seismic forces introduced into the superstructure. However, due to the artificially capped demand in the design displacement spectrum following EC8 (Figure 10.12, dashed lines), there is no longer a trade-off, and there is theoretically no limit to the effectiveness of the isolation that can be obtained. A sensible starting point, therefore, may be to target an effective period of 3 s, which is the upper limit of the range allowed by EC8 for simplified linear analysis. The only disadvantage of targeting a flexible isolation system is that, for a given gravity load capacity, this means a taller bearing, and likely a higher unit cost.

Inverting EC8 Equation 10.1, the lower bound effective stiffness of the isolation system is given by

$$K_{eff} = M \left(\frac{2\pi}{T_{eff}} \right)^2 = 9{,}945 \, \text{t} \times \left(\frac{2\pi}{3.0 \, \text{s}} \right)^2 = 43{,}620 \, \text{kN/m},$$

which is allocated amongst the 48 isolators, to give 908.8 kN/m, or 0.9088 kN/mm per isolator.

The effective stiffness is defined in Cl. 10.9.2(2) of EC8 as the secant value of the stiffness at the total design displacement. Since the force–displacement response of lead-rubber bearings is approximately bilinear, the effective stiffness is displacement dependent. Therefore, according to Cl. 10.9.2(4), an iterative procedure should be adopted until the difference between assumed and calculated d_{dc} does not exceed 5%. Since this is evaluated on d_{dc}, it does not include the amplification factors γ_x and δ_x. Therefore, the effective stiffness should be evaluated at a displacement of 166 mm. This also applies to the value of effective viscous damping.

Product catalogues of isolation device suppliers may report effective stiffness values at the design isolator displacement (i.e. 231 mm in this case, rather than 166 mm), and therefore, for non-linear devices, these values need to be adjusted based on the system displacement. This should be carried out based on literature relationships for equivalent viscous damping, as discussed in Section 10.5.2.1.

Therefore, to summarise the design requirements for the isolators so far, we have

- Axial capacity of at least 5,250 kN under zero lateral displacement.
- Axial capacity of at least 3,240 kN at design displacement.
- Effective damping of around 20%.

- Design displacement of at least 231 mm.
- K_{eff} per isolator of around 0.909 kN/mm (but no lower, as then the upper limit on period would be exceeded), evaluated at a displacement of 166 mm.

In a real project, the following other requirements could also apply:

- Sufficient stiffness (or supplemental lock-up devices) under small displacements such that wind does not result in excessive movement.
- Design displacement possibly limited by non-structural considerations such as building services.
- Limits to expected residual deformation at the end of a design earthquake ground motion.
- There would be little benefit to reducing the seismic forces applied to the superstructure below the applied wind forces, and therefore this could limit the value in designing for long effective periods.

Consulting the product catalogue of a supplier, it was determined that a lead-rubber bearing with rubber diameter at least 750 mm would be required to resist the gravity loads. Larger isolators were also considered, but the device with properties summarised in Table 10.1 gave the lowest effective stiffness at the bearing displacement, and was therefore selected for the design.

An iterative procedure was adopted whereby the isolation system displacement was calculated, the corresponding effective stiffness and effective damping of the isolators were evaluated, giving a new effective period and a new estimate of system displacement, and so on. The process converged on a design displacement of 152 mm, and an effective isolation period comprising 48 of these devices of

$$T_{eff} = 2\pi\sqrt{\frac{M}{K_{eff}}} = 2\pi\sqrt{\frac{9{,}945\,\text{t}}{48 \times 2{,}250\,\text{kN/m}}} = 1.91\,\text{s},$$

which satisfies the limits for the application of the simplified linear analysis procedure, and is actually slightly less than the corner period of 2.0 s.

The vertical stiffness of the isolators is also reported in Table 10.1. Simplified linear analysis to EC8 imposes two requirements on the vertical stiffness of the isolation system, given in Equation 10.24. The ratio of the vertical and horizontal stiffness of the isolation system is

Table 10.1 Properties of isolation device selected for building design

Diameter of bearing	750 mm
Total rubber thickness	135 mm
Total height	297 mm
Lead plug diameter (back-calculated from yield force)	140 mm
Design displacement	270 mm
Max force @ $d_{db} = 0$	5,350 kN
Max force @ d_{db} = design displacement	3,240 kN
K_{eff} @ design displacement	1.80 kN/mm
K_{eff} @ 143 mm	2.25 kN/mm
Vertical stiffness	1,533 kN/mm
Effective damping @ 143 mm	25.6%

$$\frac{K_v}{K_{eff}} = \frac{48 \times 1{,}533}{48 \times 2.25} = 681 \geq 150.$$

The fundamental period in the vertical direction is

$$T_v = 2\pi\sqrt{\frac{M}{K_v}} = 2\pi\sqrt{\frac{9{,}945\,\text{t}}{48 \times 1{,}533{,}000\,\text{kN/m}}} = 0.073\,\text{s} \leq 0.1\,\text{s}.$$

Therefore, both the requirements on vertical stiffness are satisfied.

Note that the isolation system design presented in this section was predominantly con-trolled by the required strength of the isolators under gravity loading. Therefore, an alter-native, more efficient, design could be envisaged in which a portion of the gravity load was carried on slider surfaces, allowing smaller, more flexible, lead-rubber bearings to be specified. This design would also have to take into account the friction on the sliding interface in the calculation of effective period and damping. It is likely that an effective period closer to the 3.0 s value initially targeted in the design could be achieved using this approach.

10.6.3.1.2 Design assuming corner period (T_D) increased

In this example, the effective period determined for the isolation system did not exceed the corner period, and therefore the design also applies for the case where the corner period is increased to more realistic values. Therefore, the isolation device specified in Table 10.1 is also adequate for this case.

If the alternative suggestion of a hybrid bearing and slider isolation system were adopted, higher effective periods could be achieved, and the decision would have to be made as to whether the value of T_D from EC8 is used without modification, or if a more realistic, higher value is adopted.

10.6.3.2 Design of superstructure

The design of the isolation system was essentially a displacement-based procedure, in which the spectral displacement was evaluated directly against the displacement capacity of the bearing devices (with modification for accidental torsion and increased reliability required for seismic isolation). The design of the superstructure to EC8 is a force-based procedure, taking into account the reduction in seismic forces expected due to the period elongation and energy dissipation in the isolation system.

Applying the simplified linear analysis procedure, the horizontal forces at level j are calcu-lated according to Equation 10.21. The effective period and damping value should be evalu-ated at the actual displacement expected in the isolation system, as discussed previously. The spectral acceleration is

$$S_e(T_{eff}, \xi_{eff}) = 3.0\,\text{m/s}^2 \times 1.15 \times \sqrt{\frac{10}{5+25.6}} \times 2.5 \times \left[\frac{0.6\,\text{s}}{1.91\,\text{s}}\right] = 1.55\,\text{m/s}^2.$$

These seismic forces may be divided by a behaviour factor of up to 1.5 for design of the superstructure. The resulting design seismic forces are shown in Table 10.2, for the whole

Table 10.2 Total lateral forces for base-isolated building design

Level		Total lateral forces (kN)	
		Total	GL1 only
8		613	99.7
7		1,021	166.0
6		1,021	166.0
5		1,021	166.0
4		1,021	166.0
3		1,021	166.0
2		1,021	166.0
1		1,740	282.8
0		1,795	291.7
Base shear (kN)	Incl L0	10,277	1,670.0
	Not incl L0	8,482	1,378.3
Base moment (MNm)		126.6	20.6

building, and for the frame on gridline 1. The forces for GL1 were increased by a factor of 1.3 to account for accidental torsion (as in the previous design examples).

Comparing the total lateral forces with those for different frame types in Table 3.5, it is apparent that the total base shear is comparable to a steel MRF, and less than for a concrete MRF and dual system. This comparison does not include the lateral force on the ground level (L0), which is not carried by the superstructure. The base moment is decreased with respect to these different frame types, as the uniform acceleration distribution results in a centre of seismic force around halfway up the building, rather than around 2/3 of the way up for a triangular force distribution.

Comparing the forces derived for GL1 for the steel CBF example in Table 6.8, it is apparent that the design forces have been reduced, except for the forces applied in the lower levels (due to the constant acceleration distribution). Similarly, the base shear and base moment have also been reduced significantly.

These comparisons of design forces, though instructive, do not take into account the primary benefit of seismic isolation. The design forces in Table 10.2 were evaluated for a q factor of 1.5, corresponding to nominally elastic response in the superstructure. The design forces for the other lateral systems took benefit from higher values of q, associated with the development of ductility in structural elements. In the isolated building, almost all of the deformation and energy dissipation is concentrated in the 48 engineered devices, rather than distributed over the entire building. Isolators performing up to their design displacement should not require repair or replacement following the design ground motion, and therefore it should be possible to reoccupy and use the building immediately following a rare earthquake event with minimal repair cost.

The lack of ductility demand in the structure is reflected in EC8 Cl. 10.10(3) and (4), which allow ductility class L to be adopted for isolated steel, concrete or steel-concrete composite buildings, and explicitly does not require capacity design, global or local ductility checks to be satisfied. In the present example, this means that braces in compression can also be taken into account (which was not allowed for DCM and DCH designs), which further reduces the demand on individual braces. Furthermore, limitations on allowable cross-section class and member slenderness associated with higher ductility classes do not need to be satisfied. Finally, beam and column actions are significantly

Table 10.3 Axial forces in the braces for the
seismic combination

Storey	Element no.	Section	N_{Ed} (kN)
1	69	$200 \times 100 \times 12.5$	±486.0
	70	$200 \times 100 \times 12.5$	±444.6
2	71	$150 \times 100 \times 12.5$	±417.5
	72	$150 \times 100 \times 12.5$	±266.1
3	73	$150 \times 100 \times 8.0$	±320.0
	74	$150 \times 100 \times 8.0$	±254.6
4	75	$150 \times 100 \times 6.3$	±266.4
	76	$150 \times 100 \times 6.3$	±202.3
5	77	$150 \times 100 \times 5.0$	±204.4
	78	$150 \times 100 \times 5.0$	±173.0
6	79	$150 \times 100 \times 4.0$	±151.3
	80	$150 \times 100 \times 4.0$	±120.4
7	81	$120 \times 60 \times 8.0$	±83.6
	82	$120 \times 60 \times 8.0$	±80.9
8	83	$120 \times 60 \times 3.6$	±32.2
	84	$120 \times 60 \times 3.6$	±29.7

reduced, as they do not need to be adjusted for the development of over-strength in the braces.

The axial forces in the braces are summarised in Table 10.3, noting that braces now act in both tension and compression, and therefore values are positive and negative. Members were sized to EC3 provisions for buckling resistance (EC3 Section 6.3). Note that brace sizes are significantly smaller than for the CBF example in Table 6.10. As noted earlier, these would also need to be checked for wind loading in a full design.

Designing for DCL means that other framing members do not require capacity design provisions to be taken into account. Beams, columns and connections can be designed directly for the forces from the analysis, rather than considering the development of brace over-strength. The analysis model showed that gravity load combinations govern their design, and therefore the preliminary column and beam sizes given in Section 10.6.2 may be retained for the base-isolated structure. This is in contrast to the non-isolated CBF case where these were increased from preliminary sizes to accommodate seismic demand.

Note also that according to EC8 Cl. 4.3.3.5.2, the vertical component of the seismic action should always be taken into account for base-isolated structures if the reference vertical acceleration, a_{vg}, is greater than 2.5 m/s². In this example, a_{vg}, is 2.7 m/s², so vertical input should be considered. This was ignored in the calculation of the peak vertical load in the isolation system in Section 10.6.3.1, but would generally be included in the analysis model by applying 30% of the vertical design seismic action according to the load combinations given by Equations 4.20 and 4.21 in EC8. In applying the vertical ground motion to the superstructure, the design forces may be reduced by a q factor up to 1.5 (Cl. 3.2.2.5(6)).

10.7 CONCEPTS OF SUPPLEMENTAL DAMPING

An alternative (or complementary) approach to reducing the seismic demand of a structure is to distribute passive energy dissipation throughout, rather than concentrating flexibility and damping in a single interface. From an energy perspective, seismic isolation works by reducing the earthquake energy that reaches the structure, while distributed supplemental damping dissipates a sufficiently large portion of earthquake energy in damping devices (Christopoulos and Filiatrault, 2006). In many ways, a distributed damping strategy is no different from that adopted in conventional seismic design, where allowing material non-linearity in ductile mechanisms reduces the design demand, and capacity design principles are used to ensure that energy dissipation is not concentrated in just a few elements. However, in common with seismic isolation, a supplemental damping strategy has the advantage that energy dissipation is primarily confined to engineered devices, which can be replaced if necessary following a large earthquake event, rather than in structural elements, whose inspection, repair or replacement can be prohibitively expensive.

For a single degree of freedom (SDOF) system, the advantages of supplemental damping are evident from a typical earthquake response spectrum evaluated for multiple damping levels. It is clear from Figure 10.13 that increasing the viscous damping from a typically assumed value of 5% (which is, in any case, unrealistically high for many structures, particularly tall buildings; see Smith et al., 2010), to 10% or 20% can significantly reduce the seismic demand. On the other hand, increasing damping further from 20% to 30% has a smaller impact. For practical damping systems, the achievable viscous damping is between 5% and 30%, with the upper limit a function of the structural configuration, and the load paths that are engaged by the dampers.

As alluded to above, the inherent viscous damping in buildings is not well understood nor quantified. An advantage of inserting engineered devices into the structure is that a reliable viscous damping can be achieved, with known dependence (if any) on frequency and amplitude of loading.

For multiple degree of freedom (MDOF) systems, dampers are typically distributed over the height of the building, to share the energy dissipation over multiple devices, to effectively engage dampers in multiple modes of response and to provide redundancy in case one or more devices do not perform as anticipated.

The configuration of dampers depends on both architectural considerations, and on the type of device employed. Figure 10.14 shows some typical arrangements. The diagonal brace arrangement in Figure 10.14a is appropriate where devices are designed to dissipate energy under axial loading, such as for hydraulic viscous dampers, buckling restrained braces and

Figure 10.13 Pseudo-acceleration response spectra with 5%, 10%, 20% and 30% viscous damping, for Tabas 1978 earthquake, Dayhook recording station.

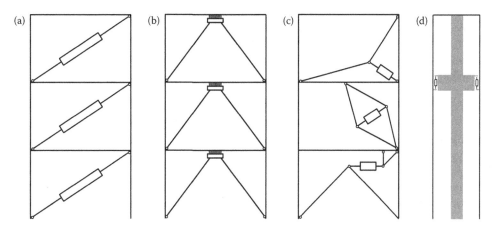

Figure 10.14 Configurations for supplemental dampers. (a) Diagonal bracing; (b) chevron bracing; (c) examples of geometrical amplification configurations (top: toggle brace, middle: scissor jack, bottom: horizontal brace); (d) damped outrigger system for tall buildings.

some friction-based and viscoelastic dampers. The chevron arrangement in Figure 10.14b is appropriate when devices dissipate energy in shear, such as many hysteretic dampers. See Section 10.8 for a description of these damping devices.

Sigaher and Constantinou (2003) pointed out that the diagonal arrangement in Figure 10.14a is not optimal, in that it is possible to achieve larger damper strokes under the same level of storey drift. Some possible configurations to provide geometrical amplification are shown in Figure 10.14c. This can be necessary for stiff frames, where design damper strokes may be very small, in which case the need to overcome friction, tolerance and elastic deformation in the connections, and compressibility of fluid (in fluid viscous dampers) means that damping is limited or negligible. For more information on geometrical amplification, refer to Sigaher and Constantinou (2003).

Not all supplemental damping systems involve distributing devices over the full height of the building. The damped outrigger concept, for example, is a variant of a core and outrigger system, commonly used in tall buildings to increase the lateral stiffness (Smith and Willford, 2007). Instead of connecting the outrigger rigidly to the perimeter framing, thereby engaging the perimeter framing and increasing the stiffness significantly, the outrigger is connected via a fluid viscous damper, as shown in Figure 10.14d. As the building vibrates, the outrigger semi-rigidly transforms the rotation of the core into relative vertical movement at the perimeter of the building, which engages the viscous damper, dissipating energy. The concept was employed on the St Francis Shangri-La Place development in Manila, the Philippines (Smith and Willford, 2007). The dampers, provided by FIP Industriale, were designed with a high damping exponent (see Section 10.8.2) for low velocities, and with a pressure relief valve to limit the forces transmitted under extreme seismic excitation.

As with isolation, the material savings in the structure that can be realised by introducing supplemental dampers must be balanced with the cost of the devices themselves, as well as their testing, installation, maintenance and inspection. Dampers could be used in conjunction with a limited ductility (DCL) design of the structure, assuming that the seismic forces can be brought down to a level that can be economically designed for elastically. Alternatively, they could be used to complement energy dissipation obtained from the development of a ductile yielding mechanism, satisfying the design requirements of DCM or DCH in the relevant chapter of EC8. In the latter case, the various load paths should be carefully considered, as yielding of the structure will significantly reduce its stiffness, and this may 'fuse out' the dampers, reducing their effectiveness.

10.8 DAMPING DEVICES

As with the isolation systems discussed in Section 10.3, a number of supplemental damping devices are available for seismic protection of buildings. A brief discussion is provided in the following, but the reader is referred to Christopoulos and Filiatrault (2006) for more details of the systems.

Damping systems may be grouped into displacement-dependent and velocity-dependent – the former comprising devices based on material (typically steel) hysteresis and those based on friction, and the latter comprising hydraulic viscous and viscoelastic dampers. Self-centring devices, which may typically be considered displacement-dependent, are considered separately in the following. Tuned mass dampers are not considered here, as they are typically used for wind applications only, and are of limited effectiveness for seismic protection (Carr, 2005).

10.8.1 Displacement-dependent devices

Displacement-dependent devices are those that dissipate energy based on the relative displacement across the device independently of the frequency of excitation. The most common among these devices are those which use the hysteretic energy dissipation associated with the cyclic yielding of metals, and those that use friction to dissipate energy. Most displacement-dependent devices exhibit elastic-perfectly-plastic or bilinear hysteresis, as shown in Figure 10.9. For friction-based devices, the initial stiffness is very large.

For metallic hysteretic dampers, the primary concern is developing a configuration that exhibits dependable, repeatable cyclic hysteresis and is not susceptible to low cycle fatigue. Therefore two important considerations are (1) determining a geometry of metallic elements that effectively distributes the yielding over as large a volume of material as possible, and not concentrating plasticity in one area; and (2) preventing buckling of elements.

For steel hysteretic elements based on yielding in flexure, the first objective is met by shaping the element such that its profile matches the bending moment diagram. If this is achieved, the element will yield approximately simultaneously throughout the whole height of the member. This is the basis of the added damping–added stiffness (ADAS) systems, and triangular ADAS (TADAS) systems, which have profiles matching a fixed-fixed, and fixed-pin bending moment diagram under constant shear force, respectively. Several steel plates are placed in parallel to share the energy dissipation, thereby providing redundancy. These may be placed in the chevron configuration shown in Figure 10.14b to engage the lateral movement of the braces with respect to the beam. Since no gravity load is transmitted from the beam, the only axial load in the steel elements is due to second order non-linear geometrical effects, and therefore buckling is unlikely to be a concern.

An alternative approach to satisfying the two considerations discussed above is represented by the buckling restrained brace (BRB). BRBs comprise steel brace elements, encased in concrete or other material to prevent them from buckling, and allowing them to yield in both tension and compression. The interface between steel and concrete is lubricated to ensure that no axial load is transmitted to the concrete, and so that stress concentrations in the steel caused by local bonding are prevented.

Steel is not the only material used for metallic dampers. Lead is an attractive alternative, for the reasons discussed in Section 10.3.1. One system making use of the hysteretic behaviour of lead is the lead extrusion device (LED), which resembles a fluid viscous damper (see next section), but involves the extrusion of lead through a constricted tube. This device exhibits very stable hysteretic response.

Several devices make use of energy dissipation in friction. Generally, the sliding surfaces are carefully controlled to ensure that a predictable friction coefficient can be obtained. Coulomb friction is essentially rigid-perfectly-plastic, and depends only on the preload normal to the surface. Unlike friction-based isolation devices (Section 10.3.2), generally friction dampers do not bear the weight of the building, and therefore pre-tensioning must be used to provide a controlled normal preload and slip force. Friction damping devices are discussed in detail by Christopoulos and Filiatrault (2006); they do not appear to have been used on any European projects to date.

10.8.2 Velocity-dependent devices

Velocity-dependent devices dissipate energy based on the relative velocity across them, and are therefore frequency-dependent. Fluid viscous dampers are purely velocity dependent, while dampers based on viscoelastic materials are both displacement and velocity dependent.

Fluid (hydraulic) viscous dampers comprise fluid, typically silicon based, forced through a narrow orifice under pressure. Depending on the fluid characteristics and orifice dimensions, the force–velocity (F_d–v) response can be linear or non-linear, given by the following expressions, respectively:

$$F_d = Cv, \quad \text{and} \quad F_d = C_{NL}v^\alpha. \tag{10.25}$$

Note that the units of the linear damper coefficient, C, are force/velocity (e.g. kNs/m), while those of the non-linear coefficient C_{NL} depend on the exponent, α. Typical values for α are between 0.2 and 2.0. Values less than 1.0 are most useful for seismic applications, as damper forces are reduced at high velocities, and the energy dissipated per cycle for the same stroke is greater (Christopoulos and Filiatrault, 2006).

Fluid dampers are purely velocity-dependent and do not transmit forces under slowly applied loading. Devices based on viscoelastic material, on the other hand, have a displacement-dependent component in addition to their velocity dependence. Their response is described by the following expression:

$$F_d = Kd + Cv, \tag{10.26}$$

where d and v are the displacement and velocity across the device, respectively, and K and C are stiffness and damper constants that depend on the constitutive properties of the material and the geometry of the material within the device.

Viscoelastic material damping is typically defined in terms of the loss factor, η, which is related to the other constants by the following expression:

$$\eta = \frac{C\omega}{K}, \tag{10.27}$$

where ω is the excitation frequency. Note that for purely viscous devices, the stiffness term, K, is zero, and therefore the loss factor is undefined.

10.8.3 Self-centring devices

Traditional seismic design assumes that the peak response of a structure during a strong earthquake ground motion is the most relevant parameter for assessing the structure's performance. While this may be approximately true for assessing the possibility of collapse (and even in this case, energy-based considerations are relevant for cyclic degradation and low-cycle fatigue), it does not consider in what state the structure will be left at the end of the ground shaking. For many structures, the costs of damage repair and/or downtime following an earthquake may be severe, even if the structure has performed 'well' with respect to the design intent of maintaining life safety.

There is a growing recognition in the earthquake engineering community that traditional approaches to earthquake resistant design may not be optimal from the point of view of the residual deformation in the structure after the earthquake. Although there are strategies available at the design stage to mitigate potential residual deformations (Pettinga et al., 2007), a more direct approach is to employ a system that inherently dissipates energy but returns to a residual displacement of zero when load is removed. These systems are referred to as 'self-centring'.

Aside from some systems already discussed that may be considered self-centring – rocking and fluid viscous dampers – and some of the mitigation strategies that may be incorporated into conventional seismic design (Pettinga et al., 2007), self-centring systems may be classified into two main categories: (1) those that use stand-alone self-centring devices, such as the energy dissipating restraint (EDR) and ring springs, as the key source of energy dissipation, and (2) those that make use of unbonded post-tensioning to provide a restoring force and allow joints to open and close during an earthquake event.

The latter works on exactly the same principle as rocking, discussed in Section 10.3.4, and, in fact, many of the systems that have been developed using this technology also make use of rocking at the base. The first such systems were developed for pre-cast reinforced concrete as part of the PRESSS program at the University of California, San Diego, in the 1990s (Priestley et al., 1999), but variants on the theme have been continually developed throughout the world in the last one and a half decades. Instead of the restoring force being provided by the weight of the structure, it is provided by the force in the post-tensioning. On its own, this leads to non-linear elastic force–displacement response; energy dissipation can be added by connecting joints with mild steel rods or other damping devices, without compromising the self-centring behaviour.

Christopoulos and Filiatrault (2006) provide an excellent review of self-centring technologies, although, as with any new field, new developments appear regularly in the technical literature.

10.9 NUMERICAL ANALYSIS AND DESIGN OF DAMPED SYSTEMS

Chapter 10 of EC8 is explicitly not intended for the design of distributed damping systems (Cl. 10.1(4)). Chapter 18 of the US Standard ASCE 7-10 contains significantly more developed procedures for the design of structures with damping systems, and this is a useful resource provided that the reader keeps in mind the usual dangers of mixing codes. The following is a brief set of design criteria, based primarily on the work of Christopoulos and Filiatrault (2006), consistent with the EC8 rules for design of isolated structures. These criteria are intended to be used for preliminary design only, and the final design should be verified with time history analysis that explicitly accounts for the force–velocity and/or force–displacement of the devices used.

10.9.1 Viscous and viscoelastic dampers

Christopoulos and Filiatrault (2006) present an analysis of the modal response of an elastic MDOF system with added linear viscoelastic dampers, assuming that the damping matrix corresponding to the added dampers is classical; i.e. that it is orthogonal with respect to the mode shape vectors, as are the mass and stiffness matrices. Since dampers are often arranged in a configuration in which they connect adjacent storeys, such as in Figure 10.14a, it may sometimes be assumed that the added damping matrix is proportional to the stiffness matrix, and therefore retains the mode shape orthogonality. Figure 10.15a is a schematic illustration of the model, in which a spring representing the structure is in parallel with a viscoelastic spring-damper (k_2 and c_2), neglecting the mass of the dampers and the inherent damping of the structure, which is small compared to the damping added. Note, however, that Figure 10.15a shows an SDOF model, while the results are applicable to MDOF systems, provided that the damping matrix is classical.

Christopoulos and Filiatrault (2006) show that the modal damping ratio of the ith mode of this system is given by

$$\xi_i = \frac{\eta_i}{2}\left(1 - \frac{\tilde{T}_i^2}{T_i^2}\right), \tag{10.28}$$

where $\tilde{T}_i$ is the ith period of the system incorporating the additional elastic stiffness provided by the dampers, T_i is the ith period of the undamped system, and η_i is the loss factor of the viscoelastic material responding at a period $\tilde{T}_i$.

For design of structures equipped with viscoelastic dampers, Christopoulos and Filiatrault (2006) propose that the designer selects a target damping ratio in the fundamental mode of the structure, ξ_1, and dampers are provided based on the available viscoelastic material such that the added stiffness at storey i, $\bar{k}_i$, is proportional to the stiffness of ith storey of the undamped structure, k_i. Then, the added stiffness and the loss factor of the dampers are selected such that

$$\bar{k}_i = \frac{2\xi_1 k_i}{\eta - 2\xi_1}. \tag{10.29}$$

From this, the properties of the viscoelastic material and the geometry of the damper configuration can be determined. Refer to Christopoulos and Filiatrault (2006) for more details.

Figure 10.15 Spring-damper models for numerical analysis of supplemental viscous or viscoelastic damping.

For purely viscous dampers (i.e. in Figure 10.15a, $k_2 = 0$), the loss factor is undefined, and Equations 10.28 and 10.29 cannot be used. Christopoulos and Filiatrault (2006) propose an alternative design procedure based on modal analysis, in which viscous dampers are replaced with fictitious springs. To retain the classical damping assumption, the stiffness of each fictitious spring is set to be proportional to the storey stiffness, with a value given by

$$\hat{k}_0 = \frac{2\pi}{T_1} C_L, \tag{10.30}$$

where C_L is the damping coefficient of the damper.

The authors then show that the required period of the fictitiously braced structure for a given target viscous damping ratio, ξ_1, is given by

$$\hat{T}_1 = \frac{T_1}{\sqrt{2\xi_1 + 1}}, \tag{10.31}$$

where T_1 is the initial undamped period of the structure. This result only applies as long as the classical damping assumption holds; this will be discussed further below.

For design, the following steps are carried out:

1. The initial undamped period, T_1, is evaluated using a structural analysis program.
2. A target damping ratio, ξ_1, is selected.
3. The required fundamental period of the fictitiously braced structure is evaluated from Equation 10.31.
4. Trial fictitious springs are introduced at each storey with stiffness proportional to the storey stiffness, and another structural analysis is carried out to determine the trial damped period, $\hat{T}_{1,tr}$. Since the stiffness is proportional to the square of the period, the required spring stiffness of spring n can be calculated from the trial stiffness of spring n in a single iteration by scaling them by the following factor:

$$\frac{\hat{k}_0^n}{\hat{k}_{0,tr}^n} = \frac{1}{1 - \left((\hat{T}_1^2 - \hat{T}_{1,tr}^2) / (\hat{T}_1^2 - T_1^2) \right)}. \tag{10.32}$$

5. Calculate the damping coefficient of each damper from Equation 10.30. Other properties of the damper, such as maximum force and stroke, are determined from non-linear time history analysis.

For non-linear dampers, the required damping coefficient, C_{NL}, can be evaluated based on matching the energy dissipated per cycle in the linear damper. For non-linear dampers with exponents in the range $0.2 \leq \alpha \leq 1$, the following expression gives an approximate value for the non-linear damping coefficient:

$$\frac{C_{NL}}{C_L} \approx \frac{\sqrt{\pi}}{2} (\omega X_0)^{1-\alpha}, \tag{10.33}$$

where X_0 is the peak displacement amplitude, and ω is the excitation frequency. The units of C_{NL} depend on the damping exponent, and therefore consistent units must be used in Equation 10.33.

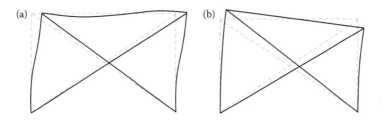

Figure 10.16 (a) Shear and (b) flexural deformation of one-storey MRF with diagonal dampers.

The methods of Christopoulos and Filiatrault (2006) are appropriate for preliminary design of viscoelastic and viscous damping systems, and they stress that they should always be verified with time history analysis. One weakness of the methods is their reliance upon the classical damping assumption, which, even for simple structures, breaks down as the target damping ratio increases. Almost all structural configurations exhibit multiple load paths, rather than the simple spring with stiffness k_1 representing the structural response in Figure 10.15a. For example, consider the simple single-storey moment resisting frame connected with a diagonal damper illustrated in Figure 10.16. The frame can deform in global shear (flexural and shear deformation in beam and columns), as shown in Figure 10.16a, and global flexure ('push-pull' axial deformation in columns), as shown in Figure 10.16b. Only the former is restrained by the diagonal damper; the latter would require dampers placed vertically in the columns, similar to the rocking South Rangitikei rail bridge and Maison Hermès examples discussed in Section 10.3.4. Michael Willford (2010) developed an unpublished design method based on the system shown in Figure 10.15b, which takes into account alternative load paths in real structures, and consequently the upper limits on damping that can be added based on damping one load path only.

10.9.2 Friction and hysteretic dampers

Damping devices based on dissipating energy in friction or in material hysteresis are similar, in that they are both displacement-dependent (rather than velocity-dependent like the viscous and viscoelastic devices discussed in the previous section). They can also usually be described by the same hysteresis models, defined by a stiff or rigid initial stiffness, followed by yield or slip and reduced stiffness, and large hysteresis loops under cyclic loading. The primary design parameter of interest is the slip/yield force to specify for each device.

Christopoulos and Filiatrault (2006) summarise the design method of Filiatrault and Cherry (1990), based on the concept of optimum hysteretic design spectra. The method is only applicable to buildings of up to 10 storeys, with dampers arranged regularly, such as in Figure 10.14a, is not applicable in the near-field of a major fault where pulse-like ground motions may be expected, and values of certain parameters are also limited, as discussed below. Nevertheless, for buildings where these criteria are met, it provides a quick and simple preliminary design method for friction and hysteretic dampers that can be verified and fine-tuned with time history analysis.

The method is based on optimising a 'relative performance index' which is a dimensionless index based on energy dissipated in the damped structure compared with that in the undamped structure. A structural model is analysed both with and without dampers, to give fundamental periods T_b and T_u. Note that the former includes the actual stiffness of the braces, pre-yield or slip. The authors recommend that braces are sized such that they are as stiff as is economically feasible, with an upper limit of T_b/T_u of 0.4 for the damping to be effective.

Through parametric analyses, Filiatrault and Cherry determined the following expression for the optimum total shear force, V_0, required to activate or yield all the dampers in the structure, as a ratio of the total weight, W:

$$\frac{V_0}{W} = \frac{a_g}{g} Q\left(\frac{T_b}{T_g}, \frac{T_b}{T_u}, N_f\right), \tag{10.34}$$

where a_g is the peak ground acceleration, and Q is an empirical, dimensionless function of the three variables shown. N_f is the number of storeys, with $N_f \leq 10$. T_g is the predominant period of the site, which is perhaps the most difficult parameter to determine, especially for a structural engineer. For $T_g/T_u \geq 1$, the value of Q is relatively insensitive to T_g.

Values of the empirical function, Q, are presented in terms of optimum hysteretic design spectra, reproduced from Christopoulos and Filiatrault (2006) in Figure 10.17. They also provide the expressions in parametric form. The design spectra are piecewise linear, exhibiting a very different dependence on T_g/T_u for values less than and greater than unity.

Aside from the limitations already discussed, the optimum hysteretic design spectra are valid only within the following range of parameters: $0.2 \leq T_b/T_u \leq 0.8$, $0.2 \leq T_b/T_g \leq 0.8$, $0.05 \leq a_g/T_g \leq 0.8$ (the units of the final inequality are not clear from the reference, and the example given in Christopoulos and Filiatrault (2006) seems to fall outside this range whether a_g is in units of g or m/s²).

When the total slip or yield shear force has been determined from Equation 10.34, this is distributed over all the braces in the structure. Christopoulos and Filiatrault (2006) recommend the total shear is divided evenly among all the braces, and this is likely to be the simplest design solution for most structures. Final checks of the design would be to ensure that the slip or yield capacity of each damper is sufficiently high to not slip or yield under gravity

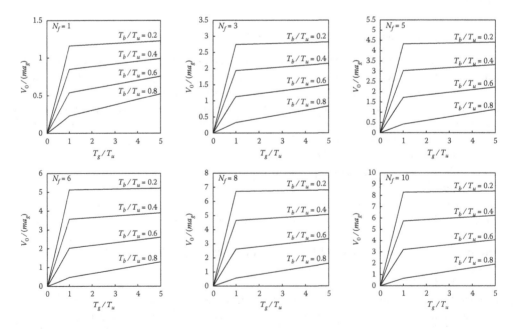

Figure 10.17 Optimum hysteretic damping design spectra for structures ranging from $N_f = 1$ to 10 storeys. (Adapted from Christopoulos, C. and Filiatrault, A. 2006. *Principles of Passive Supplemental Damping and Seismic Isolation*, IUSS Press, Pavia, Italy).

or wind loading, and that the brace capacities are sufficient to develop the slip or yield force in the dampers under ultimate conditions.

10.9.3 Other considerations for application with Eurocode 8

Although supplemental damping is not covered by EC8, some guidelines for preliminary design have been given in the preceding sections. Until provisions are developed in future versions of EC8, time history analysis should always be used to verify these designs. For consistency with the provisions for seismic isolation, the following recommendations should also be adopted:

- Damper strokes should be multiplied by a reliability factor at least as high as that used for seismic isolation, $\gamma_x = 1.2$.
- Maximum damper strokes should be determined based on lower bound damper properties (damper coefficients, loss factors, slip force or yield force).
- Maximum damper forces and forces in the rest of the structure should be determined based on upper bound damper properties.
- Generally, elastic or near-elastic response in the main structural elements should be targeted. To this end, behaviour factors of 1.0 for the dampers, and up to 1.5 for the rest of the structure should be used. DCL provisions may be adopted for the rest of the structure.

REFERENCES

Almufti, I., Motamed, R., Grant, D.N. and Willford, M. 2015. Incorporation of velocity pulses in design ground motions for response history analysis using a probabilistic framework. *Earthquake Spectra*, 31(3), 1647–1666.

Almufti, I. and Willford, M. 2014. The REDi™ rating system: A framework to implement resilience-based earthquake design for new buildings. *Proceedings of the 10th U.S. National Conference on Earthquake Engineering*, Anchorage, Alaska.

Carr, J. 2005. Seismic response of buildings equipped with Tuned-Mass Dampers. *MSc Project*, Department of Civil, Structural and Environmental Engineering, University of Buffalo.

Chopra, A.K. 2011. *Dynamics of Structures: Theory and Applications to Earthquake Engineering*. 4th ed. Prentice Hall, Englewood Cliffs, NJ.

Christopoulos, C. and Filiatrault, A. 2006. *Principles of Passive Supplemental Damping and Seismic Isolation*. IUSS Press, Pavia, Italy.

Field, C.J., Mole, A. and Arkinstall, M. 2005. The use of advanced computer simulation in structural design. *Ninth International IBPSA Conference*, Montreal, Canada.

Filiatrault, A. and Cherry, S. 1990. Seismic design spectra for friction damped structures. *Journal of Structural Engineering, ASCE*, 116(5), 1334–1355.

Grant, D.N., Blandon, C.A. and Priestley, M.J.N. 2005a. Modelling inelastic response in direct displacement-based design. *Report ROSE 2005/01*, IUSS Press, Pavia, Italy.

Grant, D.N., Fenves, G.L. and Auricchio, F. 2005b. Modelling and analysis of high-damping rubber bearings for the seismic protection of bridges. *Report ROSE 2005/03*, IUSS Press, Pavia, Italy.

Grant, D.N., Fenves, G.L. and Whittaker, A.S. 2004. Bidirectional modelling of high-damping rubber bearings. *Journal of Earthquake Engineering*, 8(Special Issue 1), 161–185.

Grant, D.N., Padilla, D. and Greening, P.D. 2011. Orientation dependence of earthquake ground motion and structural response. In *Protection of Built Environment against Earthquakes*, M. Dolšek, ed., Springer, the Netherlands, 352pp.

Huang, W., Fenves, G.L., Whittaker, A.S. and Mahin, S.A. 2000. Characterization of seismic isolation bearings for bridges from bi-directional testing. *Proceedings of the 12th World Conference on Earthquake Engineering*, The New Zealand Society for Earthquake Engineering, Auckland, New Zealand, pp. 2047–2048.

Huang, Y., Whittaker, A.S. and Luco, N. 2008. Maximum spectral demands in the near-fault region. *Earthquake Spectra*, 24(1), 319–341.

Iervolino, I. and Cornell, C.A. 2008. Probability of occurrence of velocity pulses in near-source ground motions. *Bulletin of the Seismological Society of America*, 98(5), 2262–2277.

Iwan, W.D. and Gates, N.C. 1979. Estimating earthquake response of simple hysteretic structures. *Journal of the Engineering Mechanics Division, ASCE*, 105(EM3), 391–405.

Kamai, R., Abrahamson, N. and Graves, R. 2014. Adding fling effects to processed ground-motion time histories. *Bulleting of the Seismological Society of America*, 104(4), 1914–1929.

Kelly, J.M. 1990. Base isolation: Linear theory and design. *Earthquake Spectra*, 6(2), 223–244.

Mosqueda, G., Whittaker, A.S. and Fenves, G.L. 2004. Characterization and modeling of friction pendulum bearings subjected to multiple components of excitation. *Journal of Structural Engineering*, 130(3), 433–442.

Pettinga, D., Christopoulos, C., Pampanin, S. and Priestley, N. 2007. Effectiveness of simple approaches in mitigating residual deformations in buildings. *Earthquake Engineering and Structural Dynamics*, 36(12), 1763–1783.

Priestley, M.J.N., Calvi, G.M. and Kowalsky, M.J. 2007. *Displacement-Based Seismic Design of Structures*. IUSS Press, Pavia, Italy.

Priestley, M.J.N., Seible, F. and Calvi, G.M. 1996. *Seismic Design and Retrofit of Bridges*. Wiley, New York.

Priestley, M.J.N., Sritharan, S., Conley, J.R. and Pampanin, S. 1999. Preliminary results and conclusions from the PRESSS five-storey precast concrete test building. *PCI Journal*, 44(6), 42–67.

Sigaher, A.N. and Constantinou, M.C. 2003. Scissor-jack-damper energy dissipation system. *Earthquake Spectra*, 19(1), 133–158.

Skinner, R.I., Robinson, W.H. and McVerry, G.H. 1993. *An Introduction to Seismic Isolation*. Wiley, New York.

Smith, R., Merello, R. and Willford, M. 2010. Intrinsic and supplementary damping in tall buildings. *Proceedings of the ICE – Structures and Buildings*, 163(2), 111–118.

Smith, R.J. and Willford, M.R. 2007. The damped outrigger concept for tall buildings. *The Structural Design of Tall and Special Buildings*, 16(4), 501–517.

Thompson, A.C.T., Whittaker, A.S., Fenves, G.L. and Mahin, S.A. 2000. Property modification factors for elastomeric seismic isolation bearings. *Proceedings of the 12th World Conference on Earthquake Engineering*, Auckland, New Zealand.

Wen, Y.K. 1976. Method for random vibration of hysteretic systems. *Journal of the Engineering Mechanics Division*, ASCE, 102, 249–263.

Willford, M. R. 2010. Theoretical basis for scheme design of viscous and visco-elastic damped structures. (Unpublished manuscript, 2010).

Shallow foundations

Gopal S. P. Madabhushi, Indrasenan Thusyanthan,
Zygmunt Lubkowski and Alain Pecker

CONTENTS

11.1 INTRODUCTION

11.1.1 Dynamic properties of structures

Practical and comprehensive seismic design methods for both shallow and deep foundations are not fully established as yet. The 1990s have seen the publication of a series of important papers on the seismic design of shallow foundations. However, several important gaps between theory and observation are yet to be bridged. In 1987, Dowrick could write that an authoritative rationale for the design of deep foundations did not exist. Progress has been made in the intervening years but substantial further work is required fully to establish seismic design methods both for shallow and deep foundations (e.g. Pappin 1991; Pender, 1996). It is, therefore, pleasing to state that EN1998-5:2004 provides one of the most comprehensive codes of practice for addressing seismic foundation design.

In order to develop robust and reliable design methods good observations of field performance are indispensable. Despite the problems of making such observations on foundations, evidence has accumulated from major earthquakes over the past few decades. Many of these, from Alaska in 1964 to the Hyogoken-Nambu (Kobe) earthquake of 1995, have provided evidence of the highly damaging effects of liquefaction and lateral spreading for both shallow and deep foundations (Ross et al., 1969; Tokimatsu and Asaka, 1998). The unusual soft clay conditions of Mexico City (1985) gave rise to a range of foundation failures rarely reported elsewhere (Mendoza and Auvinet, 1988; Zeevaert, 1991). In general, spread footings properly designed for static loadings have been observed to perform adequately under seismic loading although cases of significant settlement have been reported (Richards et al., 1993). In contrast, the poor performance of raked or battered piles in bridge abutments and jetties has been highlighted in several earthquakes including Loma Prieta in 1989 (Seed et al., 1991).

Current trends suggest that displacement-based design methodologies will come in time to play a major role in seismic foundation design, but their full development is yet to come.

11.1.2 Overview of soil behaviour

The ground presents the earthquake engineer with major challenges. Its behaviour often falls far short of that which would be desirable for the support of structures in earthquakes. However, the engineer is obliged to work with the ground, modifying its performance where necessary, to produce economic and safe foundations. The ground properties of particular interest when considering the seismic design of foundations are its stiffness, damping and strength. In addition, its density and degree of over-consolidation provide important indicators of its likely behaviour under seismic excitation.

The stiffness of both granular and cohesive soils is highly non-linear with stiffness reducing with increasing strain once a threshold strain has been exceeded. While the stiffness of soils reduces with increasing strain, the hysteretic or material damping increases. It should also be noted that the stiffness and the hysteretic damping of soils are generally not dependent on the frequency of the loading.

The non-linear behaviour of soils poses significant problems to the designer for assessing the foundation response to both static and dynamic loadings. Close to the foundation structural element (e.g. the footing or pile), strains may be high and the ground response soft. However, at more remote locations the strains will be small and the ground behaviour will be stiff. The response of the foundation to loading will represent an integration of both the near field and the far field strains (see Jardine et al., 1986). This makes the use of linear elastic solutions difficult except at very small load levels where the strains may be largely below the threshold values at all locations. At higher loadings, the choice of a representative stiffness and damping values requires considerable care.

The stiffness and strength of soils are dependent on the effective stresses in the ground. The effective stresses are directly affected by the pore water pressures. These can vary both as a result of fluctuations in groundwater levels and as a result of stress changes in the ground. For saturated loose sands or normally to lightly over-consolidated clays, cyclic shear stresses produce increases in pore water pressure and progressive losses of strength and stiffness.

The designer must not forget that natural ground is a heterogeneous material created by complex and often non-uniform processes. The information provided by a normal ground investigation will strictly relate only to a tiny fraction of the volume of ground, which will affect the behaviour of structural foundations. The influences of foundation construction imposed loadings and of the seismic excitation itself will alter the ground's behaviour. Hence the designer requires a good understanding of the processes involved and a suitably cautious attitude.

11.1.3 Soil stiffness

The response of the foundation as a system under earthquake loading is strongly affected by the stiffness profile and depth of the foundation stratum. These affect the fundamental frequency of the stratum, which, in turn, affects the amplification of bedrock motions to the surface and the foundation system damping characteristics. Three stiffness profiles are shown in Figure 11.1. The fundamental frequencies of the strata are given by

Constant stiffness with depth:

$$f = \frac{0.25 \, v_s}{H} \tag{11.1}$$

Figure 11.1 Typical stiffness profiles for foundation strata. (a) Constant stiffness (typical of overconsolidated clay). (b) Parabolic stiffness (typical of sand). (c) Linear stiffness (typical of soft clay).

Parabolic stiffness with depth:

$$f = \frac{0.22\, v_s}{H} \tag{11.2}$$

Linearly increasing stiffness:

$$f = \frac{0.19\, v_s}{H} \tag{11.3}$$

where v_s = the shear wave velocity at the base of the stratum (Depth = H).

Additional results for an arbitrary increasing with depth stiffness profile are given by Pecker (2005).

The stiffness, damping and strength characteristics of the soil column affect the transmission of seismic motions from the bedrock to the ground surface. Various methods are used to assess the seismic motions at the foundation level. Direct measurements may be available from the site or another site with similar characteristics. A word of caution is necessary here because measurements are only valid if the range of induced strains in the soil profile is representative of the design situation. Code requirements may be used for the class of site (usually based on depth of deposit and shear wave velocity) and site location. Alternatively, the ground motions may be calculated using one-dimensional wave propagation codes such as SHAKE (Schnabel et al., 1972), SIREN (Pappin, 1991) or DESRA-2 (Lee and Finn, 1978), and where the geometry is more complex, 2-D wave propagation codes such as FLUSH (Lysmer et al., 1975) or DYNA3D.

Section 4.2.3 of EN1998-5 provides a table that gives average soil damping and reduction factors (Table 11.1) for shear wave velocity and shear modulus, which can be used in the absence of specific measurements.

The maximum acceleration that can be transmitted through a soil stratum in a seismic event is limited by the shear strength of the soil. For dry non-cohesive soils, the peak acceleration is given by

$$k_{h\,\mathrm{lim}} = \tan\phi \tag{11.4}$$

Application of this relationship indicates limiting horizontal accelerations of between 0.5 g and 0.8 g for dry granular soils. In the case of loose saturated deposits, the effects of cyclic

Table 11.1 Average soil damping ratios and reduction factors

Ground acceleration ratio, αS (g)	Damping ratio	$v_s/v_{s,max}$	G/G_{max}
0.10	0.03	0.90 (±0.07)	0.80 (±0.10)
0.20	0.06	0.70 (±0.15)	0.50 (±0.20)
0.30	0.10	0.60 (±0.15)	0.36 (±0.20)

Note: $v_{s,max}$ is the average v_s value at small strain (<10⁻⁵), not exceeding 360 m/s.

G_{max} is the average shear modulus at small strain.

Through the ± one standard deviation ranges the designer can introduce different amounts of conservatism, depending on such factors as stiffness and layering of the soil profile. Values of $v_s/v_{s,max}$ and G/G_{max} above the average could, for example, be used for stiffer profiles, and values of $v_s/v_{s,max}$ and G/G_{max} below the average could be used for softer profiles.

loading are likely to lead to liquefaction at acceleration levels significantly below the limit given in Equation 11.4. Once liquefaction occurs, the liquefied horizon substantially reduces the transmitted peak acceleration. However, it may be noted that several cycles of earthquake loading are likely to occur prior to the general onset of liquefaction. The appropriate strength to use for saturated granular materials (loose or dense) is discussed to some extent in Pecker (2005). For dense saturated sands, values in excess of those given by Equation 1.4 may be achieved.

For cohesive soils, the equivalent relationship for limiting horizontal acceleration is

$$k_{h\,\lim} = \frac{s_u}{\gamma h} \qquad (11.5)$$

where
s_u = the undrained shear strength at depth h
γ = the average bulk unit weight of the soil

The non-linear stress–strain response of soils result in different amplifications of the bedrock motion through the soil column, depending on the magnitude of the earthquake motions (Idriss, 1990). The greatest amplification of bedrock accelerations occurs at low peak acceleration levels. As the peak acceleration level increases for larger earthquakes, so the amplification of the soil column decreases and becomes less than unity at high bedrock acceleration levels. Observations from soft soil sites suggest that the cross-over from amplification to de-amplification occurs at bedrock accelerations of about 0.3–0.5 g (Mohammadioun and Pecker, 1984, Idriss, 1990, Suetomi and Yoshida, 1998, Kokusho and Matsumoto, 1998).

11.1.4 Soil strength

The value of the soil strength parameters applicable under static undrained conditions may generally be used. As in EN1997 these are characteristic strength parameters, which are defined as a cautious estimate of the value affecting the occurrence of the limit state. Further discussion on selection of characteristic soil parameters can be found in Frank et al. (2004).

For cohesive soils, the appropriate strength parameter is the undrained shear strength c_u, adjusted for the rapid rate of loading and cyclic degradation effects under the earthquake loads when such an adjustment is needed and justified by adequate experimental evidence. For cohesionless soil, the appropriate strength parameter is the cyclic undrained shear strength $\tau_{cy,u}$, which should take the possible pore pressure build-up into account.

Alternatively, effective strength parameters with appropriate pore water pressure generated during cyclic loading may be used. For rocks the unconfined compressive strength, q_u, may be used.

EN1998-5 requires that a partial factor (γ_M) is applied to the material properties c_u, $\tau_{cy,u}$ and q_u. These are denoted as γ_{cu}, $\gamma_{\tau cy}$ and γ_{qu}, and those for tan φ' are denoted as $\gamma_{\varphi'}$. The recommended values are $\gamma_{cu} = 1.4$, $\gamma_{\tau cy} = 1.25$, $\gamma_{qu} = 1.4$, and $\gamma_{\varphi'} = 1.25$ (EN 1998-5:2004 pg 13).

11.2 SITING REQUIREMENTS

11.2.1 General

The primary cause of building damage has been identified as ground shaking, however in most earthquakes, the overall damage to buildings is caused by more than one hazard. The principal secondary cause of building damage is ground failure, which can be divided

Figure 11.2 Examples of poorly sited structures. (a) Fault Rupture – Luzon, Philippines 1990. (b) Lique-faction – Adapazari, Turkey 1999. (c) Slope Instability – Pacific Palisades, Los Angeles, 1994.

into five elements namely fault rupture, topographic amplification, slope instability, lique-faction and shakedown settlement (Bird and Bommer, 2004).

Section 4 of EN1998-5:2004 requires that these earthquake phenomena are identified and hence they can be minimised. Figure 11.2 shows just a few examples where a failure to assess these phenomena has impinged on the performance of structures during a major earthquake. By ensuring these potential hazards at a site are identified, the designer can then take appropriate actions to minimise those hazards.

11.2.2 Active faults

Section 4.1.2 of EN1998-5 states that buildings of importance classes II, III and IV (i.e. all buildings except agricultural buildings) should not be sited in the immediate vicinity of active tectonic faults.

No minimum distance requirement between a building and an active fault is quoted. Requirements in countries such as New Zealand, Russia and the United States range from about 15 to 200 m.

In areas of high seismicity, the code requires geological investigations to be carried out for important structures near active tectonic faults, in order to determine the hazard in terms of ground rupture and severity of ground shaking.

For structures that are not critical to public safety the absence of movement in the Late Quaternary (last 10,000 years) may be used to define non-active faults.

11.2.3 Slopes

Structures adjacent to slopes may be subject to two different phenomena, first slope instability and second topographic amplification.

11.2.4 Slope instability

As part of a natural process, slopes undergo a process of landsliding in order to reduce their slope angle and to re-establish equilibrium. This process takes place in a variety of forms such as soil creep, cambering and rotational slips. These have different effects on structures, but the degree to which they affect a structure will also depend on the foundations of the structure itself.

Where instability is shallow, for example, where there is soil creep or flow sliding, the foundations of the structure may displace unless constructed beneath the plane along which slipping is occurring, as shown in Figure 11.3a.

Where the foundations are at depth, the structure should remain stable and may help stabilise the local area, as shown in Figure 11.3b. However, material may eventually build up

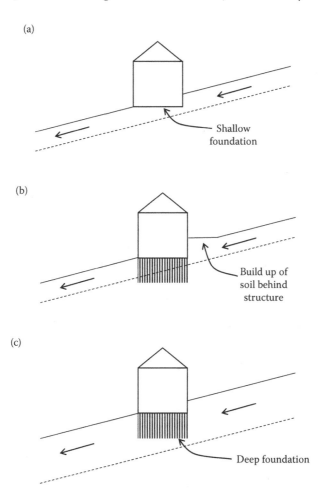

Figure 11.3 Vulnerability of structures to landslide hazard. (a) Shallow landslide, shallow foundations = unstable. (b) Shallow landslide, deep foundations = stable. (c) Deep landslide, deep foundations = unstable.

behind the structure and apply a horizontal pressure to it. The possible effects of this should be taken into consideration.

Where deep landslides occur, there is a greater risk of severe structural damage, with structures being translated downhill or undermined by the slip, as shown in Figure 11.3c. While previous deep landslides may be stabilised to some extent, for example by incorporating drainage measures and/or re-grading, construction in these areas should be avoided where possible.

The code recommends the pseudo-static method of analysis to determine the degree of slope instability.

11.2.5 Topographic amplification

Earthquake ground motion experienced near the top of a slope or ridge is often greater than the ground motion felt on level ground, assuming similar soil conditions. This phenomenon is referred to as the topographical effect and is understood to be a function of the height and inclination of the topography and the wavelength of the ground motion. Topographical effect studies were first conducted after the 1971 San Fernando earthquake in California. The amplitude of ground motions at the crest of hills were two to three times greater than amplitudes measured at the base of the hills (Finn et al. 1995). Similar results have been reported from Matsuzaki in Japan, Northridge in California and many other parts of the world.

Section 4.1.3.2 of EN1998-5:2004 requires that, for structures erected near slopes, the amplification factor (S_T) should be determined and applied to the seismic action derived in Section 3.2.2 of EN1998-1:2004. Simple guidelines on assessing determining S_T are given in Annex A of the code. The following limits on this effect should be noted:

The amplification factor should, in general, only be applied to long ridges and cliffs of height greater than about 30 m.

For average slope angles of less than about 15°, the topography effects may be neglected.

It should also be noted that ground motions are also strongly influenced by sub-surface topography, though this is not considered explicitly in EN1998. A major factor contributing to the amplification of ground motion and increased damage in alluvium filled valleys and basins is interpreted to be the generation of surface waves at the valley edges and the reflection of these waves back and forth through the alluvium in-filling the valleys. This phenomenon is referred to as the basin effect (Borcherdt and Glassmoyer, 1992; Finn et al. 1995). Basin effects were interpreted to have significantly influenced the characteristics of the earthquake ground motion and location of major damage centres in the Los Angeles and San Fernando Basins during recent earthquakes (Borcherdt and Glassmoyer, 1992). The most common manifestations of the basin effect are an increase in the duration and a shift to lower frequencies that are more damaging for taller structures during the earthquake strong ground shaking.

11.3 LIQUEFACTION

11.3.1 Effect of soil liquefaction on structures

Liquefaction is a process by which non-cohesive or granular sediments below the water table temporarily lose strength and behave as a viscous liquid rather than a solid when subjected to strong ground shaking during an earthquake. Typically, saturated, poorly graded, loose, granular deposits with a low fine content are most susceptible to liquefaction.

Liquefaction does not occur at random, but is restricted to certain geological and hydrological environments, primarily recently deposited sands and silts in areas with high ground water levels. Dense and more clayey soils, including well compacted fills, and older deposits (Pleistocene deposits, Youd and Perkins, 1978) have low susceptibility to liquefaction.

The liquefaction process itself may not necessarily be particularly damaging or hazardous. For engineering purposes, it is not the occurrence of liquefaction that is of importance, but the capability of the process and associated hazards to cause damage to structures. The adverse effects of liquefaction can be summarised as follows:

- *Flow failures:* Completely liquefied soil or blocks of intact material ride on a layer of liquefied soil. Flows can be large and develop on moderate to steep slopes.
- *Lateral spreads:* Involve lateral displacement of superficial blocks of soil as a result of liquefaction of a subsurface layer. Spreads generally develop on gentle slopes and move toward a free face such as an incised river channel or coastline.
- *Ground oscillation:* Where the ground is flat or the slope too gentle to allow lateral displacement, liquefaction at depth may disconnect overlying soils from the underlying ground, allowing the upper soil to oscillate back and forth in the form of ground waves. These oscillations are usually accompanied by ground fissures and fracture of rigid structures such as pavements and pipelines.
- *Loss or reduction in bearing capacity:* Liquefaction is induced when earthquake shaking increases pore water pressures, which, in turn, causes the soil to lose its strength and hence bearing capacity.
- *Settlement:* Soil settlement may occur as the pore-water pressures dissipate and the soil densifies after liquefaction. Settlement of structures may occur due to the reduction in bearing capacity or due to the ground displacements noted above.
- *Increased lateral pressure on retaining walls:* Occurs when the soil behind a wall liquefies and so behaves as a 'heavy' fluid with no internal friction.
- *Flotation of buried structures:* Occurs when buried structures such as tanks and pipes become buoyant in the liquefied soil.

Other manifestations of liquefaction, such as sand boils, can also occur and may pose a risk to structures, particularly through loss or reduction in bearing capacity and settlement.

11.3.2 Liquefaction potential

Section 4.1.4 of EN1998-5 describes the requirements for assessing liquefaction potential. Furthermore, it provides a normative methodology in Annex B. It should, however, be noted that there have been numerous developments in liquefaction assessment methodologies in recent years (e.g. Seed et al., 2003; Boulanger and Idriss, 2004) and the methods described in the code may be potentially unconservative, especially for materials with high fines content. It is therefore recommended that an expert should be employed to carry out liquefaction assessment.

A liquefaction susceptibility evaluation should be made when the soil includes extended layers of thick lenses of loose sand (with or without silt/clay fines), beneath the water table and when the water table level is close to the ground surface. EN 1998-5 recommends that the shear stress approach is applied. In this method, the horizontal shear stresses generated by the earthquake are compared with the resistance available to prevent liquefaction. In Annex B of EN 1998-5, a set of liquefaction potential charts can be found for a magnitude $M_s = 7.5$ earthquake. The shear stresses 'demand' are expressed in terms of a cyclic stress ratio (CSR), and the 'capacity' in terms of a cyclic resistance ratio (CRR).

The CRR is assessed based on corrected SPT blowcount using the empirically derived liquefaction charts which are shown schematically for silty sand in Figure 11.4. These charts compare CRR (τ/s'_{v0}), with corrected SPT blowcount ($N_{1(60)}$). In Figure 11.4, the dependence of liquefaction potential on the % fines content in the silty soil is also seen by comparing the three lines. For a given corrected SPT blow count, clean sands with fines content of <5% liquefy more easily compared to silty sands with greater % of fines content. The procedure for correcting the field N values to obtain the corrected $N_{1(60)}$ is explained later in Section 11.3.3.

The CSR is assessed by first calculating the cyclic shear stress (τ_e) using Equation 11.6.

$$t_e = 0.65\,\alpha S \sigma_{v0} \tag{11.6}$$

where
   α is the ratio of the design ground acceleration on type A ground
   a_g, to the acceleration of gravity g
   S is the soil factor and σ_{v0} is the overburden pressure

It must be pointed out that Equation 11.6 is conservative because it neglects the stress reduction factor with depth (r_d).

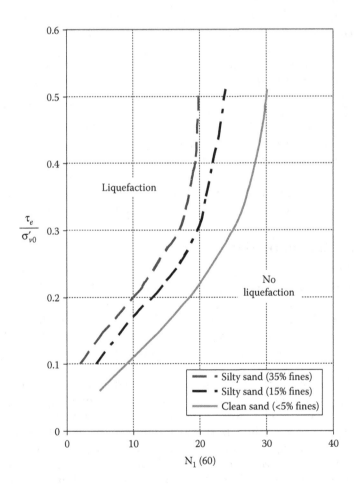

Figure 11.4 Liquefaction assessment using corrected SPT values.

This expression may not be applied for depths larger than 20 m. A soil shall be considered susceptible to liquefaction whenever $CRR > \lambda \times CSR$, where λ is recommended to be 0.8, which corresponds to a factor of safety of 1.25.

If soils are found to be susceptible to liquefaction, mitigation measures such as ground improvement and piling (to transfer loads to layers not susceptible to liquefaction) should be considered to ensure foundation stability.

The use of pile foundations alone should be considered with caution due to the large forces induced in the piles by the loss of soil support in the liquefiable layers, and to the inevitable uncertainties in determining the location and thickness of such layers.

For buildings on shallow foundations, liquefaction evaluation may be omitted when the saturated sandy soils are found at depths greater than 15 m.

11.3.3 Design example on determination of liquefaction potential

In this section, we shall outline the liquefaction assessment for site A. The foundations for the hotel building can take the form of shallow foundation provided that the chosen site does not pose a major risk of liquefaction. In other words, liquefaction potential of the chosen site should be low. The design of shallow foundations will be considered in this chapter. However, certain sites where there is significant liquefaction risk, pile foundation may be preferred. The design of pile foundation will be considered in Chapter 9. In either case, it is important to carry out an assessment of liquefaction potential for any building site. The method for carrying out such an assessment Site 'A' is shown in this section. As explained in Chapter 4, site 'A' has loose sand layers below water table. The bore hole data obtained from site investigation is presented in Figure 11.5 along with the strength parameters and the water table.

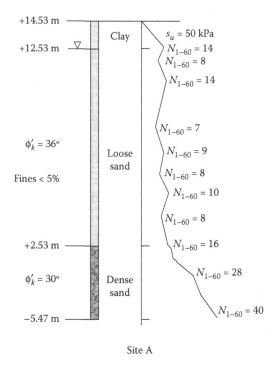

Site A

Figure 11.5 Bore hole data from site A.

11.3.3.1 Check for liquefaction

N_{SPT} from a field SPT test are to be normalised as given below to obtain $N_1(60)$.
In the present question, this has already been done and values of $N_1(60)$ are given.

$$N_1(60) = N_{SPT} \sqrt{\frac{100}{\sigma'_{v0}}} \frac{ER(\%)}{60} \sigma'_{v0} \text{ in kPa} \tag{11.7}$$

where ER = the ratio of the actual impact energy to the theoretical free-fall energy.

In Europe, a value of $ER = 70\%$ is commonly used. However, it is recommended that as much as possible, measurements of ER should be made at the start of the site investigation as the values for ER vary significantly from one equipment to another and even from one operator to another.

Liquefaction hazard may be neglected when $\alpha S < 0.15$ and at least one of the following is satisfied (see p. 16, 17 EC 8 - part 5).

1. The sands have clay content >20% with PI >10
2. Sands have slit content >35% and $N_1(60)$ >20
3. Sands are clean and $N_1(60)$ >30

Seismic shear stress τ_e

$$\tau_e = 0.65 \cdot \alpha \cdot S \cdot \sigma_{v0} \quad \text{for depths <20 m} \tag{11.8}$$

For present case, $\alpha = 0.3$, $S = 1.15$. Also take the saturated unit weight of the clay and sand to be 20 kN/m³. M_s is surface wave magnitude. For $M_s = 6$, from table B.1 = >CM is 2.2 Pg 34, EC 8 - part 5.

In Table 11.2, the calculations for the seismic shear stress and its normalisation with the effective vertical stress and CM factor are presented for various depths. Note that EC8 - Part 5 Clause 5.1.4(11) requires liquefaction factor of safety check. Normalised data is put into Figure B.1 in Annex B of EC 8 - part 5.

From the plot in Figure 11.6 we can determine that liquefaction is possible in the loose sand layer.

11.3.4 Conclusion

The loose sand layer from elevation +12.53 to +2.53 (i.e. 10 m of sand layer) is susceptible to liquefaction.

11.4 SHALLOW FOUNDATIONS

11.4.1 Overview of behaviour

The performance of shallow or spread foundations subject to seismic loading can be considered as consisting of several modes (see Figure 11.7). The long-term static loading will have produced some foundation displacement (1). For relatively small seismic loadings, most foundations will respond in an essentially linear elastic manner (2). As the loading increases towards the ultimate dynamic capacity, non-linear soil responses become significant and

Table 11.2 Calculation of seismic shear stress with depth

Depth (m)	$N_{1(60)}$	Total stress (kPa)	Effective stress (kPa)	Seismic shear stress (kPa)	Seismic shear stress/ Effective stress	$(\tau_e/\sigma'_{v0})/CM$
1	5	20	20	4.49	0.2243	0.10
2	5	40	40	8.97	0.2243	0.10
3	14	60	50	13.46	0.2691	0.12
4	8	80	60	17.94	0.2990	0.14
5	10	100	70	22.43	0.3204	0.15
6	14	120	80	26.91	0.3364	0.15
7	10	140	90	31.40	0.3488	0.16
8	7	160	100	35.88	0.3588	0.16
9	9	180	110	40.37	0.3670	0.17
10	8	200	120	44.85	0.3738	0.17
11	10	220	130	49.34	0.3795	0.17
12	8	240	140	53.82	0.3844	0.17
13	12	260	150	58.31	0.3887	0.18
14	16	280	160	62.79	0.3924	0.18
15	20	300	170	67.28	0.3957	0.18
16	28	320	180	71.76	0.3987	0.18
17	35	340	190	76.25	0.4013	0.18
18	40	360	200	80.73	0.4037	0.18

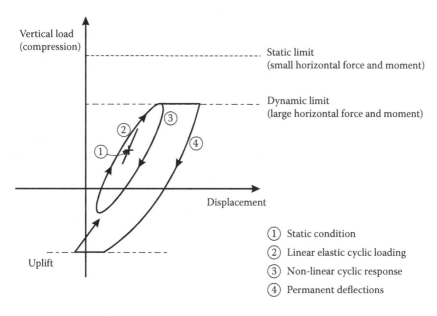

Figure 11.6 Determination of liquefaction potential.

the foundation response may be affected by partial uplift (3). The ultimate capacity of the foundation will be significantly influenced by the dynamic loadings imposed, with transient horizontal loads and moments acting to reduce the ultimate vertical capacity. For transient loadings that exceed yield, permanent displacements may occur (4). Lateral loading may generate sliding with larger sliding displacements accumulating if the transient horizontal

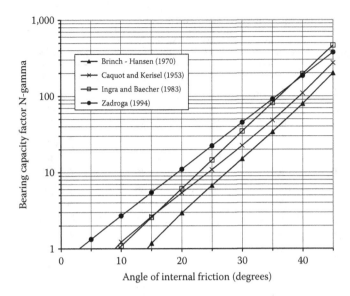

Figure 11.7 Conceptual response of spread foundation to seismic loading.

loading is biased in one direction. Uplift and rocking behaviour may result in permanent rotations while bearing capacity failure will lead to settlement, translation and tilt.

In addition to the transient and permanent deformations that arise from loads transmitted through the structure into the foundation, additional displacements may arise from ground movements imposed on the foundation. In this class of behaviour are settlements arising from densification of the soil, the effects of liquefaction and lateral spreading.

Historically, seismic foundation design has aimed to avoid yield of the foundation material. Fixed base assumptions have often been made for the structural analysis and the foundation design has attempted to produce this behaviour. However, the recent trend has been to recognise that limited foundation displacements (both transient and permanent) may absorb substantial energy and allow significant economies in construction. Practical design methodologies have been developed to enable implementation of this approach particularly in American and New Zealand practice. EN1998:2004 does not explicitly discuss displacement based geotechnical design.

11.4.2 Ultimate capacity of shallow foundations

EN1998-5 requires the ultimate seismic capacity of footings to be assessed for the onset of sliding and bearing capacity 'failure'. These modes of behaviour are considered in the following sections.

11.4.3 Sliding

The friction resistance for footings on cohesionless deposits above the water table, F_{Rd}, may be calculated from the following expression:

$$F_{Rd} = N_{Ed} \frac{\tan\delta}{\gamma_M} \qquad (11.9)$$

where

> N_{Ed} = the design normal force on the horizontal base
> δ = interface friction angle
> γ_M is the partial factor (1.25 for tan δ).

For cohesive soils the equivalent relationship is

$$F_{Rd} = \frac{s_u A}{\gamma_M}$$ (11.10)

where

> A = plan area of foundation
> s_u = undrained strength
> γ_M is the partial factor (1.4 for s_u).

Most foundations are embedded and derive additional resistance to sliding by mobilising passive resistance on their vertical faces. For some classes of foundation (e.g. bridge abutments), this resistance provides a major contribution to their performance. However, the mobilisation of full passive resistance requires significant displacements, which may amount to between 2% and 6% of the foundation's depth of burial (see e.g. Martin and Yan (1995)). Such displacements may exceed the maximum allowable values for the structure and hence the foundation design may incorporate only a proportion of the full passive resistance.

EN1998-5 requires that to ensure no failure by sliding on a horizontal base, the following expression must be satisfied:

$$V_{Ed} \le F_{Rd} + E_{Rd}$$ (11.11)

where E_{Rd} = design lateral resistance from earth pressure, not exceeding 30% of the full passive resistance

11.4.4 Bearing capacity

11.4.4.1 Static bearing capacity

Bearing capacity formulae for seismic loading are generally related to their static counterparts. For the static case:

$$q = cN_c s_c + 0.5\gamma B N_\gamma s_\gamma + p_0 N_q s_q$$ (11.12)

where

> q = ultimate vertical bearing pressure
> c = cohesion
> γ = soil density
> B = foundation width
> p_0 = surcharge at foundation level
> N_c, N_γ, N_q = bearing capacity factors
> s_c, s_γ, s_q = shape factors

Table 11.3 Formulations for bearing capacity factors

$N_q = e^{\pi\tan\phi} \tan^2(45 + (\phi/2))$	Terzaghi and Peck (1967)
$N_c = (N_q - 1)\cot\phi$	Terzaghi and Peck (1948)
$N_\gamma = 2(N_q + 1)\tan\phi$	Caquot and Kerisel (1953), API (2000)
$N_\gamma = 2(N_q - 1)\tan\phi$	EN1997-1 (2004)
$N_\gamma = \exp(-1.646 + 0.173\phi)$	Strip footing – Ingra and Baecher (1983)
$N_\gamma = 0.657\exp(0.141\ \phi)$	Strip footing – Zadroga (1994)

Closed form solutions exist for N_c and N_q but not for N_γ. Thus while the factors N_c and N_q have widely accepted definitions, a considerable range of solutions have been proposed for N_γ based on approximate numerical studies or on experimental results. A selection of the suggested values are presented in Table 11.3 and plotted in Figures 11.8 and 11.11.

Inclined loading is incorporated into the bearing capacity equation either by incorporating inclination factors into each term of Equation 11.13 or by direct modification of the bearing capacity factors. Thus

$$q = cN_c s_c i_c + 0.5\gamma BN_\gamma s_\gamma i_\gamma + p_0 N_q s_q i_q \tag{11.13}$$

where i_c, i_γ, i_q = inclination factors.

Various proposals for the inclination factors are shown in Table 11.4. While there may not be unanimity on the precise formulation of the various inclination factors, the key issue they all indicate is that the ultimate vertical capacity of a foundation is severely reduced by relatively modest horizontal loading.

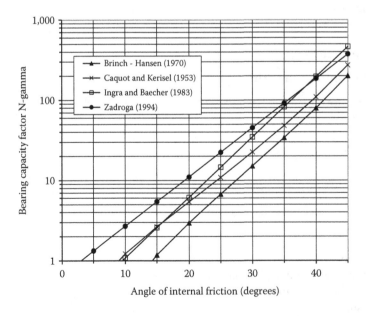

Figure 11.8 Published relationships between $N\gamma$ and ϕ for static loading.

Table 11.4 Published relationships for inclination factors

Inclination factor	EN1997-1 (2004)	Vesic (1975)
i_c undrained	$0.5\left(1-\sqrt{\dfrac{H}{As_u}}\right)$	$1-\dfrac{mH}{As_u N_c}$
i_c drained	$\dfrac{\left(i_q N_q -1\right)}{\left(N_q -1\right)}$	$\dfrac{\left(i_q N_q -1\right)}{\left(N_q -1\right)}$
i_q	$\left(1-\dfrac{0.7H}{V+Ac\cot\phi}\right)^3$	$\left(1-\dfrac{H}{V+Ac\cot\phi}\right)^m$

Note: H = horizontal loading (parallel to B); V = vertical loading; A = plan area of foundation = BL; s_u = undrained shear strength.

$$m = \frac{2+B/L}{1+B/L}$$

A moment acting on the foundation is treated by defining an effective foundation width B′. The horizontal and vertical loads are applied to the effective foundation. B′ is defined as follows:

$$e = \frac{M}{V} \quad \text{and} \quad B' = B - 2e \tag{11.14}$$

where M = applied moment.

11.4.4.2 Seismic bearing capacity

Significant earthquake events substantially reduce the ultimate bearing capacity of spread footings due principally to the following effects:

- The imposition of transient horizontal loads and moments arising from the inertia of the supported structure.
- Inertial loading of the foundation material.
- Changes in the strength of foundation materials due to rapid cyclic loading.

In addition, the soil strata which comprise the foundation may act to limit the maximum seismic accelerations that can be transmitted to the foundation level.

Several solutions have recently been published for bearing capacity that take account of inertia effects in the foundation material. The methods due to Sarma and Iossifelis (1990), Budhu and Al-Karni (1993) and Shi and Richards (1995) are all based on Equation 11.11 with modified bearing capacity factors that incorporate the effects of load inclination and inertia in the foundation. Thus the seismic bearing capacity may be expressed as

$$q = cN_{cE}s_c + 0.5\gamma BN_{\gamma E}s_\gamma + p_0 N_{qE}s_q \tag{11.15}$$

where
q = vertical component of the ultimate bearing pressure
$N_{cE}, N_{\gamma E}, N_{qE}$ = seismic bearing capacity factors

While this expression appears suitable for the evaluation of shallow foundation behaviour on either granular or cohesive soil, some caution is required. The rate of loading applied by seismic events is sufficiently high to cause the response of a saturated granular stratum to be essentially undrained beneath the footing. The undrained strength of sand under such loadings is not well understood.

Returning to Equation 11.15, Sarma and Iossifelis (1990) and Budhu and Al-Karni (1993) assume that the horizontal loading applied to the foundation by the structure is given by

$$H = k_h V \tag{11.16}$$

For many real foundations subjected to seismic loading, this condition will not be satisfied. For foundations of base isolated structures or bridge piers with sliding bearings, the applied horizontal loadings may be greatly reduced. Conversely, many structures will amplify the applied base accelerations, leading to horizontal loadings much higher than those suggested by Equation 11.16. Phase differences between the ground accelerations and those of the structure complicate the assessment of the appropriate horizontal load. For a more comprehensive discussion on these issues and other limitations, refer to Pecker (1994).

Shi and Richards (1995) have assessed the effects of a range of horizontal loadings on the seismic bearing capacity. They define the horizontal load as

$$H = f k_h V \tag{11.17}$$

where f = a shear transfer factor.

Solutions have been presented as the ratios of the static to the dynamic bearing capacity factors for cases where the shear transfer factor is 0, 1 or 2 (see Figure 11.9). The solution obtained by Shi and Richards for a shear transfer factor of unity agrees closely with those obtained by Sarma and Iossifelis. It may be noted from Figure 11.9 that inertia effects within the foundation material have negligible effect on N_{cE} ($\phi = 0$) while $N_{\gamma E}$ and N_{qE} are substantially affected even in cases where the horizontal loading imposed by the foundation remains at its static value.

Annex F of EN1998-5 presents an alternative method for assessing bearing capacity of strip, shallow foundations. The results are based on a long-term European research program, including field evidence, analytical and numerical solutions and few experimental results (Pecker and Salençon, 1991, Dormieux and Pecker, 1995, Salençon and Pecker, 1995a,b, Auvinet et al., 1996, Paolucci and Pecker, 1997a,b, Pecker, 1997). The stability against seismic bearing failure of a shallow foundation may be checked with the following inequality:

$$\frac{(1-e\bar{F})^{c_T}}{\bar{N}^a[(1-m\bar{F}^k)^{k'}-\bar{N}]^b} + \frac{(1-f\bar{F})^{c'_M}(\gamma\bar{M})^{c_M}}{\bar{N}^c[(1-m\bar{F}^k)^{k'}-\bar{N}]^d} - 1 \le 0 \tag{11.18}$$

where for a footing of dimensions width B and length L

$$\bar{N} = \frac{\gamma_{Rd}N_{Ed}}{N_{max,tot}} \tag{11.19}$$

$$\bar{V} = \frac{\gamma_{Rd}V_{Ed}}{N_{max,tot}} \tag{11.20}$$

$$\bar{M} = \frac{\gamma_{Rd}M_{Ed}}{BN_{max,tot}} \tag{11.21}$$

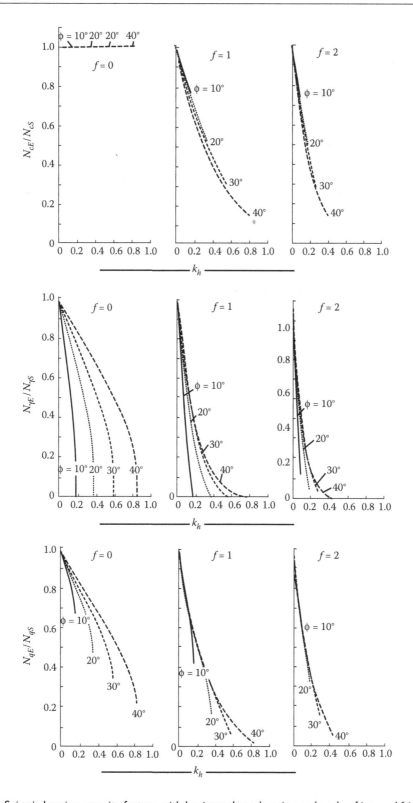

Figure 11.9 Seismic bearing capacity factors with horizontal acceleration and angle of internal friction.

Table 11.5 Values of parameters used in Equation 11.18

Purely cohesive soils				Purely cohesionless soil			
Parameter	Value	Parameter	Value	Parameter	Value	Parameter	Value
A	0.70	k	1.22	a	0.92	k	1
B	1.29	k'	1.00	b	1.25	k'	0.39
C	2.14	c_t	2.00	c	0.92	c_t	1.14
D	1.81	c_m	2.00	d	1.25	c_m	1.01
E	0.21	c'_m	1.00	e	0.41	c'_m	1.01
F	0.44	β	2.57	f	0.32	β	2.9
M	0.21	γ	1.85	m	0.96	γ	2.8

Table 11.6 The model partial factor γ_{Rd}

Soil type	γRd
Medium-dense to dense sand	1
Loose dry sand	1.15
Loose saturated sand	1.5
Non-sensitive clay	1
Sensitive clay	1.15

where N_{Ed}, V_{Ed} and M_{Ed} are the design action effects at the foundation level, and the rest of the numerical parameters in Equations 11.18 through 11.21 depend on the type of soil and are given in Tables 11.5 and 11.6.

11.4.4.2.1 Purely cohesive soils

The ultimate bearing capacity of the foundation under a vertical centred load N_{max} is given by Equation 11.22.

$$N_{max} = (\pi + 2)\frac{s_u}{\gamma_{cu}}B \tag{11.22}$$

where
   s_u is the undrained shear strength of the soil
   γ_{cu} is the partial factor for the undrained shear strength

The dimensionless soil inertia $\bar{F}$ is given by Equation 11.23.

$$\bar{F} = \frac{\rho a_g S B}{s_u} \tag{11.23}$$

where
   ρ is the unit mass of the soil
   a_g is the design ground acceleration on Type A ground, given by $a_g = \gamma_I a_{gR}$

a_{gR} is the reference Peak Ground Acceleration
γ_I is the importance factor, depending on the building importance
S is the soil factor.

The following constraints apply to the general bearing capacity expression in Equation 11.18.

$$0 < \bar{N} \le 1, |\bar{V}| \le 1 \tag{11.24}$$

11.4.4.2.2 Purely cohesionless soils

The ultimate bearing capacity of the foundation under a vertical centred load N_{max} is given by the following equation:

$$N_{max} = 0.5\rho g \left(1 \pm \frac{a_v}{g}\right) B^2 N_\gamma \tag{11.25}$$

where
a_v is the vertical ground acceleration, given by $a_v = 0.5\, a_g\, S$
N_γ is the bearing capacity factor, given by Equation 11.26.

$$N_\gamma = 2\left[\tan^2\left(45° + \frac{\phi'_d}{2}\right)e^{\pi \tan\phi'_d} + 1\right]\tan\phi'_d \tag{11.26}$$

where ϕ'_d is the design shearing resistance angle given by Equation 11.27.

$$\phi'_d = \tan^{-1}\left(\frac{\tan\phi'}{\gamma_{\phi'}}\right) \tag{11.27}$$

where ϕ' is the shearing resistance angle.
The dimensionless soil inertia $\bar{F}$ is given by Equation 11.28.

$$\bar{F} = \frac{a_g S}{g \tan\phi'_d} \tag{11.28}$$

The following constraints apply to the general bearing capacity expression in Equation 11.18:

$$0 < \bar{N} \le (1 - m\bar{F})^{k'} \tag{11.29}$$

where k' is a coefficient from Table 11.5.
The previous formulation has been recently extended to circular foundations on homogeneous and heterogeneous foundations by Chatzigogos et al. (2007).

11.5 SEISMIC DISPLACEMENTS

In cases where the transient seismic loadings exceed the available foundation resistance, permanent displacements will occur. The accelerations at which displacement commences are termed threshold accelerations. In many cases, the peak earthquake accelerations can exceed the threshold values by a substantial margin with minimal foundation displacement occurring. Though EN1998-5 generally requires that foundations remain elastic, for foundations above the water table, where the soil properties remain unaltered and the sliding will not affect the performance of any lifelines connected to the structure, a limited amount of sliding may be tolerated.

Designing on the basis of allowable deflections can result in significant economies by comparison to alternative 'elastic' design approaches. However, a cautious approach is required to the assessment of seismic displacements because modest variations in design parameters can result in substantial variations in displacements.

11.5.1 Sliding displacements

The principles whereby permanent seismic displacements can be calculated were set out by Newmark (1965) in his Rankine lecture. These are illustrated in Figure 11.10 for a block subjected to a rectangular acceleration pulse. The method considers that the block accelerates with the ground until threshold acceleration (N_g) is reached. The ground acceleration continues to rise to peak acceleration (A_g) but the acceleration of the block is limited by the shear capacity of the base to a value of N_g. The equations of motion give the velocity of the block and the ground and their relative displacement.

The Newmark analysis may be used directly to calculate the sliding displacement of a foundation provided that design acceleration time histories are available and the threshold acceleration for sliding has been established.

In many instances, design acceleration time histories will not be available for routine foundation design. Several authors have used the Newmark approach combined with earthquake acceleration records to derive 'design lines' relating sliding displacements to the ratio of threshold to peak accelerations (N/A). Notable examples are those of Franklin and Chang (1977), Richards and Elms (1979), Whitman and Liao (1985) and Ambraseys and Menu (1988).

The Ambraseys and Menu relationships are shown in Figure 11.11 for various probabilities of exceedance. It may be noted that both unsymmetrical (one-way) sliding and symmetrical (two-way) sliding have been considered. Significant differences between the two cases only arise when the peak acceleration is more than twice the threshold (i.e. $N/A < 0.5$). The Ambraseys and Menu database included earthquakes of M_S 6.4–7.7.

11.5.2 Shakedown settlement

Settlements under cyclic loads should be assessed when extended layers or thick lenses of loose, unsaturated cohesionless materials exist at a shallow depth. Excessive settlements may also occur in very soft clays because of cyclic degradation of their shear strength under ground shaking of long duration. If the settlements caused by densification or cyclic degradation appear capable of affecting the stability of the foundations, consideration should be given to ground improvement methods.

Dynamic settlement can be estimated using empirical relationships between volumetric strain, SPT N-values (corrected for overburden), and the cyclic shear strain. For example, the approach developed by Tokimatsu and Seed (1987) is based on relationships between the volumetric strain, the cyclic shear strain and SPT N-values. The peak shear strain computed from

Figure 11.10 Sliding displacement for a block with a rectangular base acceleration pulse. (After Newmark, N.M. 1965, *Geotechnique*, 15(2), 139–160).

the one-dimensional response analysis and the SPT corrected N-value at that point are entered into the Tokimatsu and Seed chart (see Figure 11.12) to yield the volumetric strain. The total settlement can then be obtained by integrating these volumetric strains as a function of depth.

11.5.3 Foundation horizontal connections

Tie beams should be provided between all foundations, except for ground type A (rock), or on ground types A and B (stiff soil) in areas of low seismicity.

The tie beams should be designed to withstand an axial force, considered in both tension and compression, equal to

$$\pm 0.3\, \alpha S N_{Ed} \text{ for ground type } B \tag{11.30}$$

$$\pm 0.4 \alpha S N_{Ed} \text{ for ground type } C \tag{11.31}$$

$$\pm 0.6 \alpha S N_{Ed} \text{ for ground type } D \tag{11.32}$$

where N_{Ed} = mean value of the design axial forces of the connected vertical elements.

Figure 11.11 Displacement versus N/A for various probabilities of exceedance.

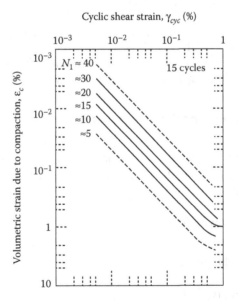

Figure 11.12 Assessment of volumetric strain. (Adapted from Tokimatsu, K. and Seed, H.B. 1987, *Journal of Geotechnical Engineering, ASCE*, 113(8), 861–878).

11.6 DESIGN EXAMPLE ON A SHALLOW FOUNDATION: PAD FOUNDATION

11.6.1 Sites

There are four sites (A, B, C and D) that are available for construction of the hotel as stated in Chapter 4 (Section 4.3).

The design ground acceleration is taken as $a_{gR} = 0.3\ g$.

The building importance factor for the hotel is taken as $\gamma_I = 1$ in this example.

Preliminary site investigation was carried out at all the sites. Borehole data and SPT and field vane shear tests were carried out at each site. This information is assimilated in Figure 11.13.

Site selected for the hotel:

Choose 'Site C' for shallow foundation design

Reasons:

1. Good, dense sand layer with 16 m thickness with high SPT numbers, overlying stiff clay
2. Angle of internal friction is 36°
3. Above the water table

In a practical design situation, site D may also be considered, at least as an initial candidate and the design calculations may be carried out using the undrained shear strength of the stiff clay with appropriate partial factors as outlined in Section 11.4.4. However, in this example, only site C will be considered.

Ultimate limit state (ULS) design:

1. Failure by sliding
2. Bearing capacity failure

(see Section 5.4 Verification and dimensioning criteria in EC8).

11.6.2 Design of pad foundation

The most heavily loaded columns in the hotel are along C and D lines separated by only 3 m spacing, so let us consider a combined PAD foundation for these two column bases.

As an initial guess, choose a 10 m × 4 m pad foundation, located 1 m below ground level to support the two columns along C and D lines shown in the plan view.

The worst loading occurs on two columns in a 4 m bay along C and D on the plan of the building. These loads are obtained in the structural design example (Chapters 3 and 5) with due consideration to the capacity design aspects and are shown in Table 11.7.

Using the above, we can obtain the following design loads:

Design vertical load, $N_{Ed} = 6,840\ \text{kN}$
Design horizontal, shear load, $V_{Ed} = 1,652\ \text{kN}$
Design moment load, $M_{Ed} = 4,493\ \text{kN m}$
Consider F.6 (EC8 on page 43)

$$N_{max} = \frac{1}{2}\rho g\left[1\pm\frac{a_v}{g}\right]B^2 N_\gamma$$

Figure 11.13 Bore hole data at the four sites from site investigation.

Table 11.7 Loads on the foundation from the columns

	Column C	Column D	Total design loads on the pad foundation
Axial load	5,978 kN	862 kN	6,840 kN
Shear load	826 kN	826 kN	1,652 kN
Moment load	2,405 kN m	2,088 kN m	4,493 kN m

The mass density of the sand = 1,650 kg/m³ (unit weight of 16.19 kN/m³).

$$a_v = 0.5 \cdot S \cdot a_g$$

Referring to Tables 3.2 and 3.1 in EC8, choose soil factor $S = 1.15$. Also

$$a_g = \gamma_I \cdot a_{gR}$$

where γ_I is the building importance factor, which is taken as unity for this hotel building. From design brief, $a_{gR} = 0.3\ g$;

$$a_v = 0.5 \times 1.15 \times 0.3\ g = 0.1725\ g$$

$$N_{max} = \frac{1}{2} 1,650 \times 9.81[1 \pm 0.225]B^2 N_\gamma$$

The friction angle for the sand needs to be reduced using the γ_m factor obtained before (see EC8 Page 23). $\gamma_m = 125$

$$\phi'_d = \tan^{-1}\left(\frac{\tan\phi_k}{\gamma_m}\right)$$

$$\phi'_d = \tan^{-1}\left(\frac{\tan 36°}{1.25}\right) = 30°$$

The bearing capacity factor can be calculated using

$$N_\gamma = 2\left[\tan^2\left(45 + \frac{\phi_d}{2}\right)e^{\pi\tan\phi_d} + 1\right]\tan\phi_d$$

$$N_\gamma = 38.8 \times \tan 30° = 22.4$$

$$N_{max} = \frac{1}{2} 1,650 \times 9.81 \times 0.8275 \times 22.4 \times B^2$$

$$N_{max} = 150\,B^2\ \text{kN/m}$$

$$N_{max,tot} = 150 \times B^2 \times L \text{ kN}$$

Substituting the dimensions of the footing (10 m × 4 m), we get

$$N_{max} = 15{,}000 \text{ kN/m}$$

and

$$N_{max,tot} = 60{,}000 \text{ kN}$$

11.6.3 Failure against sliding

Design friction resistance for footing above water table:

$$F_{Rd} = N_{Ed} \frac{\tan\delta_d}{\gamma_m}$$

Choose γ_m value from EC8 (Equation 5.1, EC8, p. 23)

$$\gamma_m = 1.25 \text{ and } \delta_d = \phi_k$$

Use Equation 5.2 (EC8, p. 23)

$$V_{Ed} \leq F_{Rd} + E_{Rd}$$

V_{Ed} is design horizontal shear force.
E_{Rd} is design lateral resistance can be up to 30% of passive resistance according to EC8.

Sliding resistance:
Angle of internal friction for this sand:

$$\phi'_k = 36°$$

$$F_{Rd} = 6{,}840 \times \frac{\tan 36°}{1.25}$$

$$F_{Rd} = 3{,}975.6 \text{ kN}$$

Therefore,

$$V_{Ed} \leq F_{Rd} \ (1{,}652 \text{ kN} \leq 3975.6 \text{ kN})$$

So, this satisfies sliding check. Note that in this example we did not have to use 30% of passive resistance clause in EC8 here, but if needed, we could estimate 30% of passive resistance for the footing, once we established the depth of the foundation below ground level.

11.6.4 Verification of bearing capacity

Now calculate $\bar{N}$, $\bar{V}$ and $\bar{M}$ using equation F.2 (see EC8 - Part 5, p. 42).
   Choose $\gamma_{Rd} = 1$, using Table F.2 (see EC8, p. 44).

$$\bar{N} = \frac{\gamma_{Rd} N_{Ed}}{N_{max,tot}} = \frac{1 \times 6,840 \text{ kN}}{60,000 \text{ kN}} = 0.114$$

$$\bar{V} = \frac{\gamma_{Rd} V_{Ed}}{N_{max,tot}} = \frac{1 \times 1,652 \text{ kN}}{60,000 \text{ kN}} = 0.0275$$

$$\bar{M} = \frac{\gamma_{Rd} M_{Ed}}{B N_{max,tot}} = \frac{1 \times 4,493 \text{ kNm}}{10 \times 60,000 \text{ kNm}} = 0.0075$$

Calculate $\bar{F}$ using F.7 (EC8, p. 43)

$$\bar{F} = \frac{a_g S}{g \tan \phi'_d}$$

$$\bar{F} = \frac{0.3 g \times 1.15}{g \tan 30°} = 0.598$$

Check using F.8 (EC8, p. 43)

$$0 \le \bar{N} \le (1 - m\bar{F})^{k'}$$

Constants m and k' are to be chosen appropriately from Table F.1 (EC8 - Part 5, page 44)
Values of $m = 0.96$ and $k' = 0.39$.

$$0 \le 0.114 \le (1 - 0.96 \times 0.598)^{0.39}$$

$$0 \le 0.114 \le 0.7169$$

So check is satisfied.
Check for bearing capacity failure (EC8 - Part 5, equation F.1):

$$\underbrace{\frac{(1 - e\bar{F})^{C_T} (\beta \bar{V})^{C_T}}{(\bar{N})^a [(1 - m\bar{F}^k)^{k'} - \bar{N}]^b}}_{I} + \underbrace{\frac{(1 - f\bar{F})^{C_m} (\gamma \bar{M})^{C_m}}{(\bar{N})^c [(1 - m\bar{F}^k)^{k'} - \bar{N}]^d}}_{II} - 1 \le 0$$

The constants for dense sand may be chosen from EC8 - Part 5 as shown in Table 11.8:
Constants: *Dense Sand*
Substitute for the values and check the inequality is satisfied.

Table 11.8 Parameters for dense sand

Parameter	Value
a	0.92
b	1.25
c	0.92
d	1.25
e	0.41
f	0.32
m	0.96
k	1
k'	0.39
C_T	1.14
C_M	1.01
C'_M	1.01
β	2.9
γ	2.8

$$\frac{(1-0.41\times0.598)^{1.14}(2.9\times0.0275)^{1.14}}{0.114^{0.92}[(-0.96\times0.598)^{0.39}-0.114]^{1.25}}=0.56375 \tag{I}$$

$$\frac{(1-0.32\times0.598)^{1.01}(2.8\times0.0075)^{1.01}}{0.114^{0.92}[(-0.96\times0.598)^{0.39}-0.114]^{1.25}}=0.22628 \tag{II}$$

Inequality is therefore $(I+II-1)$;

$$0.564+0.226-1\leq0$$

$$-0.2099\leq0$$

So, check is satisfied. The PAD foundation is safe against bearing failure. Recall that the design in EC8 is based on partial safety factors and therefore no other global safety factor needs to be applied.

Check the plan view of the building (see Figure 11.14). Columns along C and D lines are separated by 3 m spacing. Columns along B and E are separated by 20 m spacing. The required dimensions for the PAD foundation designed above are 10 m × 4 m.

11.7 DESIGN EXAMPLE ON A SHALLOW FOUNDATION: RAFT FOUNDATION

11.7.1 Design of raft foundation

A pad foundation was designed for individual columns along lines C and D of the hotel plan in the previous section. This section demonstrates the design of a raft foundation for the hotel. The design loads for the raft foundation are obtained from structural analysis.

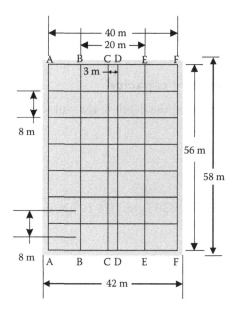

Figure 11.14 Plan view of the hotel showing the location of the pad foundation.

For this design example, we will use the structural loads for a steel frame design outlined earlier. (*Note*: The total mass of the structure will be somewhat larger when the structural design is based on concrete, but the following calculations can be easily repeated taking into account the increased mass). For a raft foundation design, we need to consider loads from all columns. Also we need to multiply the loads by 7, as there are seven bays in the building. The loads are summed up and shown below:

Design vertical load $N_{Ed} = 80.522$ MN
Design moment load $M_{Ed} = 414.4$ MN m
Design horizontal, shear load $V_{Ed} = 21.5$ MN

11.7.2 Failure against sliding

Sliding Resistance:
   Angle of internal friction for this sand:

$$\phi_k = 36°$$

$$F_{Rd} = 80.5 \times 10^3 \times \frac{\tan 36°}{1.25}$$

$$F_{Rd} = 46,802 \text{ kN}$$

Therefore,

$$V_{Ed} \le F_{Rd}(21.5 \text{ MN} \le 46.8 \text{ MN})$$

So, satisfies sliding check.

[**Note:** As in the case of pad foundation, we did not have to use 30% of passive resistance clause in EC8 here, but if needed we could estimate 30% passive resistance for the footing, once we established depth of the foundation below ground level.]

II.7.3 Verification of bearing capacity

Site C has purely cohesionless soil at the depth of the pad foundation.

Consider F.6 (EC8 on p. 43)

$$N_{max} = \frac{1}{2}\rho g \left[1 \pm \frac{a_v}{g}\right] B^2 N_\gamma$$

The mass density of the sand = 1,650 kg/m³ (16.19 kN/m³)

$$a_v = 0.5 \cdot S \cdot a_g$$

Referring to Tables 3.2 and 3.1 in EC7, choose soil factor $S = 1.15$. Also,

$$a_g = \gamma_I \cdot a_{gR}$$

where γ_I is the building importance factor, which is taken as unity for this hotel building. From design brief, $a_{gR} = 0.3$ g;

$$a_v = 0.5 \times 1.15 \times 0.3g = 0.1725\,g$$

$$N_{max} = \frac{1}{2}1,650 \times 9.81[1 \pm 0.1725]B^2 N_\gamma$$

The friction angle for the sand needs to be reduced using the γ_m factor obtained before (see EC8 Page 23). $\gamma_m = 1.25$

$$\phi'_d = \tan^{-1}\left(\frac{\tan\phi_k}{\gamma_m}\right)$$

$$\phi'_d = \tan^{-1}\left(\frac{\tan 36°}{1.25}\right) = 30°$$

The bearing capacity factor can be calculated using;

$$N_\gamma = 2\left[\tan^2\left(45+\frac{\phi_d}{2}\right)e^{\pi\tan\phi_d}+1\right]\tan\phi_d$$

$$N_\gamma = 38.8 \times \tan 30° = 22.4$$

$$N_{max} = \frac{1}{2} \times \frac{1,650}{1,000} \times 9.81 \times 0.8275 \times 22.4 \times B^2$$

$$N_{max} = 150\,B^2 \text{ kN/m}$$

If we assume that the Raft foundation is going to be a '$B \times L$' foundation we can use, as before, the following equation

$$N_{max,tot} = 150 \times B^2 \times L \text{ kN}$$

Choose a 42 m × 58 m raft foundation, located 1 m below ground level to support the whole building shown in the plan view with a 1m extension beyond the plan area.

$$N_{max} = 264,600 \text{ kN/m}$$

$$N_{max,tot} = 15,346,800 \text{ kN}$$

Now calculate $\bar{N}$, $\bar{V}$ and $\bar{M}$ using equation F.2 (see EC8 - Part 5, p. 42).
Choose $\gamma_{Rd} = 1$, using Table F.2 (see EC8, p. 44).

$$\bar{N} = \frac{\gamma_{Rd} N_{Ed}}{N_{max,tot}} = \frac{1 \times 80,522 \text{ kN}}{15,346,800 \text{ kN}} = 0.0052$$

$$\bar{V} = \frac{\gamma_{Rd} V_{Ed}}{N_{max,tot}} = \frac{1 \times 21,500 \text{ kN}}{15,346,800 \text{ kN}} = 0.0014$$

$$\bar{M} = \frac{\gamma_{Rd} M_{Ed}}{B N_{max,tot}} = \frac{1 \times 41,4500 \text{ kNm}}{42 \times 15,346,800 \text{ kNm}} = 0.00064$$

Use $\bar{F}$ value calculated before as this will not change with the loading;

$$\bar{F} = 0.598$$

Use the table of constants for dense sand from Section 11.6.4 (Table 11.8) used for the bearing capacity verification calculations for the pad foundation.
Check using F.8 (EC8, p. 43)

$$0 \le \bar{N} \le (1 - m\bar{F})^{k'}$$

Constants m and k' are taken from Table 11.8 for 'dense sand' from before, as these do not change.
Values of $m = 0.96$ and $k' = 0.39$.

$$0 \le 0.0052 \le (1 - 0.96 \times 0.598)^{0.39}$$

$$0 \le 0.0052 \le 0.7169$$

So, check is satisfied.
Check for bearing capacity failure (EC8 - Part 5, equation F.1):

$$\underbrace{\frac{(1 - e\bar{F})^{C_T} (\beta \bar{V})^{C_T}}{(\bar{N})^a [(1 - m\bar{F}^k)^{k'} - \bar{N}]^b}}_{I} + \underbrace{\frac{(1 - f\bar{F})^{C_m} (\gamma \bar{M})^{C_m}}{(\bar{N})^c [(1 - m\bar{F}^k)^{k'} - \bar{N}]^d}}_{II} - 1 \le 0$$

Figure 11.15 Plan view of the hotel showing the raft foundation.

Substitute for the values and check the inequality is satisfied.

$$\frac{(1-0.41\times0.598)^{1.14}(2.9\times0.0014)^{1.14}}{0.0052^{0.92}\left[(1-0.96\times0.598)^{0.39}-0.0052\right]^{1.25}} = 0.263264844 \tag{I}$$

$$\frac{(1-0.32\times0.598)^{1.01}(2.8\times0.00064)^{1.01}}{0.0052^{0.92}\left[(1-0.96\times0.598)^{0.39}-0.0052\right]^{1.25}} = 0.262201923 \tag{II}$$

Inequality is therefore $(I + II - 1)$;

$$0.263 + 0.262 - 1 \leq 0$$

$$-0.475 \leq 0$$

So, check is satisfied. The raft foundation is safe against bearing failure.

Hence a raft foundation with dimensions of 42 m × 58 m located 1 m below the ground level is suitable for the hotel. A plan view of the raft foundation that extends 1m the plinth area of the hotel is shown in Figure 11.15.

REFERENCES

Ambraseys, N.N. and Menu, J.M. 1988. Earthquake-induced ground displacements. *Earthquake Engineering and Structural Dynamics*, 16, 985–1006.

API. 2000. Recommended practice for planning, designing and constructing fixed offshore platforms—working stress design, RP2A-WSD, 21st edn. American Petroleum Institute, Washington.

Auvinet, G., Pecker, A. and Salençon, J. 1996. Seismic bearing capacity of shallow foundations in Mexico city during the 1985 Michoacan earthquake. In: *Proceedings of the 11th World Conference on Earthquake Engineering*, Acapulco, pp. 143–151.

Bird, J.F. and Bommer, J.J. 2004. Earthquake losses due to ground failure. *Engineering Geology*, 75(2), 147–179.

Borcherdt, R. and Glassmoyer, G. 1992. On the characterisation of local geology and their influence on ground motions generated by the Loma Prieta earthquake in the San Francisco Bay Region, California. *Bulletin of the Seismological Society of America*, 82, 603–641.

Boulanger, R.W. and Idriss, I.M. 2004. *Evaluating the Potential for Liquefaction or Cyclic Failure of Silts and Clays*. Report No. UCD/CGM 04–01, Center for Geotechnical Modeling, University of California, Davis.

Brinch Hansen, J. 1970. A revised and extended formula for bearing capacity. Bulletin of the Danish Geotechnical Institute No. 28.

Budhu, M. and Al-Karni, A. 1993. Seismic bearing capacity of soils. *Geotechnique*, 43(1), 181–187.

Caquot, A. and Kerisel, J. 1953. Sur le terme de surface dans le calcul des foundations en milieu pulverulent. In: *Proceedings 3rd International Conf. on Soil Mechanics and Geotechnical Engineering*, Zurich, Vol. 1, pp. 336–337.

Chatzigogos, C.T., Pecker, A. and Salençon, J. 2007. Seismic bearing capacity of a circular footing on a heterogeneous cohesive soil. *Soils and Foundations*, 47(4), 783–797.

Dormieux, L. and Pecker, A. 1995. Seismic bearing capacity of a foundation on a cohesionless soil. *Journal of Geotechnical Engineering Division, ASCE*, 121(3), 300–303.

Dowrick, D. 1987. *Earthquake Resistant Design*, Wiley, New York.

EN 1997-1 2004. Eurocode 7: Geotechnical design, Part 1: General rules. CEN European Committee for Standardisation.

EN 1998-1 2004. Eurocode 8: Design of structures for earthquake resistance, Part 1: General rules, seismic actions and rules for buildings. CEN European Committee for Standardisation.

EN 1998-5 2004. Eurocode 8: Design of structures for earthquake resistance, Part 5: Foundations, retaining structures and geotechnical aspects. CEN European Committee for Standardisation.

Finn, L., Iai, S. and Matsunaga, Y. 1995. The effects of site conditions on ground motions. In: *Tenth European Conference Earthquake Engineering*, Duma (ed.), Balkema, Rotterdam, pp. 23–31.

Frank, R., Bauduin, C., Driscoll, R., Kavvadas, M., Krebs Ovesen. N., Orr, T. and Schuppener, B. 2004. Designers' Guide to EN1997-1 – Eurocode 7: Geotechnical design-general rules. Thomas Telford.

Franklin, A.G. and Chang, F.K. 1977. Earthquake resistance of earth and rockfill dams; permanent displacements of embankments by Newmark sliding block analysis. Miscellaneous Paper S.71.17, U.S. Army Corps of Engineers Waterways Experimental Station, Vicksburg.

Idriss, I.M. 1990. Response of soft soil sites during earthquakes. In: *Proc. H.B.Seed Memorial Symposium*, Berkley, pp. 273–289.

Ingra, S.T. and Baecher, G.B. 1983. Uncertainty in bearing capacity of sands. *Journal of Geotechnical Engineering ASCE*, 109(1), 899–914.

Jardine, R.J., Potts, D.M., Fourie, A.B. and Burland, J.B. 1986. Studies of the influence of non-linear stress-strain characteristics in soil–structure interaction. *Geotechnique*, 36(3).

Kokusho, T. and Matsumoto, M. 1998. Nonlinearity in site amplification and soil properties during the 1995 Hyogoken-Nambu Earthquake. *Soils and Foundations*, Special Issue No. 2, September, 1–9.

Lee, M.K.W. and Finn W.D.L. 1978. DESRA-2, dynamic effective stress response analysis of soil deposits with energy transmitting boundary including assessment of liquefaction potential. Soil Mech. Series No. 38. Dept. of Civil Eng., University of British Columbia, Vancouver, B.C.

Lysmer, J., Udaka, T., Tsai, C.-F. and Seed, H.B. 1975. FLUSH a computer program for approximate 3-D analysis of soil-structure interaction problems, EERC, 75–30.

Martin, G.R. and Yan, L. 1995. Modelling passive earth pressure for bridge abutments. Earthquake-Induced Movements and Seismic Remediation of Existing Foundations and Abutments, Geotechnical Special Publication No. 55, 1–16, American Society of Civil Engineers.

Mendoza, M.J. and Auvinet, G. 1988. The Mexico earthquake of September 19, 1985–Behaviour of building foundations in Mexico City. *Earthquake Spectra*, 4(4), 835–852.

Mohammadioun, B. and Pecker, A. 1984. Low frequency transfer of seismic energy by superficial soil deposits and soft rocks. *Earthquake Engineering and Structural Dynamics*, 12, 537–564.

Newmark, N.M. 1965. Effects of earthquakes on dams and embankments. *Geotechnique*, 15(2), 139–160.

Paolucci, R. and Pecker, A. 1997a. Seismic bearing capacity of shallow strip foundations on dry soils. *Soils and Foundations*, 37(3), 95–105.

Paolucci, R. and Pecker, A. 1997b. Soil inertia effects on the bearing capacity of rectangular foundations on cohesive soils. *Engineering Structures*, 19(8), 637–643.

Pappin, J.W. 1991. Design of foundations and soil structures for seismic loading. In: *Cyclic Loading of Soils*, O'Reilly, M.P. and Brown S.F. (eds). Blackie, London, pp. 306–366.

Pecker, A. and Salençon, J. 1991. Seismic bearing capacity of shallow strip foundations on clay soils. In: *Proceedings CEE-Mexico International Workshop*, CENAPRED, 22–26 April, pp. 287–304.

Pecker, A. 1994. Seismic design of shallow foundations. In: *State of the Art, 10th European Conference on Earthquake Engineering*, Vienna, Vol. 2, pp. 1001–1010.

Pecker, A. 1997. Analytical formulae for the seismic bearing capacity of shallow strip foundations. *Seismic behaviour of Ground and Geotechnical Structures*, Seco e Pinto (ed.), Balkema, Rotterdam.

Pecker, A. 2005. Maximum ground motions in probabilistic seismic hazard analyses. *Journal of Earthquake Engineering*, 9(4), 1–25.

Pender, M.J. 1996. Earthquake resistant design of foundations. *Bulletin of New Zealand National Society for Earthquake Engineering*, 29(3), 155–171.

Ross, G.A., Seed, H.B. and Migliaccio, R.R. 1969. Bridge foundation failure in Alaska earthquake. *Proceedings of ASCE, Journal of Soil Mechanics and Foundations Div.*, 95(SM4), 1007–1036.

Richards, R. and Elms, D.G. 1979. Seismic behaviour of gravity retaining walls. *Journal of Geotechnical Engineering Division ASCE*, 105, No. GT4, pp. 449–464.

Richards, R., Elms, D.G. and Budhu, M. 1993. Seismic bearing capacity and settlements of shallow foundations. *Journal of Geotechnical Engineering ASCE*, 119, 662–674.

Salençon, J. and Pecker, A. 1995a. Ultimate bearing capacity of shallow foundations under inclined and eccentric loads-Part I: Purely cohesive soil without tensile strength. *European Journal of Mechanics/A*, 3, 349–375.

Salençon, J. and Pecker, A. 1995b. Ultimate bearing capacity of shallow foundations under inclined and eccentric loads-Part II: Purely cohesive soil. *European Journal of Mechanics/A*, 3, 377–396.

Sarma, S.K. and Iossifelis, I.S. 1990. Seismic bearing capacity factors of shallow strip footings. *Geotechnique*, 40(2), 265–273.

Schnabel, P.B., Lysmer, J. and Seed, H.B. 1972. SHAKE – A Computer Program for Earthquake Response Analysis of Horizontally Layered Sites, Report EERC 72–12, Earthquake Engineering Research Centre, University of California, Berkeley.

Seed, R.B., Dickensen, S.E. and Idriss, I.M. 1991. Principal geotechnical aspects of the Loma Prieta Earthquake. *Soils and Foundations*, 31(1), 1–27.

Seed, C., Moss, K., Wu, P., Riemer, S., Bray, K. and Faris, A. 2003. Recent advances in soil liquefaction engineering: A unified and consistent framework. In: *26th Annual ASCE LA Geotechnical Seminar, Keynote Presentation*, HMS Queen Mary, Long Beach, California, April, pp. 48–59.

Shi, X. and Richards, R. Jr. 1995. Seismic bearing capacity with variable shear transfer. *Earthquake-Induced Movements and Seismic Remediation of Existing Foundations and Abutments*, Geotechnical Special Publication No. 55, 17–32, American Society of Civil Engineers.

Suetomi, I. and Yoshida, N. 1998. Nonlinear behavior of surface deposit during the 1995 Hyogoken-Nambu earthquake. *Soils and Foundations*, Special Issue No. 2, September, 11–22.

Terzaghi, K. and Peck, R.B. 1948. *Soil Mechanics in Engineering Practice*. Wiley, New York; Chapman and Hall, London.

Terzaghi, K. and Peck, R.B. 1967. *Soil Mechanics in Engineering Practice*. 2nd edn, Wiley, New York.

Tokimatsu, K. and Asaka, Y. 1998. Effects of liquefaction-induced ground displacements on pile performance in the 1995 Hyogoken-Nambu Earthquake. *Soils and Foundations*, Special Issue No. 2, September, 163–177.

Tokimatsu, K. and Seed, H.B. 1987. Evaluation of settlements in sand due to earthquake shaking. *Journal of Geotechnical Engineering ASCE*, 113(8), 861–878.

Vesic, A.S. 1975. Bearing capacity of shallow foundations. In *Foundation Engineering Handbook*, Chapter. 3, Winterkorn and Fang (ed.), Van Nostrand Reinhold, Amsterdam.

Whitman, R.V. and Liao, S. 1985. Seismic design of gravity retaining walls. Miscellaneous Paper GL-85-1, U.S. Army Corps of Engineers Waterways Experimental Station, Vicksburg.

Youd, T.L. and Perkins, D.M. 1978. Mapping of liquefaction-induced ground failure potential. *Journal of Geotechnical Engineering ASCE*, 104, GT4, 433–446.

Zadroga, B. 1994. Bearing capacity of shallow foundations on non-cohesive soils. *Journal of Geotechnical Engineering ASCE*, 120(11), 1991–2008.

Zeevaert, L. 1991. Seismosoil dynamics of foundations in Mexico City earthquake, 19 September, 1985. *Journal of Geotechnical Engineering ASCE*, 117(3), 376–428.

Pile foundations

Gopal S. P. Madabhushi and Robert May

CONTENTS

12.1 INTRODUCTION

Pile foundations are widely used both onshore and offshore to transfer heavy structural loads to competent load-bearing soil strata or bed rock. Geotechnical engineers are called upon to design deep foundations when the shallow layers of soils beneath the building are either unable to support the loads imposed by the superstructure on the shallow foundations or if the shallow layers may become unstable due to the cyclic shear stresses induced by the earthquake loading. Under such circumstances, it is imperative to look for pile foundations that transfer the load from the superstructure to more firm and stable soil strata at deeper levels or onto bed rock. In this chapter, the seismic design of pile foundations is considered in the light of the EC8 provisions as well as some of the current research findings. It is perhaps helpful if some of the well-known examples of failures of pile foundations during or following an earthquake loading are considered first.

12.1.1 Examples of pile foundation failures following earthquake loading

Although pile foundations are widely used in the regions of high seismicity around the world, there are a number of examples where the pile foundations have failed during strong earthquake events. Such failures can cause either collapse of the superstructure or cause excessive settlements and rotations.

During the 1964 Niigata earthquake, the Showa bridge collapsed (as shown in Figure 12.1a). Figure 12.1b shows one of the piles that was extracted during the post earthquake investigation, while Figure 12.1c shows the schematic diagram of the collapsed spans (Figure 12.1a–c).

The Showa bridge collapse was attributed to many causes. For example, Hamada (1992) proposed that lateral spreading of the soil following liquefaction (see Figure 12.1c) has caused large displacements at the pile heads and resulted in the dislodging of the spans. Bhattacharya et al. (2005a) have proposed that buckling of the piles in liquefied sands could have caused the collapse of the Showa bridge. The Showa bridge collapse is not a unique event. There have been many other failures involving pile foundations.

More recently, the Harbour Master's building at Kandla port suffered a rotation of about 11° from the vertical following the Bhuj earthquake of 2001 as shown in Figure 12.2. The pile foundations supporting this building have suffered differential settlement. Similarly, Tokimatsu et al. (1997) describe the failure of a three-storied building supported on pile foundations during the 1995 Kobe earthquake as shown in Figure 12.3. They suggest that the failure of the quay wall allowed the seaward movement of the soil that caused the pile foundations to fail. Lateral spreading of soil subjects the pile to additional loading. The piles need to be adequately designed to sustain these additional lateral loads.

Figure 12.1 (a) Collapse of the Showa bridge. (b) Excavated pile. (After Hamada, M. 1992. *Case Studies of Liquefaction and Lifeline Performance during Past Earthquakes* NCEER, NY.) (c) Collapsed spans of the Showa bridge. (After Takata et al. 1965. *Damage to bridges in Niigata earthquake. Report no. 125-5*, Public Works Research Institute. (In Japanese.))

12.1.2 Lessons learnt from pile foundation failures

Pile foundations seem to suffer from earthquake loading for a variety of reasons. A comprehensive list of pile foundations of various structures that have performed poorly was compiled by Bhattacharya et al. (2004). The load-bearing soil strata into which the piles are transferring the load may change their character under strong cyclic loading. In addition, the piles have to bear the inertial and kinematic loading described in detail later in Section 12.4.1. In many cases, the pile foundation failures appear to be associated with 'liquefaction' of the ground to some depth around the piles. Similarly presence of a non-liquefied

Figure 12.2 Rotation of a tall masonry building on pile foundations during the Bhuj earthquake.

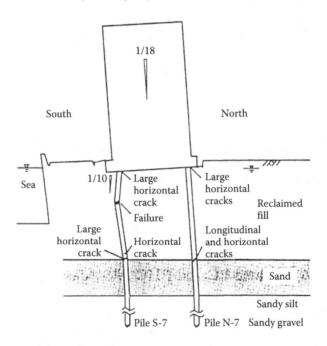

Figure 12.3 Failure of piles in a three-storied building in 1995 Kobe earthquake. (Adapted from Tokimatsu, K. et al. 1997. *Journal of Structural Engineering AIJ (Japan)*, 495, 95–100).

layer such as stiff clay overlying a liquefiable layer appears to cause additional loading particularly if the ground is on a slope. Research has shown that slopes as gentle as 1–3° can result in lateral spreading of liquefied soil and that of any non-liquefied soil crust overlying the liquefied soil, Haigh et al. (2000).

12.1.3 Eurocode 8 provisions

The normal static design of pile foundations must be carried out under the provisions of Eurocode 7. In addition, EC8 recommends that the liquefaction potential of all the soil layers at a given site be carefully determined based on the SPT tests conducted at the site. It is also suggested that careful consideration of any additional loading on the piles and pile caps that may arise due to the lateral spreading of the soil, particularly in the presence of a non-liquefiable soil strata present overlying a liquefiable layer. In addition, where liquefaction is anticipated, it is suggested that the strength of the liquefied soil must be ignored.

12.2 PILE FOUNDATION DESIGN UNDER STATIC LOADING

The static design of the pile foundations has to be carried out in accordance with Eurocode 7. A procedure is outlined below for cohesionless soils. Similar approach can be used for cohesive soils with suitable modification.

The pile capacity can be determined as a combination of the base capacity and the shaft capacity.

$$\underset{\text{Pile capacity}}{Q} = \underset{\text{Base capacity}}{Q_b} + \underset{\text{Shaft capacity}}{Q_s} \tag{12.1}$$

12.2.1 Base capacity

The base capacity depends on the bearing capacity of the soil at the pile tip level. It can be calculated using;

$$Q_b = q_b \cdot A_b \tag{12.2}$$

where A_b = base area of pile shaft.

$$q_b = \sigma'_v \cdot N_q \tag{12.3}$$

The bearing capacity factor N_q for a deep foundation can be obtained using the chart shown in Figure 12.4. These types of charts were originally proposed by Berezantzev et al. (1961) and subsequently modified by several researchers.

12.2.2 Shaft capacity

The shaft capacity is obtained by estimating the shear stress generated along the shaft, which can be calculated as

$$\tau_s = K_s \cdot \sigma'_v \cdot \tan\delta \tag{12.4}$$

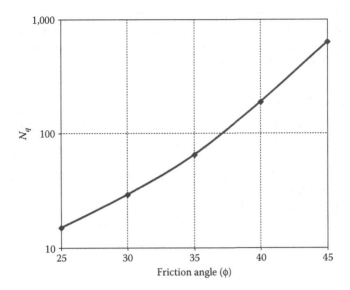

Figure 12.4 Bearing capacity factor N_q for deep foundations.

where K_s depends on type of pile and installation (driven or cast in-situ piles), σ'_v is the effective stress at the elevation where shear stress is being calculated and δ is the friction angle between the pile and the soil.

For driven piles, $K_s \geq 1$, so choose $K_s = 1$ conservatively.

In order to obtain the shaft capacity due to skin friction, we need to integrate the shear stress over the surface area of the pile using the following equation:

$$Q_s = 2\pi r \times \int_0^L \tau_s \qquad (12.5)$$

where r is the pile radius and L is the length of the pile.

The overall pile capacity can be determined by adding the base capacity and shaft capacity. It must be noted that where multiple piles are present, the pile spacing must be nominally 2–3 pile diameters. Closer the spacing of the piles, lower the efficiency of the pile group.

In addition, the piles need to be designed following the procedure outlined in Eurocode 7 taking into account appropriate National Annex. This aspect is outlined in the design example presented in Section 12.7.

12.3 LIQUEFACTION EFFECTS ON PILE FOUNDATIONS

Soil liquefaction is the association of phenomena like piping, boiling, mud volcanoes, etc. that lead to severe loss of strength in loose saturated soils. It is well known that loose sandy soils and sandy silts are particularly vulnerable from a liquefaction point of view. In Sections 8.3.1 through 8.3.3 of Chapter 8, we have seen how to determine whether a given site is susceptible to liquefaction by using in-situ tests such as SPT or CPT. EC8 requires the assessment of a site to determine its vulnerability to liquefaction.

Pile foundations can suffer the effects of soil liquefaction around them in a number of ways. These are considered in detail next.

12.3.1 Buckling of piles in liquefiable soils

Piles are slender columns that are supported by surrounding soil. Normally pile foundations do not suffer buckling except when placed in very soft soils and are carrying large axial loads. As discussed in Section 12.1, the load-carrying mechanism is via base capacity and skin friction. The horizontal stresses generated around the surface area of the pile provide the skin friction and also offer lateral support by acting like closely spaced struts.

During earthquake loading, if the soil suffers liquefaction (as determined using the EC8 – Part 5 procedure outlined in Chapter 8, Section 8.3.3) then the lateral support (and the skin friction) may be lost. According to EC8 – Part 5 the strength of any soil layer that liquefies must be ignored. In addition, if the piles are carrying large axial loads, then they may become vulnerable to buckling failure. Recent research at Cambridge (Bhattacharya et al., 2004, 2005b) has shown that pile buckling as a possible mechanism of failure if the following conditions are satisfied:

The pile is fully end bearing, that is, the pile tip is socketed into the bed rock.
If the pile is carrying a relatively large axial load compared to the Euler buckling load of an equivalent column.

The Euler buckling load can be calculated quite easily using the following equation:

$$P_E = \frac{\pi^2 EI}{L_e^2} \tag{12.6}$$

where EI is the flexural rigidity of the pile and L_e is the equivalent length of the pile.

12.3.1.1 Fixity condition and equivalent pile length

Equivalent length (L_e) depends on the end conditions of the pile. For example the top of the pile is almost always is connected to a pile cap. This provides a rotational restraint. However, the pile cap may be able to 'sway' laterally especially if there is laterally spreading non-liquefied crust present around the pile cap. So at the top of the pile, we generally expect a rotational restraint but not a translational restraint. At the base of the pile, if the pile tip is socketed into the bed rock to sufficient depth, then there will be both rotational and translational restraints. These conditions will yield the equivalent length (L_e) to be the length of the pile in the liquefied soil. On the other hand, if the top of the pile is free to rotate and translate and base of the pile is fixed in both rotation and translation, then L_e will be twice the length of the pile in the liquefied soil. This is demonstrated in Figure 12.5.

Similarly, we can easily obtain equivalent lengths for other end conditions by considering the buckling mode shape of the pile. It must be pointed out that the Euler buckling load is very sensitive to the equivalent length.

Once the equivalent length (L_e) is determined and the pile's flexural rigidity EI is known, the Euler buckling load can be calculated.

12.3.1.2 Slenderness ratio

The concept of slenderness ratio of the pile can also be used to check the pile design for any possible buckling. Slenderness ratio may be defined simply as

$$\kappa = \frac{L_e}{r} \tag{12.7}$$

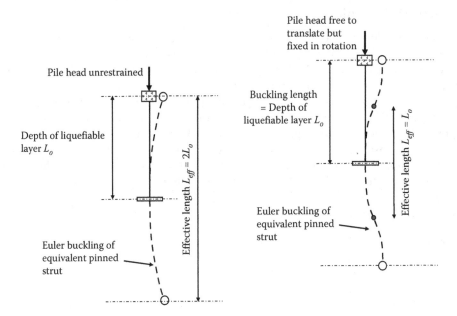

Figure 12.5 Buckling mode shape and effective length.

where r is the minimum radius of gyration of the pile section given by $\sqrt{I/A}$. I is the second moment of area about the weakest axis and A is the cross-sectional area of the pile. For a tubular pile the minimum radius of gyration can be estimated as 0.35 times the outside diameter of the pile.

The slenderness ratio κ can be used to quickly check the vulnerability of the pile to buckle, if liquefaction potential is high, that is, liquefaction of soil around the pile is to be expected based on methodology explained in Sections 8.3.1 through 8.3.3 of Chapter 8. If the slenderness ratio κ is very much less than 50, the piles can be considered generally safe from buckling as the pile would behave more as a short column rather than a long column. For slenderness ratio κ greater than 50, buckling of piles need to be considered in view of the axial load anticipated on the pile, end conditions of the pile etc. as described earlier.

12.3.1.3 Critical load on piles

Normally, the applied axial load must be less than the Euler buckling load by a factor of 5 or more. Euler buckling load calculation is for 'idealised' situations where the load is completely concentric to the axis of the pile and the pile does not have any imperfections. Any deviation from these conditions can result in a large drop in the estimated buckling load. Therefore it is prudent to have a large factor between the Euler buckling load and the design axial load on the pile.

Similarly, it must be noted that the above simple calculation does not account for the moment loading applied on the pile cap. Of course, the presence of a moment load on the pile cap again reduces the buckling load. In addition, any lateral displacement of the pile cap due to lateral spreading of the soil following liquefaction can induce additional P–Δ effects. Also any errors in the pile alignment due to pile wander during installation will reduce the buckling load. Due consideration must be given to these factors.

12.3.2 Lateral spreading of sloping ground

One of the side effects of liquefaction of soil is that the sloping ground start to move in the down slope direction. This is often termed as 'lateral spreading'. Recent earthquakes such as the 921 Ji-Ji earthquake in Taiwan and the Bhuj earthquake in India provided many examples of lateral spreading of ground. In Figures 12.6 through 12.9, some examples of lateral spreading are presented. In Figure 12.6, the lateral spreading of a slope past a bridge pier is seen. Clearly such a lateral spread will generate large lateral forces due to the passive pressures generated in the upslope soil wedge. Unrestrained, the soil would spread down the slope as seen in Figure 12.7. River banks as seen in Figure 12.7 often exhibit tension crack parallel to the river as the whole slope tries to spread into the river following earthquake induced liquefaction.

Similar damage was also seen during the Bhuj earthquake of 2001 in India. In Figure 12.8, the lateral spreading that occurred next to a railway line that serviced the bulk material transportation port of Navlakhi in Gujarat is seen. This led to serious disruption to the port operations and a large section of the railway had to be relayed. In Figure 12.9, the lateral

Figure 12.6 Lateral spreading past a bridge pier at the new Taichung bridge, Taiwan.

Figure 12.7 Lateral spreading of slopes of a river bank in Taichung, Taiwan.

Figure 12.8 Lateral spreading next to a railway track at the Navalakhi port in Gujarat, India.

Figure 12.9 Lateral spreading of the downstream slope of an earth dam in Gujarat, India.

spreading that occurred on the downstream slope of an earth dam in Gujarat is seen. Again piles are often used to stabilise the upstream and downstream slopes of earth dams. Such piles need to resist the large lateral forces created by the soil passive pressures once the whole slope is subjected to lateral spreading.

12.3.2.1 Slope angle

As the liquefied soil has very little shear resistance, by definition, it is likely that even gently sloping ground are liable to suffer lateral spreading. Based on the dynamic centrifuge

Figure 12.10 Non-liquefiable soil crust on a liquefiable soil layer.

tests carried out at Cambridge, Haigh et al. (2000) concluded that ground sloping even at 3–6 degrees will suffer lateral spreading. Similar results were reported by other researchers at RPI (Dobry et al., 2003) and UC, Davis (Brandenberg et al., 2005). The amount of lateral spreading suffered by sloping ground is usually in the order of several meters. If the excess pore water pressures generated in the liquefied ground are retained for several tens of seconds, it is possible to estimate the amount of lateral spreading suffered by the soil using Newmarkian style sliding block analysis with suitable modification to include effective stress on the sliding plane. Haigh et al. (2000) have shown that such calculations yield lateral spreading of several meters.

In the context of pile foundations that pass through laterally spreading soil, it is sufficient to recognise that the lateral spreading will be in the order of meters and therefore sufficient soil strains are mobilised to generate full passive earth pressures. This becomes more important when there are non-liquefied layers above the liquefied layers. Further, if these upper layers are of clayey nature with low hydraulic conductivity, then they exacerbate the problem by helping the liquefied layer to retain the excess pore water pressure for longer durations.

12.3.2.2 Presence of non-liquefied crust layers and their effects

The laterally spreading ground layers can impose additional loading on pile foundations passing through them. More importantly, it is possible that non-liquefied layers exist above a liquefied layer which can also start to move laterally riding on the liquefied layer as illustrated in Figure 12.10. Such non-liquefied soil crust can apply large lateral loads on pile foundations passing through them. In some cases, the passive earth pressure exerted by the non-liquefied crust can dominate the lateral loading on the pile foundations making the lateral loading applied by the liquefied layer on the piles to be relatively small. Dobry et al. (2003) recently proposed that for simplified design the lateral loading generated by the liquefied layer can be ignored provided that the passive earth pressures generated by the non-liquefied crust are accounted for.

In Section 12.7.5 of this chapter a simplified methodology is included to estimate the loading imposed by laterally spreading ground on the pile foundation. This can be used to estimate the 'upper bound' of the lateral load that can be expected to act on the pile foundation.

12.4 COMPARISON OF STATIC AND DYNAMIC PERFORMANCE REQUIREMENTS OF PILE FOUNDATIONS

As explained in Section12.2, the static design of pile must be carried out according to the guidelines provided in Eurocode 7 and its provisions. However, it is important to compare the performance requirements of pile foundations under static and dynamic loading.

12.4.1 Kinematic and inertial loading

For many classes of structure, the predominant static loading on piled foundations is vertical compressive loading. Earthquake loading will impose requirements on the piles to resist significant lateral loads and moments with the further possibility of piles being required to carry tensile loads. The deformation of piles may be substantially affected by the permanent deformations of the ground in which they are embedded and, in particular, liquefaction-induced lateral spreading can impose severe damage on piled foundations as discussed in Section 12.1.1.

The loading requirements imposed by seismic events on piles require different geotechnical and structural design of these elements compared with the static equivalent. Earthquake loading differs from other forms of environmental and machinery induced cyclic loading because the in-ground motions produce pile loadings in addition to the pile loadings derived from the motion of the supported structure. The in-ground motion generates 'kinematic interaction' between the piles and the soil while the loading imposed by the structure generates 'inertial interaction' (Figure 12.11).

EC8 – Part 5 (2003) notes that bending moments due to kinematic interaction only need to be considered when all the following conditions apply:

- The ground profile is of type D, S_1 or S_2 and contains consecutive layers of sharply differing stiffness.
- The zone is of moderate or high seismicity (i.e. αS exceeds 0.1 g), and the structure is of importance class III or IV.

12.4.2 Static pile load-deflection analyses

The static load-deflection analysis of piles has developed in two principle directions, which should be seen as complementary. These methods are the Winkler spring approach in which the pile is modelled as a beam supported by a series of independent springs and the elastic continuum approach in which an elastic pile is considered to be embedded in an elastic soil continuum (Figure 12.12).

Figure 12.11 Kinematic and inertial interaction.

Figure 12.12 Alternative models for pile load – deflection analyses. (a) Pile head fixity indeterminate. (b) Pile head fixed, pinned or M-spring. (c) Pile head fixed or pinned.

The different approaches have different strengths and weaknesses:

- The Winkler spring method allows the non-linear loading response of the soil to pile deflection to be easily incorporated through the use of non-linear p-y or t-z curves. These springs can be modified to incorporate the effects of imposed ground movements around the piles. In addition, complex soil profiles can also be accommodated in a straightforward manner. However, the springs do not account for the effects of soil movement at one location on soil movements at adjacent locations. This limits the reliability of the empirical methods used to derive p-y and t-z curves and makes the analysis of pile groups difficult with this method.
- The elastic continuum approach is more satisfactory from a theoretical standpoint as the stress and strain fields in the soils around the pile are correctly analysed. This makes the technique suitable for the analysis of the interaction of piles in pile groups. However, the available solutions are predominantly linear-elastic and based on rather simple soil profiles.

Solutions for single piles under static loading are given by Poulos and Davis (1980) with further solutions by Davies and Budhu (1986), Budhu and Davies (1987, 1988) and Gazetas (1991a,b). These are summarised by Pender (1993) for the Winkler spring and elastic models for a variety of stiffness distributions. The strengths of both methodologies can be harnessed by using the Winkler spring model to refine the soil stiffnesses selected for horizontal and vertical elastic analysis of single piles. The refined parameters from the single pile analysis may then be employed in an elastic analysis of the pile group.

For static lateral loading of an elastic pile embedded in an elastic soil, the displacement, u, and rotation, q, of the pile head are given by

$$u = f_{uH}H + f_{uM}M \tag{12.8}$$

Figure 12.13 Idealised soil stiffness profiles. (a) Constant stiffness (typical of over-consolidated clay). (b) Parabolic stiffness (typical of sand). (c) Linear stiffness (typical of soft clay).

$$\theta = f_{\theta H} H + f_{\theta M} M \qquad (12.9)$$

where
H is the horizontal load
M is the moment
$f_{uH}, f_{uM}, f_{\theta H}, f_{\theta M}$ are flexibility coefficients
with $f_{\theta H} = f_{uM}$

The pile head flexibility coefficients for the three soil stiffness profiles given in Figure 12.13 may be expressed as shown in Table 12.1.

The matrix of pile head flexibility coefficients can be inverted to obtain the matrix of pile head stiffness coefficients K_{HH}, K_{HM}, K_{MH} and K_{MM}. These can be employed to define horizontal and rotational springs, which reproduce the pile head response, thus

$$\begin{bmatrix} K_{HH} & K_{HM} \\ K_{MH} & K_{MM} \end{bmatrix} = \frac{1}{(f_{uH}f_{\theta M} - f_{uM}^2)} \begin{bmatrix} f_{\theta M} & -f_{uM} \\ -f_{\theta H} & f_{uH} \end{bmatrix} \qquad (12.10)$$

where $K_{HM} = K_{MH}$

Table 12.1 Pile head flexibility coefficients for static loading

Flexibility coefficient	Soil stiffness profile		
	Constant	Parabolic	Linear
f_{uH}	$\dfrac{1.3}{E_s D}\left(\dfrac{E_p}{E_s D}\right)^{-0.18}$	$\dfrac{2.14}{E_{sD}D}\left(\dfrac{E_p}{E_{sD}}\right)^{-0.29}$	$\dfrac{3.2}{mD^2}\left(\dfrac{E_p}{E_{sD}}\right)^{-0.333}$
$f_{\theta H} = f_{uM}$	$\dfrac{2.2}{E_s D^2}\left(\dfrac{E_p}{E_{sD}}\right)^{-0.45}$	$\dfrac{3.43}{E_{sD}D^2}\left(\dfrac{E_p}{E_{sD}}\right)^{-0.53}$	$\dfrac{5.0}{mD^3}\left(\dfrac{E_p}{E_{sD}}\right)^{-0.556}$
$f_{\theta M}$	$\dfrac{9.2}{E_s D^3}\left(\dfrac{E_p}{E_{sD}}\right)^{-0.73}$	$\dfrac{12.16}{E_{sD}D^3}\left(\dfrac{E_p}{E_{sD}}\right)^{-0.77}$	$\dfrac{13.6}{mD^4}\left(\dfrac{E_p}{E_{sD}}\right)^{-0.778}$

Note: m = rate of increase of stiffness with depth.

$$K_h = \frac{K_{HH}K_{MM} - K_{HM}^2}{K_{MM} - eK_{HM}} \text{ (kN/mm)} \tag{12.11}$$

$$K_\theta = \frac{K_{HH}K_{MM} - K_{HM}^2}{K_{HH} - K_{HM}/e} \text{ (kNm/mrad)} \tag{12.12}$$

where $e = M/H$.

The variation in stiffness of the soil in the analyses considered here can be taken in an idealised fashion as shown below. These would be reasonably good approximations for the soil types indicated in Figure 12.13 if the soil layer is homogeneous. However, when the soil strata under consideration have distinct layers, suitable approximations have to be made.

12.4.3 Dynamic pile load-deflection analyses

12.4.3.1 General behaviour

Summaries of the methods used to assess the responses of piles and pile groups to seismic loading are provided by Gazetas (1984), Novak (1991) and Pender (1993). Numerical studies indicate that the response of a pile shaft under seismic loading can be considered in three zones:

1. *The near surface zone*: This zone extends to approximately eight pile diameters beneath the soil surface and is dominated by inertial loading effects.
2. *An intermediate zone*: This zone exists between the near surface and deep zones and is influenced by both inertial and kinematic effects.
3. *The deep zone*: This zone is below 12–15 pile diameters form the surface and is dominated by kinematic effects.

The effective length of pile, L_{ad}, which participates in the inertial response may be determined for elastic soil profiles (Gazetas, 1984) as a function of the stiffnesses of the pile and the soil and the pile diameter. With reference to the idealised soil profiles shown in Figure 12.13:

Constant stiffness with depth: $L_{ad} = 2D\left(\dfrac{E_p}{E_{sD}}\right)^{0.25}$ \hfill (12.13)

Parabolic stiffness with depth: $L_{ad} = 2D\left(\dfrac{E_p}{E_{sD}}\right)^{0.22}$ \hfill (12.14)

Linear increasing stiffness: $L_{ad} = 2D\left(\dfrac{E_p}{E_{sD}}\right)^{0.20}$ \hfill (12.15)

where
  D = diameter of pile
  E_p = Young's modulus of the pile
  E_{sD} = Young's modulus of soil at depth

These active lengths are somewhat greater than the equivalent lengths, which can be determined for piles under static loading. Field studies such as those by Hall (1984) and Makris et al. (1996) on instrumented piled structures under significant levels of seismic loading show that the stiffness of the pile group tends to decrease significantly as the number of load cycles increases. This is due to effects such as a decrease in soil stiffness as shear-induced pore water pressures increase in the near surface zone and the development of gapping around the top of the pile shafts. These effects will increase the effective or active length of the piles to be considered in the inertial loading response.

The effective pile length concept is useful for differentiating between 'long' and 'short' piles. For 'long' piles, an increase in length does not affect the horizontal response to iner-tial loading. 'Short' piles, of length less than Lad, exhibit a softer response to inertial load, which is a function of pile length.

12.4.3.2 Pile flexibility

As an alternative to the method suggested above the flexibility of the pile can be determined using the following procedure. The elastic length of pile can be determined using the fol-lowing equation.

$$\text{Elastic length of pile, } T = \left(\frac{E_p I_p}{k} \right)^{0.2} \tag{12.16}$$

where $E_p I_p$ is the flexural stiffness of the pile and, k is gradient of the soil modulus may vary 200–2000 kN/m^3 · k takes the value of 2,000 kN/m^3 for loose saturated conditions. Thus elastic length of the pile is determined as a function of the relative pile–soil stiffness.

Using the value of T calculated above, the Z_{max} is calculated as follows:

$$Z_{max} = \frac{L_p}{T} \tag{12.17}$$

If $Z_{max} > 5$, the pile is considered to be flexible, that is, its behaviour is not affected by the length, and collapse is always caused by a flexural failure, with formation of a plastic hinge. The pile is semi-flexible if $5 > Z_{max} > 2.5$, and the pile is rigid if $Z_{max} < 2.5$.

Piles that are classified as flexible will 'move' with the surrounding soil and therefore would attract the inertial shear load imposed by the superstructure during earthquake loading. Rigid piles, on the other hand, will attract significant soil load, as the piles stay in position and the soil would exert passive pressures on either side of the pile in alternative load cycles. This additional lateral load applied by the soil must be considered in the pile design.

12.5 KINEMATIC RESPONSE

12.5.1 Classical approach

It is convenient to analyse the kinematic response of the pile or pile group separately from the inertial response. The kinematic response at depth may be used assess the structural requirement of the pile in the intermediate and deep zones. The kinematic response of the pile head is an input into the inertial response analysis.

In the deep zone, the presence of piles has little effect on the ground motion or natural frequency of the stratum. The pile and soil motions are likely to be practically coincident for frequencies up to at least 1.5 times the natural frequency (f_n) of the stratum. This observation is of practical significance as the deflected shape of the pile can be obtained from a 1-D equivalent linear shear wave propagation analysis. Having obtained the deflected shape of the pile its bending moments and shear forces may readily be determined. Larkin (1986) and Makris et al. (1996) discuss field studies that make useful observations on this mode of behaviour. It should be noted that substantial bending moments may be induced in piles at the levels of interfaces between zones of appreciably different stiffness. EC8 – Part 5 (2003) requires piles to remain elastic, though under certain conditions they are allowed to develop plastic hinges at their heads. The regions of plastic hinging should be designed according to EC8 – Part 1, clause 5.8.4.

In order to perform the inertial response analysis, the kinematic pile head response is required. Numerical studies indicate that the kinematic response derived for single piles is applicable to pile groups and that the kinematic interaction between the soil and a free-headed pile is conservative if applied to a fixed head pile.

Pender (1993) describes an approximate technique based on Gazetas (1984), which may be used to evaluate the kinematic response of the pile head. Firstly the free field response at the top of the soil column is determined at a point remote from the pile group. The horizontal amplitude of the free field motion is u_0. Then a frequency dependent horizontal interaction factor, I_u, is determined such that

$$I_u = \frac{u_p}{u_o} \tag{12.18}$$

where u_p is the horizontal amplitude of the pile head motion.

For constant stiffness with depth:
$$F = \left(\frac{f}{f_n}\right)\left(\frac{E_p}{E_{sD}}\right)^{0.30}\left(\frac{L}{D}\right)^{-0.5} \tag{12.19}$$

For parabolic stiffness with depth:
$$F = \left(\frac{f}{f_n}\right)\left(\frac{E_p}{E_{sD}}\right)^{0.16}\left(\frac{L}{D}\right)^{-0.35} \tag{12.20}$$

For linear increasing stiffness:
$$F = \left(\frac{f}{f_n}\right)\left(\frac{E_p}{E_{sD}}\right)^{0.10}\left(\frac{L}{D}\right)^{-0.4} \tag{12.21}$$

where
    f = response spectrum frequency considered
    f_n = natural frequency of stratum
    E_p = Young's modulus of pile
    E_{sD} = Young's modulus of soil at depth D
    L & D = length and diameter of pile.

Table 12.2 Coefficients for horizontal kinematic interaction factor

Coefficient	Soil stiffness profile		
	Constant	Parabolic	Linear
A	0	3.64×10^{-6}	-6.75×10^{-5}
B	0	-4.36×10^{-4}	-7.0×10^{-3}
C	-0.21	6.0×10^{-3}	3.3×10^{-2}

Using the appropriate equation, values of F are calculated for discrete frequencies across the frequency range of interest (e.g. 0.5 to 40 Hz). Corresponding values of I_u are calculated from the following expression (Gazetas, 1984):

$$I_u = aF^4 + bF^3 + cF^2 + 1.0 \tag{12.22}$$

with a minimum value of $I_u = 0.5$.

The coefficients a, b and c in the above equation are given in Table 12.2.

The interaction factors produced by this procedure are strictly applicable only to a Fourier spectrum. However, approximate results can be obtained by applying the interaction factors directly to the free field spectral acceleration versus frequency response spectrum. The horizontal spectral acceleration of the pile head is simply obtained by multiplying the free field acceleration by the value of I_u for each frequency considered.

The study of spectral acceleration responses produced by the above procedure shows that the piles damp the higher frequency excitation seen in the free field. The extent of that damping depends on the ratio of pile to soil stiffness and particularly on the soil stiffness profile. The linear increasing stiffness profile produces damping at lower frequencies than the other stiffness profiles. Pender (1993) observes that the response of instrumented piles in earthquakes tends to that of the linear increasing stiffness profile even if the SI data suggests a constant or parabolic profile. This is considered to be due to softening of the soil close to the surface under seismic loading.

The studies undertaken by Gazetas (1984) show that the rotational interaction factor is sufficiently small to be neglected.

12.5.2 Kinematic loading induced by laterally spreading soil

A more common problem is when pile foundations are used to transfer load from the superstructure through a shallow layer of clay and through a liquefiable soil layer into firmer soil stratum or bed rock below. If the ground at such a site is sloping (even with very small slope angles), earthquake-induced liquefaction may cause the clay layer to spread laterally, thereby inducing large lateral loads on the piles and pile cap. Recent research at Cambridge (Haigh and Madabhushi, 2005) and at Renessealer Polytechnic Institute, New York (Dobry et al., 2003) and University of California, Davis (Brandenberg et al., 2005) have looked at the loading applied by non-liquefied crust onto the pile cap and piles due to lateral spreading.

In such situations, Dobry et al. (2003) propose that in pile design, the lateral load from the non-liquefied crust plays the most important role and the contribution of the resistance offered by liquefied soil can be ignored.

Thus the lateral load applied by a clay crust can be determined by using shallow foundation bearing capacity factors. For example, if the clay crust overlying the liquefied layer had

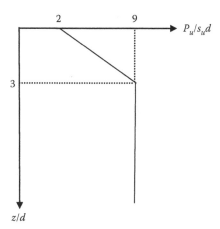

Figure 12.14 Variation in normalised lateral load with normalised depth.

an undrained strength of S_u then the lateral pressure applied on the pile cap and piles in this region can simply be calculated using

$$q = (\pi + 2) \cdot S_u \tag{12.23}$$

For other soil types, the above expression can be suitably modified.

The lateral deflection of the pile cap and rotation of the pile cap and pile heads can be determined under the action of the lateral load induced by q above over the resisting surface area (sides of the pile cap and the portion of the piles in this region). The above expression is based on upper bound theorem of plasticity and therefore should provide a safe bound.

Alternatively, the upper bound theory by Murff and Hamilton (1993) for lateral resistance can also be used in cohesive soil (shown below in Figure 12.14 where P_u is lateral resistance). This expression allows for a more gradual change of the undrained shear strength with depth from the surface of the soil crust to deeper regions and therefore should be used if the non-liquefied crust is reasonably deep.

It must be noted that the lateral loading due to inertia from the superstructure and the kinematic loading due to the lateral spreading of the soil will not generally occur at the same time. However, for design these can be superposed which is a conservative assumption. The superposed load can be used to estimate the lateral deflection of the pile heads and their rotation.

12.5.3 Adoption of static pile head static stiffness concept

EC8 – Part 5 (2003) Annex C provides guidance on the pile head stiffness coefficients for the three types of idealised soil stiffness profiles presented in Figure 12.13. These are reproduced in Table 12.3 and can be used in preference to the flexibility coefficients presented earlier in Section 12.4.2, Table 12.1. As before, the key parameters are

E – the Young's modulus of the soil model equal to 3G
E_p – the Young's modulus of the pile material
E_s – the Young's modulus of the soil at 1 pile diameter depth
d – the pile diameter and
z – the pile depth.

Table 12.3 Static stiffness of flexible piles embedded in three soil models

Soil Model	K_{HH}/dE_s	K_{MM}/d^3E_s	K_{HM}/d^2E_s
Linear variation $E = E_s \dfrac{z}{d}$	$0.60\left(\dfrac{E_P}{E_s}\right)^{0.35}$	$0.14\left(\dfrac{E_P}{E_s}\right)^{0.80}$	$-0.17\left(\dfrac{E_P}{E_s}\right)^{0.60}$
Square root variation $E = E_s\sqrt{\dfrac{z}{d}}$	$0.79\left(\dfrac{E_P}{E_s}\right)^{0.28}$	$0.15\left(\dfrac{E_P}{E_s}\right)^{0.77}$	$-0.24\left(\dfrac{E_P}{E_s}\right)^{0.53}$
Constant $E = E_s$	$1.08\left(\dfrac{E_P}{E_s}\right)^{0.21}$	$0.16\left(\dfrac{E_P}{E_s}\right)^{0.75}$	$-0.22\left(\dfrac{E_P}{E_s}\right)^{0.50}$

Once the loads on the pile heads and the pile cap are determined as discussed in Section 12.4.3, the stiffness coefficients in Table 12.3 can be used to determine the lateral and cross pile head displacements and the pile cap rotation.

12.6 INERTIAL RESPONSE

The inertial response analysis uses the dynamic response obtained from the kinematic interaction study to assess the seismic displacements and rotations of the pile head or of the structure. The forces driving the pile head are derived from the mass and stiffness of the structure.

Typically, the structure may be simplified to a single degree of freedom system while the piled foundation is considered to have translational and rotational degrees of freedom.

12.6.1 Relative stiffness of pile–soil system

The response of the foundation to the horizontal inertial loading and moments is determined by a combination of stiffness and damping in a manner analogous to the response of a shallow foundation. While the single pile stiffness is not sensitive to frequency, the pile group interaction terms and the radiation damping are frequency dependent.

A common way of addressing the response of the single pile or group to inertial loading is by the use of the impedance concept:

$$S(\omega) = \frac{R(t)}{U(t)} \tag{12.24}$$

with $S(\omega) = K(\omega) + i\omega C$.

where
  $S(\omega)$ = impedance for mode of response (sliding, rocking etc.)
  $R(t)$ = dynamic force or moment
  $U(t)$ = dynamic displacement or rotation
  $K(\omega)$ = dynamic pile stiffness (kN/m)
  ω = frequency (rad/s)
  C = damping coefficient (kN s/m)
  $i = (-1)^{0.5}$

The impedance function is conveniently expressed as a complex variable because the damping component, being a function of velocity, is out of phase with the elastic stiffness. The damping may also be expressed as dimensionless frequency dependent coefficients, $\zeta(\omega)$, for the various modes of response where

$$\zeta(\omega) = \frac{\pi f C}{K} = \frac{\omega C}{2K} \tag{12.25}$$

This enables an alternative expression for the impedance to be developed:

$$S(\omega) = K[k(\omega) + 2\zeta(\omega)i] \tag{12.26}$$

with the impedance functions defined in above equations, any appropriate static expression for single pile or pile group loading response can be used for the dynamic loading case, substituting the complex impedance terms for their static counterparts.

Numerical studies undertaken by Gazetas (1984) show that $k(\omega)$ is approximately unity for most practical values of pile–soil stiffness ratio over the frequencies of interest and for the horizontal, rocking and vertical modes. Hence the dynamic stiffnesses for the various modes can be taken as similar to their static counterparts.

12.6.2 Damping coefficients

Values for the damping coefficients ζ are given by Gazetas (1991b) for single piles embedded in elastic soils with the stiffness profiles shown in Figure 12.13 as shown in Table 12.4. Note that ζ_{HH} is the damping due to horizontal movement under horizontal loading, ζ_{HM} refers to horizontal movement due to applied moment and ζ_{MM} refers to rotation due to applied moment.

All of the expressions in Table 12.4 apply only when $f > f_n$ for the stratum. If the exciting frequency is below the natural frequency of the stratum, then there will be no radiation damping and the damping coefficients will be the left hand term in each case.

Calculations based on the formulae in Table 12.4 have been compared with a limited amount of field data mainly derived from experiments where vibrators have been mounted on single piles (Pender, 1993). The field data suggest that the damping coefficient values obtained from these expressions under-predict actual damping by about 30%.

Using the impedance terms, the pile head behaviour may be reduced to translational and rotational springs. The inertial loading may be determined using a single degree of freedom

Table 12.4 Dimensionless pile head damping coefficients

Damping coefficient	Soil stiffness profile		
	Constant	Parabolic	Linear
ζ_{HH}	$0.8\beta + \dfrac{1.10 f D}{v_s}\left(\dfrac{E_p}{E_{sD}}\right)^{0.17}$	$0.7\beta + \dfrac{1.20 f D}{v_s}\left(\dfrac{E_p}{E_{sD}}\right)^{0.08}$	$0.6\beta + \dfrac{1.8 f D}{v_s}$
ζ_{HM}	$0.8\beta + \dfrac{0.85 f D}{v_s}\left(\dfrac{E_p}{E_{sD}}\right)^{0.18}$	$0.6\beta + \dfrac{0.70 f D}{v_s}\left(\dfrac{E_p}{E_{sD}}\right)^{0.05}$	$0.3\beta + \dfrac{1.0 f D}{v_s}$
ζ_{MM}	$0.35\beta + \dfrac{0.35 f D}{v_s}\left(\dfrac{E_p}{E_{sD}}\right)^{0.2}$	$0.22\beta + \dfrac{0.35 f D}{v_s}\left(\dfrac{E_p}{E_{sD}}\right)^{0.1}$	$0.2\beta + \dfrac{0.4 f D}{v_s}$

structural model. The equations required to solve the response of such an SDOF system are given by Wolf (1985). Useful worked examples are given by Pender (1993).

Because the impedance terms are complex numbers, the calculated displacements also have real and imaginary parts. The maximum (real) response is readily determined by applying the SRSS technique.

Calculations on the response of pile groups require the use of dynamic pile group interaction factors. These are frequency-dependent complex functions. Interaction factors for various loading directions and responses are given by Gazetas (1991b), Gazetas et al. (1991) and Makris and Gazetas (1992).

12.6.3 Combination rules

Since the response of the soil and structure will be at different natural frequencies, the combination rules given in Clause 4.3.3.5 of EC8 – Part 1 can be used to calculate the cumulative effect of kinematic and inertial interaction.

12.7 DESIGN EXAMPLE ON A PILE FOUNDATION

In this section, we shall outline the design of a pile foundation for a typical column of the building for which the seismic designs were carried out in earlier chapters. Of course, in reality, the design of pile foundations will be carried out for individual columns with the associated reductions in the pile lengths and/or pile diameters to suit the design load on the column. Here we shall only consider one typical column along the D line on the plan of the building.

Another premise that is made here is the requirement of the pile foundations. It is assumed that the building will be located at 'site A' for economic and operational reasons.

12.7.1 Configuration of the problem

In Chapter 8, the EC8 Part 5 provisions were used to determine the liquefaction potential of 'Site A'. The soil profile at this site as determined from bore hole data is presented in Figure 12.15. Based on this it was determined that this site has

- A non-liquefiable clay crust of 2 m thickness close to ground
- Liquefaction potential analysis confirms that a 10 m thick layer of loose sand underlying the clay layer is 'liquefiable' during the design earthquake event

The above ground conditions at this site would necessitate the requirement of pile foundations. The pile foundations would be required to pass through the loose sand layer and end-bearing fully into the dense sand layer.

12.7.2 Structural loading on piles

In Chapter 3, the structural analysis of the building frame is considered. Here we use the loading obtained from those analyses (using q factor of 3.9 and choosing concrete frame building, which has the more severe loading case). These loads are obtained with due consideration to the capacity design aspects and are shown below in Table 12.5. Please note that the worst loading occurs on columns along the lines C and D, with each line reaching a maximum load while the other is at a minimum.

Figure 12.15 Bore hole data from Site A.

Therefore the loading on the pile group is

Design vertical load $N_{Ed} = 5{,}978$ kN
Design moment load $M_{Ed} = 2{,}405$ kN m
Design horizontal, shear load $V_{Ed} = 826$ kN

Based on the above requirements, the following will be assumed regarding the pile foundations. Choose

- 2×2 pile group for columns along the D line
- Steel tubular driven pile
- ~15 m pile length
- Pile diameter 800 mm; pile wall thickness 20 mm
- Pile spacing = 2-D = 1.6 m
- Pile group efficiency $\eta = 70\%$ (conservatively)

Various other pile types can be considered for this application, such as concrete bored piles, pre-cast concrete driven piles or steel H-piles for example.

Table 12.5 Loading on the foundation from the columns

	Column C	Column D
Axial load	5,978 kN	862 kN
Shear load	826 kN	826 kN
Moment load	2,405 kNm	2,088 kNm

12.7.3　Static pile design

The piles are required to be designed according the provisions of Eurocode 7. Here the UK National Annex provisions are also taken into consideration.

Assumptions and simplifications:

Assume pile density is equal to soil density.
Assume moment on group is carried by couple in piles.

Individual axial pile load Q_A is given by

$$Q_A = \frac{N_{Ed}}{4} \pm \frac{M_{Ed}}{2 \times 1.6\,\text{m}}$$

$$Q_A = 1{,}495 \pm 752\,\text{kN} \quad Q_{A\max} = 2{,}247\,\text{kN}$$

Ignore shaft friction from upper clay layer.
Assume pile is plugged and can develop full end bearing capacity.

12.7.4　Axial pile design

Use BS EN1997 Design Approach DA-1.

Two combinations must be considered. In combination 1, partial factors are applied to the pile loading. In combination 2, partial factors are applied to components of the pile resistance.

Note: Refer to the UK National Annex for appropriate partial factors for pile design.

12.7.4.1　Combination 1

Partial factor sets A1 + M1 + R1 apply

From A1 adopt factor $\gamma_G = 1.35$
For M1 all material factors $\gamma_M = 1$
For R1 all resistance factors $\gamma_R = 1$

This is a simplification. Separate factors apply to permanent and transient loads.

A model factor (M_F) is also required. From the UK NA, the model factor is 1.4 if the pile has been designed from soil test data alone. If the pile capacity has been verified using a maintained load test the model factor is 1.2.

BS EN1997 is not prescriptive concerning the method of calculating the pile capacity, only requiring that the method should be one that is verified against pile load test data.

End bearing: Consider the end bearing of a pile. End bearing of pile may be calculated as

$$Q_b = N_q \sigma'_v A_b$$

where A_b = Base area of shaft, assuming fully plugged base.

The friction angle for dense sand $\phi_k = 36°$

The bearing capacity factor N_q for a deep foundation can be obtained using the following chart given by Berezantzev et al. (1961) shown in Figure 12.4.

Reading the value of N_q from Figure 12.4 for the friction angle of 36°, we get;
$N_q = 65$ from Berezantzev et al. (1961)
Assume that the pile starts from 1 m below ground level to allow for pile cap of 1 m thickness.

For a 15 m long pile, the base is at 16 m below ground level (allowing for 1 m thick pile cap). So calculate the effective stress at 16 m depth.

The pile diameter of 800 mm (assume that the pile is plugged at the base).

$$\sigma'_v = 16 \times 20 - 14 \times 10 = 180 \text{ kPa}$$
$$q_b = \sigma'_v \cdot N_q$$

$$q_b = 180 \times 65 = 11,700 \text{ kPa}$$

$$Q_b = \pi r^2 \times q_b = \frac{\pi \times 0.4^2 \times 11,700}{1,000} = 5,881 \text{ kN}$$

Shaft friction: We can ignore the 1 m of clay layer just below the pile cap in estimating the shaft capacity (conservative assumption)

$$\tau_s = K_s \cdot \sigma'_v \cdot \tan \delta$$

Assume the following:

For driven piles, $K_s \geq 1$, choose $K_s = 1$.
For the loose sand layer around the shaft, $\phi_k = 30°$
As we are using driven, smooth, steel tubular piles,

$$\delta_d = \frac{2}{3}\phi_{cvd} = \frac{2}{3} \times 30° = 20°$$

Therefore, at +12.53 elevation,

$$\tau_S = 1 \times 40 \times \tan 20° = 14.55 \cong 15 \text{ kPa}$$

At −1.47 m elevation (15 m long pile),

$$\tau_S = 1 \times 180 \times \tan 20° = 65.5 \cong 66 \text{ kPa}$$

$$Q_s = 2\pi r \times \int_0^L \tau_s$$

$$Q_s = 2\pi r \times 0.4 \times \frac{(15+66)}{2 \times 1,000} \times 15 = 1,527 \text{ kN}$$

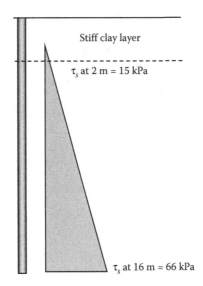

Stiff clay layer

τ_s at 2 m = 15 kPa

τ_s at 16 m = 66 kPa

Figure 12.16 Variation of shear stress along the length of the pile.

The applied design load for combination 1 is $2{,}247 \times 1.35 = 3{,}033$ kN.
(See the shear stress distribution along the length of the pile in Figure 12.16.)
[In practice, shear stress is limited to 100 kPa, $\tau_s < 100$ kPa.]
For combination 1, the total design resistance R_{cd1} is calculated as follows:

$$R_{cd1} = R_{bd1} + R_{sd1}$$

where

$$R_{bd1} = \frac{Q_b}{M_F} \quad \text{and} \quad R_{sd1} = \frac{Q_s}{M_F}$$

Therefore

$$R_{cd1} = \frac{5{,}881 + 1{,}527}{1.4} = 5{,}291 \text{ kN} > 3{,}033 \text{ kN}$$

12.7.4.2 Combination 2

Partial factor sets A2 + M1 + R4 apply.
For A2 factors on actions $\gamma_G = 1$
For M1 all material factors $\gamma_M = 1$
For R4 various factors apply to the shaft and base resistances.
For a driven pile without explicit load test verification of the SLS, the following factors
   apply (see UK National Annex):

$\gamma_b = 1.7$
$\gamma_s = 1.5$
$M_F = 1.4$ as for combination 1.

As before the unfactored end bearing and shaft resistances are given as

$Q_b = 5,881 \text{ kN}$

$Q_s = 1,527 \text{ kN}$

For combination 2, the total design resistance R_{cd2} is calculated as follows:

$R_{cd2} = R_{bd2} + R_{sd2}$

where

$$R_{bd2} = \frac{Q_b}{\gamma_b \times M_F} \quad \text{and} \quad R_{sd1} = \frac{Q_s}{\gamma_s \times M_F}$$

Therefore

$$R_{cd2} = \frac{5,881}{1.7 \times 1.4} + \frac{1,527}{1.5 \times 1.4} = 2,471 + 727 = 3,198 \text{ kN} > 2,247 \text{ kN}$$

In this calculation, combination 2 is more critical than combination 1 but nevertheless the required inequality is satisfied. In the case that the loose sand layer is subject to liquefaction, the shaft capacity in the liquefied layer will temporarily be lost. Raised pore pressures will also migrate into the dense sand reducing the effective stresses and therefore also reducing the shaft friction in this stratum. During subsequent reconsolidation of the loose sand, down-drag forces will be applied to the pile shaft. The designer will wish to take these issues into account when deciding whether the pile length is adequate.

Choose pile cap of 3 m × 3 m, with pile spacing of 2 diameters as shown in Figure 12.17.

12.7.5 Factor of safety against pile buckling

Normally piles rely on lateral resistance offered by soil. However, when the surrounding soil liquefies this lateral resistance is significantly reduced. Long, slender piles in liquefiable soils

Figure 12.17 Plan view of pile cap.

can suffer buckling failure based on recent research based on dynamic centrifuge modelling (Bhattacharya et al. 2004).

A simple check can be carried out to see if piles are vulnerable to buckling mode of failure. For Euler buckling, we know that

$$P_E = \frac{\pi^2 EI}{L_e^2}$$

where EI is the flexural rigidity of the pile and L_e is the equivalent length of the pile, which depends on the end conditions. For our case, the base of the pile is fixed with sufficient embedment length into the dense sand (5 pile diameters = $5 \times 0.8 = 4$ m). At the pile cap, the pile head has rotational fixity but not translational fixity. This yields a buckling mode shape that dictates the effective length of the pile to be equivalent to length of the pile. Ignoring the 1 m of clay above the liquefiable layer, and considering the embedment length into dense sand,

$$L_E = 1 + 10 + 4 = 15 \text{ m}$$

$E = 210$ GPa for steel; Also the steel tubular piles are assumed with outer diameter of 0.8 m and wall thickness of 20 mm.

$$I_P = \frac{\pi}{64}(D_o^4 - D_i^4) = \frac{\pi}{64}(0.8^4 - 0.76^4) = 0.00373 \text{ m}^4$$

$$P_E = \frac{\pi^2 EI_P}{L_e^2} = \frac{3.14^2 \times 210 \times 10^9 \times 0.00373}{15^2} = 34355450.66 \text{ N}$$

$$P_E = 34.35 \text{ MN}$$

This is significantly more than the applied vertical load, so the pile should be safe against buckling. However, please note that the Euler buckling does not take into account any imperfections in the pile or pile wandering during driving. Similarly any lateral displacements due to lateral spreading may cause additional P–Δ effects, which you must consider.

12.7.6 Flexibility of the pile

The elastic length of pile,

$$T = \left(\frac{E_P I_p}{k}\right)^{0.2}$$

where
$E_P I_p$ = the flexural stiffness of the pile
E_P = 210 GPa

$$I_P = \frac{\pi}{64}(D_o^4 - D_i^4) = \frac{\pi}{64}(0.8^4 - 0.76^4) = 0.00373 \text{ m}^4$$

where k is gradient of the soil modulus (k takes the value of 2,000 kN/m³ for loose saturated conditions) may vary 200–2000 kN/m³).

For the present case, $k = 5,000$ kN/m³ and length of pile, $L_p = 15$ m.

$$T = \left[\frac{E_p I_p}{k}\right]^{0.2} = \left[\frac{210 \times 10^9 \times 0.00373}{5,000 \times 10^3}\right]^{0.2} = 2.75$$

$$Z_{max} = \frac{L_p}{T} = \frac{15}{2.75} = 5.46$$

Since $Z_{max} > 5$, the pile is flexible, that is, its behaviour is not affected by the length, and collapse is always caused by a flexural failure, with formation of a plastic hinge.

[Note that the pile is semi-flexible if $5 > Z_{max} > 2.5$, and the pile is rigid if $Z_{max} < 2.5$.]

Piles that are classified as flexible will 'move' with the surrounding soil and therefore would attract the inertial shear load imposed by the superstructure during earthquake loading.

12.7.7 Lateral loading due to the clay layer below the surface

Pile cap side area $= 1 \times 3 = 3 \text{ m}^2$

Projected area of one pile for 1 m $= 1 \times 0.8 = 0.8 \text{ m}^2$

The maximum loading on the pile cap due to the laterally spreading clay layer occurs when the clay layer fails loading the pile cap.

Lateral load (upper bound) $= (\pi + 2) \times s_u \times \text{Area}$
$$= 5.14 \times 50 \times (3 + 4 \times 0.8) = 1593.4 \text{ kN}$$

Alternatively, the upper bound theory by Murff and Hamilton (1993) for lateral resistance can also be used in cohesive soil (as shown in Figure 12.14 before in which P_u is lateral resistance). Note that the calculation carried out above falls at 5.14, which is close to the average of points 2 and 9 on the x-axis of Figures 12.14 and 12.18.

In our design, let us assume that the liquefied soil zone does not offer any additional loading that contributes to the pile group movement. This is also suggested by Dobry et al. (2003) as a design approximation. However, JRA1990 suggest that a resistance of 30% of total vertical stress can be used for fully liquefied soil zones. In general, non-liquefied clay crust contributes the larger proportion of the loading on the pile cap.

From Table 12.3 for the soil model with square root variation in strength with depth, choose the normalised stiffness coefficients:

$$E = E_s \sqrt{z/d}, \quad E_p = 210 \text{ GPa}.$$

Figure 12.18 Sectional view through the pile cap.

This elastic modulus can be used directly for a solid pile. However, our pile is a tubular pile. Therefore, the elastic modulus must be reduced in proportion to the reduced second moment of area. This can be done as follows:

$$E_{p\_corrected} = \frac{E_p}{(I_{solid}/I_{tubular})}$$

$$E_{p\_corrected} = \frac{E_p}{(D_o^4/\{D_o^4 - D_i^4\})}$$

The outer diameter of the pile is 800 mm and the inner diameter is 760 mm. Substituting these, we get

$$E_{p\_corrected} = \frac{210}{\left(0.8^4/\{0.8^4 - 0.76^4\}\right)} = 38.95 \cdot GPa$$

Take $E_s = 30$ MPa for clay at one depth equivalent to one pile diameter.

[**Note:** This is quite a critical parameter that affects the estimation of lateral displacement and rotation of the pile cap. In practice, we have to determine this carefully based on laboratory data from experiments on the clay crust].

$$\frac{K_{HH}}{dE_s} = 0.79\left(\frac{E_p}{E_s}\right)^{0.28}$$

$$K_{HH} = 0.8 \times 30 \times 10^6 \times 0.79 \times \left[\frac{38.95 \times 10^3}{30}\right] = 1.41 \times 10^8 \text{ N/m}$$

$$\frac{K_{MM}}{d^3 E_s} = 0.15\left(\frac{E_p}{E_s}\right)^{0.77}$$

$$K_{MM} = 0.8^3 \times 30 \times 10^6 \times 0.15 \times \left(\frac{38.95 \times 10^3}{30}\right)^{0.77} = 5.75 \times 10^8 \text{ Nm/rad}$$

$$\frac{K_{HM}}{d^2 E_s} = -0.24 \left(\frac{E_p}{E_s}\right)^{0.53}$$

$$K_{HM} = 0.8^2 \times 30 \times 10^6 \times -0.24 \left(\frac{38.95 \times 10^3}{30}\right)^{0.53} = -2.06 \times 10^8 \text{ N/rad}$$

12.7.8 Pile cap deflection and rotation

Design horizontal, shear load $V_{Ed} = 826$ kN
Design moment load $M_{Ed} = 2{,}405$ kN m
Lateral load due to the clay layer $= 1593.4$ kN

Note that the lateral load due to the clay layer (kinematic loading) and the design shear load (inertial load) from the superstructure do not occur at the same time. Therefore, it is sufficient to consider the largest of these loads in estimating the displacement. However, let us choose to superpose both these loads as a conservative approximation.

$$\text{Horizontal force one each pile} = \frac{1593.4 + 826}{4} = 604.85 \text{ kN.}$$

$$\text{Moment load on each pile} = \frac{2{,}405}{4} = 601.25 \text{ kNm}$$

12.7.8.1 Pile cap displacement

The pile cap displacement can be calculated using the following equation (Figure 12.19):

$$\delta_{HH} = \frac{H}{K_{HH}} + \frac{M/h}{K_{HM}}$$

where h is the equivalent height at which a horizontal force will cause a moment of M. Assume that this equivalent height is 6 m corresponding to the first floor level of the building.

Figure 12.19 Horizontal and moment loading on pile cap.

$$\delta_{HH} = \frac{604.85 \times 10^3}{1.41 \times 10^8} + \frac{601.25 \times 10^3/6}{2.06 \times 10^8} = 0.004776 \text{ m}$$

So lateral displacement of pile cap will be 4.78 mm, which is very small.

12.7.8.2 Pile cap rotation

The pile cap rotation can be calculated using the following equation:

$$. \; \theta_{HH} = \frac{M}{K_{MM}} + \frac{H}{K_{HM}}$$

$$\theta_{MM} = \frac{601.25 \times 10^3}{5.75 \times 10^8} + \frac{604.85 \times 10^3}{2.06 \times 10^8} = 3.982 \times 10^{-3} \text{ rad}$$

So the pile cap rotation will be 0.228°, which is very small.

[**Note:** Although the pile cap deflection and rotation is small, note that the above calculations were based on the assumption of the elastic modulus of the soil at the depth of a one pile diameter. So the deflection in reality may be somewhat larger than what we calculated.]

REFERENCES

Bhattacharya, S., Bolton, M.D. and Madabhushi, S.P.G., 2005a. A reconsideration of the safety of piled bridge foundations in liquefiable soils. *Soils and Foundations*, 45(4), 13–26.

Bhattacharya, S., Madabhushi, S.P.G. and Bolton, M.D. 2004. An alternative mechanism of pile failure during seismic liquefaction. *Geotechnique*, 54(3), 203–213.

Bhattacharya, S., Madabhushi, S.P.G. and Bolton, M.D., 2005b. Discussion on an alternative mechanism of pile failure during seismic liquefaction. *Geotechnique*, 55(3), 259–263.

Brandenberg, S.J., Boulanger, R.W. and Kutter, B.L. 2005. Discussion of Single piles in lateral spreads: Field bending moment evaluation. *Journal of Geotechnology and Geoenvironmental Engineering*, 131(4), 529–534.

Budhu, M. and Davies, T.G. 1988. Analysis of laterally loaded piles in soft clays. *Journal of Geotechnical Engineering ASCE*, 114(1), 21–39.

Budhu, M. and Davies, T.G. 1987. Nonlinear analysis of laterally loaded piles in cohesionless soils. *Canadian Geotechnical Journal*, 24, 289–296.

Caquot, A. and Kerisel, J. 1953. Sur le terme de surface dans le calcul des foundations en milieu pulverulent. In: *Proceedings 3rd International Conference on Soil Mechanics and Geotechnical Engineering*, Zurich, Vol. 1, pp. 336–337.

Davies, T.G. and Budhu, M. 1986. Nonlinear analysis of laterally loaded piles in heavily overconsolidated clays. *Geotechnique*, 36(4), 527–538.

Dobry, R., Abdoun, T., O'Rourke, T.D. and Goh, S.H. 2003. Pile response to lateral spreads: Centrifuge modeling. *Journal of Geotechnology and Geoenvironmental Engineering, ASCE*, 129(10), 879–889.

Eurocode 7 – Part 1 1995. Geotechnical design, general rules. *CEN European Committee for Standardisation*, DD ENV 1997–1:1995.

Eurocode 8 – Part 5 2003. Design provisions for earthquake resistance of structures – foundations, retaining structures and geotechnical aspects. *CEN European Committee for Standardisation*, prEN 1998–5:2003.

Franklin, A.G. and Chang, F.K. 1977. Earthquake resistance of earth and rockfill dams; permanent displacements of embankments by Newmark sliding block analysis. *Miscellaneous Paper S.71.17*, U.S. Army Corps of Engineers Waterways Experimental Station, Vicksburg.

Gazetas, G. 1984. Seismic response of end-bearing piles. *Soil Dynamics and Earthquake Engineering*, 3(2), 82–93.

Gazetas, G. 1991a. Formulas and charts for impedances of surface and embedded foundations. *Journal of Geotechnical Engineering ASCE*, 117(9), 1363–1381.

Gazetas, G. 1991b. Foundation vibrations. In: *Foundation Engineering Handbook*, 2nd Edition, H.-Y. Fang (ed.), Van Nostrand Reinhold, Asterdam, pp. 553–593.

Gazetas, G., Fan, K., Kaynia, A. and Kausel, E. 1991. Dynamic interaction factors for floating pile groups. *Journal of Geotechnical Engineering ASCE*, 117, 1531–1548.

Haigh, S.K. and Madabhushi, S.P.G. 2005. The Effects of Pile Flexibility on Pile Loading in Laterally Spreading Slopes, invited paper. In: *ASCE-GI Special Publication on Simulation and Seismic Performance of Pile Foundations in Liquefied and Laterally Spreading Ground*, R.W. Boulanger and K. Tokimatsu (eds.), ASCE Geotechnical Special Publication No. 145, Reston, West Virginia, USA. ISBN 0-7844-0822-X, pp. 24–37.

Haigh, S.K., Madabhushi, S.P.G., Soga, K., Taji, Y. and Shamoto, Y. 2000. Lateral spreading during centrifuge model earthquakes. In: *Proceedings of Geo2000 International Conference on Geotechnical Engineering*, University of Sydney, Australia, pp. 216–222.

Hall, J.F. 1984. Forced vibration and earthquake behaviour of an actual pile foundation. *Soil Dynamics and Earthquake Engineering*, 3(2), 94–101.

Hamada, M. 1992. *Case Studies of Liquefaction and Lifeline Performance During Past Earthquakes* NCEER, NY.

Madabhushi, S.P.G., Patel, D. and Haigh, S.K. 2005. *Geotechnical Aspects of the Bhuj Earthquake, Chapter 3, EEFIT Report*. Institution of Structural Engineers, London, UK, ISBN 0901297 372.

Makris, N. and Gazetas, G. 1992. Dynamic pile-soil-pile interaction. Part II: Lateral and seismic response. *Earthquake Engineering and Structural Dynamics*, 21, 145–162.

Makris, N., Gazetas, G. and Delis, E. 1996. Dynamic soil-pile-foundation-structure interaction: Records and predictions. *Geotechnique*, 46(1), 33–50.

Murff, J.D. and Hamilton, J.M. 1993. P-Ultimate for undrained analysis of laterally loaded piles. *Journal of Geotechnical Engineering, ASCE*, 119(1), 91–107.

Novak, M. 1991. Piles under dynamic loads: State of the Art. In: *Proceedings of the 2nd International Conference on Recent Advances in Geotechnical Earthquake Engineering and Soil Dynamics*, St. Louis, Vol. 3, pp. 250–273.

Pender, M.J. 1993. A seismic pile foundation design analysis. *Bulletin of New Zealand National Society for Earthquake Engineering*, 26(1), 49–174.

Poulos, H.G. and Davis, E.H. 1980. *Pile Foundation Analysis and Design*. Wiley, New York Sons.

Suetomi, I. and Yoshida, N. 1998. Nonlinear behavior of surface deposit during the 1995 Hyogoken-Nambu Earthquake. *Soils and Foundations*, Special Issue No. 2, 11–22 September.

Takata, T., Tada, Y., Toshida, I. and Kuribayashi, E. 1965. *Damage to Bridges in Niigata Earthquake. Report no. 125-5*, Public Works Research Institute. (In Japanese.)

Tokimatsu, K. and Asaka, Y. 1998. Effects of liquefaction-induced ground displacements on pile performance in the 1995 Hyogoken-Nambu Earthquake. *Soils and Foundations*, Special Issue No. 2, September, 163–177.

Tokimatsu, K., Oh-oka Hiroshi, Satake, K., Shamoto Y. and Asaka, Y. 1997. Failure and deformation modes of piles due to liquefaction-induced lateral spreading in the 1995 Hyogoken-Nambu earthquake. *Journal of Structural Engineering AIJ (Japan)*, 495, 95–100.

Wolf, J.P. 1985. *Dynamic Soil-Structure Interaction*. Wiley, New York.

Index

Printed in the United States
by Baker & Taylor Publisher Services